Muzzled Oxen

Genevieve Rogers Grant Sadler ("Brick")

Muzzled Oxen

*Reaping Cotton and Sowing Hope
in 1920s Arkansas*

Genevieve Grant Sadler

BUTLER
CENTER
BOOKS

BUTLER
CENTER
BOOKS

The Butler Center for Arkansas Studies
Central Arkansas Library System
100 Rock Street
Little Rock, Arkansas 72201

www.butlercenter.org

First edition: September 2014

ISBN 978-1-935106-69-2
ISBN 978-1-935106-70-8 (e-book)

All photographs in this book appear
courtesy of the Sadler family.

Book design:
Studio E Books, Santa Barbara, California

LIBRARY OF CONGRESS CATALOGING-IN-PUBLICATION DATA
Sadler, Genevieve Grant, 1893–1967.
Muzzled Oxen: Reaping Cotton and Sowing Hope in 1920s Arkansas /
Genevieve Grant Sadler. — First Edition.
358 pages
ISBN 978-1-935106-69-2 (pbk. : alk. paper)
1. Sadler, Genevieve Grant, 1893–1967. 2. Cotton farmers—Arkansas—Biography
3. Arkansas—History—20th century. I. Title.
HD8039.C66A3 2014
633.5'1092—dc23
[B]
2014003199

Butler Center Books, the publishing division of the Butler Center for Arkansas Studies, was
made possible by the generosity of Dora Johnson Ragsdale and John G. Ragsdale Jr.

Printed in the United States of America

This book is printed on archival-quality paper that meets requirements of the
American National Standard for Information Sciences, Permanence of Paper,
Printed Library Materials, ANSI Z39.48-1984.

"Thou shalt not muzzle the ox when he treadeth out the corn"

—*Deuteronomy 25:4*

*Dedicated to the share-croppers
and cotton pickers of the South.*

Contents

Foreword

GENEVIEVE SADLER first wrote this true account of her seven-year sojourn in Arkansas in the 1920s in the form of letters to her mother. Those letters set out the sharp contrasts between her early life in Santa Cruz, California, among family and friends, and her new life, isolated on a cotton farm in Yell County in central Arkansas.

Genevieve's letters, which were long and detailed, provided vignettes of the lives of the people among whom she lived, especially the share-croppers and cotton pickers, both black and white, who worked on the Sadler lands and nearby. The letters included faithful quotations from conversations with her new-found acquaintances, and the lyrics of origi-nal and traditional songs she heard them sing. The letters contained clear images of the Arkansas landscape, the farms and farm operations.

Genevieve, her husband Wayne, and their two young sons traveled from Santa Cruz to Dardanelle, Arkansas, in the summer of 1920 in a Model T Ford touring car. She was 28 years old, Wayne was 29, Jimmy and Donnie were eight and seven. For the next seven years Wayne led the struggle to make a living raising cotton on land along a small stream which emptied into the Arkansas River. The land had first been cleared for farming by Wayne's grandfather, Rufus Crispinius Sadler, using slave labor.

The hard work of planting, tending and picking cotton was overbur-dened by a host of problems, including the boll weevil, malaria, lawsuits, and, finally, flood waters. In 1927 the entire Mississippi River basin, from Canada to the Gulf of Mexico, from Ohio to Colorado, experienced one of the greatest floods in its history. The Mississippi River and all of its tributaries overflowed their banks and busted their levees. The Arkansas River flooded the land for miles on either side. It put many feet of water over the Sadler farmlands, and washed our family back to California.

When Genevieve returned home, her mother turned over to her a small trunk, containing all of the hundreds of letters she had written home. Genevieve then wrote this book over a period of many years, making the letters the heart and soul of this broader narrative of her life in Arkansas.

Genevieve Grant Sadler, my mother, was born on February 1, 1893, in Armstrong, British Columbia. Her parents, Alexander Grant and Elizabeth Mary Seed, although Protestants, named her after St. Genevieve, patron saint of Paris. In about 1906, Elizabeth left Alexander in British Columbia, and moved with Genevieve and three of her siblings to Los Gatos, California, where Elizabeth operated a boarding house to support the family. Genevieve's oldest brother, Harry, remained in Canada with his father, Alexander, and later served in France with the Canadian forces in World War I.

Genevieve attended high school in Santa Cruz, California, and there she met a classmate, Anthony Wayne Sadler, who had moved to California from Yell County with his parents. Genevieve and Wayne eloped and were married on March 21, 1912. Their son James Douglas Sadler was born in Palo Alto, California, on Christmas Day in 1913.

During their courtship, Wayne gave Genevieve the nickname of "Brick," after her red hair, which she wore in long braids or coiled on her head. And Brick she became, to family and friends. I never called her "mother," even as a small child.

In early 1914, Genevieve's older brother Ellis divorced his wife, Grayce. Their daughter, Orpha, age three, remained with her mother. But for whatever reason—always a family mystery—Ellis took custody of their infant son, Donald, who was born on February 21, 1914. Ellis almost at once brought the ten-month-old Donnie to the home of Genevieve and Wayne. There he immediately became a permanent member of the Sadler family. Donnie and Jimmy grew up together, almost as twin brothers, only weeks apart in age. Genevieve and Wayne formally adopted Donald, following the death of Ellis.

Genevieve and Wayne's second child, Gareth Wayne Sadler, was born on October 10, 1924, in Yell County, when his two brothers were ten years old. His relations with them were more like those between a nephew and uncles than those between siblings. As the family left Arkansas in 1927, when Gary was three, his knowledge of Arkansas was all secondhand. Jimmy and Donnie had clear memories of their lives in the South—seven years working in the fields, riding horseback, and growing up with children of both races.

Wayne died in 1965. Two years later, Genevieve's life ended. At her graveside, a piper played "Leaves of the Forest"—for loss, "Amazing Grace"—for hope, and "Scotland the Brave"—for courage and joy. Today their descendants include Donnie's son, Paul; Gary's daughters, Martha, Anne and Sharon; and Anne's daughter, Jesse Camilla Dunne.

Genevieve's finest qualities are reflected in this book. Those which come most strongly to mind are her eager observation of life and her joyful sharing.

Gareth W. Sadler
Santa Rosa, July 4, 2012

PART I

Hearts Turned Back to Dixie

1. Vnto the Mountain of Nebo

A WAVE OF HATRED, a burning resentment, swept through me, leaving me speechless and limp as I sat in our old Ford car parked by the drug store and looked up and down the street. How I hated it all!

This feeling had begun at the time my brother-in-law remarked dryly, when I asked him if I looked all right, "Sure, but you'll find that Southern ladies don't wear pants in public." I was wearing at that time an outfit of tan riding breeches and leggings, with a tan shirt, visor cap, and gloves. To me these clothes were both comfortable and appropriate, and had served well on the long trip from California, but I changed to a green linen dress that had been crisp and new when I left home, yet was now quite wrinkled and bedraggled looking. My best hat, that I had cherished carefully in a hat bag for six weeks through as many states, I now threw away for it was bent and broken and unfit to wear.

That first timid drive, down through the town onto the main street, eyeing with a palpitating interest every face, and being in turn surveyed with either a look of indifference or a calculating stare, had ended—here. We sat under the burning heat of the sun, in a car that was hot, dusty, and overcrowded. My husband, Wayne, and his brother, Henry, sat in the front seat, and I knew by the look on their faces that they were trying hard to keep from expressing any opinion. We had been so hopeful and eager in making this trip to the Southland, and though our plans were still a little vague, we had looked forward to a place where conditions would be at least as satisfactory as those we had left behind.

The heat was excessive and my hair clung to my head. The night before, we had made our last camp on a hillside just outside of town where there was no water and little shade. Now, here in the forenoon, rumpled and nervous and disappointed, and with our clothes sticking to our backs in the damp heat and the perspiration beginning to run down our faces, we just sat and looked about.

Our two little boys, Jimmy and Donald, travel-wise and weary, sat quietly beside me, saying nothing. Donald, nearly seven, was our adopted son; his father, now dead, was a favorite brother of mine. Jimmy was about four months older than Donald. Both boys had brown hair and dark blue eyes. They were sturdy and healthy with round pink cheeks and a few freckles, and, best of all, each had a ready smile and an eager interest in all that went on.

After our long weeks of driving steadily ahead, following no daily mileage schedule, the longing to be up and going, to see what lay beyond the distant hills, was still with us. Such a feeling of subordination of one's self there had been in the eagerness to be on! The only important thing was to say, "By nightfall we shall be at the foot of that purple mountain ahead."

We had begun to feel that we would never want to stop, but would like to drive on and on, forever. After the many beautiful towns we had been through, to stop here, of all places, seemed to bring us up with a shock to the unpleasant realization that our traveling was at an end. We had become accustomed to viewing all places with the critical eyes of travelers, and what lay before us was far from attractive. This was Dardanelle.

One short dirty main street, with papers lying about. The barber busily sweeping the hair trimmings from his floor across the sidewalk and into the street where Negroes straggled by. Small stores where listless men and boys lounged in the entrances. The unpaved street was the highway for farm wagons and automobiles, and seemed to be composed of nothing but soft, deep, dry soil that sent up a cloud of dust with every passing vehicle. When it rained, as we found out later, a black sticky mud formed a mire and each car and wagon churned along through deep ruts filled with water. Nevertheless I saw women in white summer dresses and white shoes who were picking their way daintily along among the debris, obviously accepting the dirt as a matter of course. Tufts of loose gray cotton, heavy with dust, were blown against the sidewalks, and grimy bits of fluff were caught in the cracks of the screens at the store doors and in the spokes of the wheels of the heavy mule-drawn wagons that were tied to hitching posts along the street.

"Let's be going," I suggested at last. Wayne and Henry had returned to the car. They had been sauntering up and down the street trying to find someone who would remember them. Now they stood hesitating and silent, and I turned my head so they would not be able to see the shock and dull despair on my face. My nerves were quivering. I knew that if I spoke I would break down completely. This was their old home town, and though they remembered it well, yet truly they must have thought that the years that had passed would surely have brought changes and improvements. I didn't want to hear the things I felt they might say, in attempting to apologize or explain; I wanted to spare them that. I knew it would be more than I could bear, for we were now entirely committed to this situation.

As we were about to drive farther on down the street, two elderly men came out of the bank, and recognizing Wayne and Henry, greeted them warmly and stopped and chatted eagerly, evidently glad to meet again these sons of a man who had been a lifelong friend. They were most courteous in welcoming the bewildered young daughter-in-law to their little town. They pleased Jimmy and Donald, too, with their kind words, and promised to see that we met their families. I watched the people passing by while I listened to the conversation. I was amazed to hear them speak with such evident pride in the town. Why, I simply couldn't understand. Even now, in kindest judgment, all I can say is that to them, it stood for all that can be represented in the word "home."

While I sat there, I went back in my mind over all the history of my husband's family that I could recall. Invariably the talk at meal times would veer around to stories of this or that relative in Arkansas, some of whom had the most intriguing names—Candace, Delilah, Crispinius. The continued recounting of events, with greatest accuracy as to time and place and exact kinship, amused me, bored me, and amazed me. When all the relatives had been gathered together for a Sunday dinner the same kind of talk went on and on, only then it became more stimulating and more exciting. My amazement was caused by the extreme kindness and veneration with which all their ancestors were remembered. I was amused at the exactness of even trivial details and circumstances and the real insight that was displayed in describing people and events, or in the summing up of someone's traits of character. Looking back now, it seems to me that always, day after day, year after year, the common subjects of conversation so avidly entered into were of the past, seldom of the future nor even of the present. That was what bored me and made me feel as though I were living alongside of this family, not completely with them. Any attempts I made to change the channel of thought or make teasing remarks about their ancestor worship were politely ignored, and I soon learned that among all the Southerners I met, there was a fellowship that began with one's immediate family and ended with a staunch loyalty that embraced everything that pertained to the South.

Wayne and Henry had been in their teens when they last lived here. Their father, although he owned two cotton farms, lived in town, where he was a deputy sheriff.

Going further back, I recalled that Wayne's grandmother, a widow, had held in her name the dower rights to the big farm where she and her husband had settled years ago when they first entered Arkansas from Virginia. This was out in the rolling hill country, toward Fort Smith. When her youngest son, who was Wayne's father, married Nancy, the tall, dark, beautiful girl who had always lived in the "sticks," as they called the hills, he and his wife received by will the ownership to the farm lands; and they agreed to take care of the mother the rest of her life. Her other children agreed to this because their brother was to take care of "Ma."

Later on, she and Wayne's parents decided to sell the hill farm and they bought a big cotton farm down along the Arkansas River. With it they received a warranty deed from the owner. They moved into Dardanelle to live, where Wayne and Henry were born.

The river-bottom farm was worked by the customary share-crop sys-tem—a tenant leasing the land and hiring what workers were needed in the crops. Wayne told me that he could remember going down often to the farm when his grandmother was supervising the planting of the great line of pecan trees. Wayne's father made an added purchase of a piece of land down the river on the Petit Jean, a small tributary of the Arkansas. This was not only fine virgin soil, excellent for cotton, but there also was a wonderful stand of oak and hickory.

Wayne was just ready to enter high school when his family moved to California where his married sister and elder brother were living. Through the years, his father often went back to Arkansas to lease the farms to responsible people, see to the making of necessary repairs and visit his old friends. In California, the grandmother died, honored and much respected. Whether she had ever longed to return to her old home, I never heard. She had been dead several years before I met Wayne and we were married. Wayne had three years at college after leaving high school, and then his father was killed in an accident, when Jimmy was about four years old. We were all living in the country raising chickens and hogs and turkeys and apples.

The Arkansas property was now in the hands of Wayne's mother,

Grandma, as we all called her. She continued to collect the yearly rentals through her oldest son, Frank, who lived near Little Rock and who saw to the renting of the cotton farms for her by the same old share-crop system—giving the tenant-farmer a third of the cotton and a fourth of the corn.

No doubt this arrangement would have continued for many years had it not been that at this time it was discovered that the man who gave the warranty deed to Wayne's father had not been the legal owner of the land. A claimant in Arkansas had found out by some chance that at the death of a certain mother and child in the hospital, the child had lived a few minutes longer than the mother, according to the records, so the line of inheritance was thrown to this claimant. Now he was demanding through the courts an undivided one-fourth of the upper farm, together with any accretions the river may have added. Apparently he was the only one left of four possible heirs.

The court order that had been sent to Grandma and her children to come or be represented to defend their title to the land was both a shock and a challenge. My own mother, when all this was explained to her, advised that the stand be taken that ownership of the land had been held, and taxes paid, for twenty-seven years, to the open knowledge of all. In any case, the family decided that they would meet the claimant and if necessary fight the case to the State Supreme Court. Compromise— never!

At first our plan had been for Grandma and Wayne and Henry to go to Arkansas. The lawyer we consulted assured us that the lawsuit should only take about three months at most, as it was such a simple case. I was to stay behind in California until their return.

I knew Grandma was homesick and lonesome for her old friends, especially so since the death of Grandpa. Many times she talked to me of the idea of returning to Arkansas to farm, where there would be no rent to pay on the land. She kept telling me how fine everything was. How rich and deep were the river-bottom lands, and how productive. She scoffed at any mention of malaria or poor schools or lack of opportunities. To her mind, no peaches were equal to the Elbertas raised there in the hills, no strawberries had the flavor of those she had known, no apples a finer taste, no water was sweeter or purer, and no friends were truer or kinder than those she had left behind. She now accepted with eagerness this opportunity to see again the land of her birth and the places and people she loved, for the lawsuit gave an added force to all her arguments, and of course, someone had to go anyway.

Finally it was decided that Jimmy and Donald and I were to go, for Grandma had recently had a spell of sickness and did not think she would be able to stand the long trip by automobile. She would come later by train when we sent for her, after we had arrived and were settled.

I had been a silent listener, day after day, to discussions of the possibilities that lay in the raising of dairy cattle and hogs in the South, instead of the usual rural cotton crop, and of the wonders that could be worked by rotating crops and raising alfalfa on the rich sandy soil and black gumbo lands that for generations had been only surface-scratched. And it seemed such a propitious time now, with prices so high. It was 1920, and cotton was striking the ceiling at forty cents. It was the post-war boom and hopes were high and ambition rampant, for none foresaw the immediate collapse that lay ahead.

Wayne and Henry and Grandma never openly and plainly declared definite intention of staying for any length of time in Arkansas to try these new ideas in farming in the land of cotton, but I saw that their hearts were now indeed turned back to Dixie.

One day when the Ford was being overhauled in the back yard, Henry looked thoughtfully at me and then turned to Wayne and said, "Have you told her what you would do if she doesn't like the place?"

"Well," Wayne answered, "I haven't yet, but I will now." Then turning to me, he said, rather solemnly I thought, "If you are not happy there, and decide you don't want to stay, we'll come right back here. After all, we are going because it is necessary to get this lawsuit settled. That won't take long. We can make a lot of improvements. The land should be surveyed. No doubt the houses and fences need repairing. We can leave the land in better condition to rent again, anyway. We might look it over and decide to give it a trial ourselves. We'll decide that later. But just remember I'll come back here with you if you don't like it. You know Henry and I never lived on the farm ourselves. Neither did Pa. It's been a long time since we were there. It will be up to you."

At first I had been overwhelmed at the thought of being separated from Wayne even for the few necessary months that the lawsuit would require. Then as plans were made to include the children and me in the trip to Arkansas, a curious excitement filled my days. It seemed that at no time were the words spoken, "We are going to move to Arkansas to live." It seemed that such a plain statement was avoided. We were really getting ready to move there to live, yet somehow I felt that we were in the hands of fate.

I could hardly comprehend all that such a long journey, camping out all the way, would mean, but the thought of it was exhilarating. We would cross the Sierras and the Rocky Mountains and see Nevada and Utah and Colorado and Kansas, parts of the country that had been only names to me. I gradually accepted the thought of leaving California. I packed up our possessions and watched some of our furniture being crated for shipment in a kind of unthinking daze, buoyed up by anticipation.

I felt that Wayne's promise was in some way an attempt at a fair bargain and seemed to relieve his mind greatly, and I felt that I would rather rely on that than to stay behind in California. Yet it left me wondering. Now here we were at last.

When we inquired about houses to rent, we were told that the hottest months of the summer were spent, by many, up on Mount Nebo. There was one empty house in town that we could have whenever we wanted it. Some of the merchants, they said, sent their families up to the mountain for the entire summer, then they themselves would drive up after the stores closed in the afternoon, returning to town again in the coolness of the early morning, and again go up and stay over the weekends, thus escaping most of the murky heat of the river town. We decided to drive up on the mountain and rent a cabin.

Mount Nebo was about eight miles from Dardanelle, in the Magazine Mountains, which is that spur of the Ozarks running through the western part of Arkansas and on up into Missouri.

We were soon out of town. The Arkansas River could not be seen from the main street, for a row of stores and warehouses hid it from view, and our road soon left the town and wound in and out near scattered cotton farms and into the low-lying hills. It was a dirt road, studded with deep-set rocks worn smoother by the wheels and tires of wagons and cars. It twisted and turned, rising ever higher, the color of the soil gradually changing from the brown soil of the lowlands to the brick red color of the lower hill farms.

The bushes and trees looked strange to me. There were chinkapins, hickorys, straggly thin pines, and heavy clumps of sumac bushes, and I recognized the blackberry vines and huckleberry bushes; still there were many grasses and weeds that were unfamiliar.

The engine boiled frequently as the car labored up the grade. We would draw off to one side of the narrow roadway, to allow the engine to cool off while we filled the tank with water from the five-gallon can we carried.

The road followed the mountainside, and though fairly smooth, it was very steep. Other motorists had thoughtfully left large boulders along the roadside, which we often found a real necessity. When we had to stop, the rocks were placed behind the rear wheels to hold the car on the steeper slopes. Many of the cars that passed us had a small log tied handily to the back axle for the same purpose.

Finally we reached the bench. This was a natural shelf about halfway to the top of the mountain, running nearly all the way around, and wide enough, we saw, for a roadway with space occasionally for small cabins. We drove as far as we could around this bench, but found that there were no empty houses. Everywhere was abundant shade and we passed many springs of water. These were iron and sulfur springs, Wayne told me. This was such a beautiful spot that we were disappointed in not finding a place here where we could stay. Wayne and Henry took delight, at every turn of the road, in pointing out to me the old familiar springs and the large odd-shaped rocks that they remembered, for they used to come here often when they were young.

From the bench on up to the top of the mountain the going was easier. I spied many bunches of dark grapes growing on thick matted vines that sprawled over the fallen rocks. These were called Muscadines. They were tough-skinned and far surpassing in flavor any grapes I had ever tasted.

Growing thickly among the fallen logs and thick timber were wild roses, and blackberry bushes, heavy with berries. There were no trails through these woods. The thick underbrush of sumac, the low-branched chinkapin, heavy with small round edible nuts, the small scraggy pines, were tangled with a heavy growth of ferns and vines, making it difficult, one could see, to even think of wandering off the road.

In the hot, heavy air, our nostrils were filled with a combination of odors—dust from the roadbed, the spicy scent of the wild roses, and the potpourri of pine, ripe berries, and wet ferns and moss.

Mount Nebo itself was won, at last, and we were at our journey's end.

Here was a real table-land. It was nearly level, about a mile wide and two miles long and more than a thousand feet high, with trees and bushes, grass and wildflowers, growing in profusion, and fallen dead branches everywhere.

Most of the houses were grouped at one end of this mesa, although a few were scattered here and there through the woods. The houses, with but a few exceptions, were gray and unpainted, most of them consisting of two or three rooms with tiny fenced-in yards. Narrow roads wound back and forth through the woods, each house almost entirely hidden among the trees and bushes.

We rented the only empty house we could find, a three-room cabin that stood with two others off to one side of the mountain. All around us grew the small post-oak and small pine trees, and the ground was thickly carpeted with soft grass and wild strawberry and blackberry vines. Nearby lived our landlord, Turner, with his wife, and his daughter, Hallie. My husband remembered them well, for they had been neighbors years ago. Turner was a pink-faced, pompous old man, with a very benevolent attitude toward everybody. He had been a school teacher for years, and was one of those honest, old-fashioned, proud men who, having developed a fine philosophy of living for themselves, are always to be found reacting to others with the kindly benevolent spirit which they carry in their own minds. Miss Hallie taught school in town, and I found both she and her mother kind, sympathetic, and friendly. In the other house lived a widow with her married daughter and granddaughter from Little Rock. Miss Edith, as the young mother was called by everyone, was a singer who had been accustomed to entertaining the public. We were particularly drawn to her, and she in return spent much of her time with us while we stayed on Nebo.

Our house was like most of the others on the mountain, a little unpainted Southern house, with a wide hallway, open each end, running through the center from front to back. There was a narrow porch across the front. This hallway was for coolness in summer, and a breeze seemed always to draw through the opening no matter how hot the day. In winter it would be just that much colder, but then nobody stayed on Nebo in winter.

A great surprise awaited us, for Grandma, irrepressible always and becoming impatient of waiting behind, had stolen a march on us. She was already happily and comfortably settled into the mountain life.

It was such a great pleasure to me after our weeks of camping, to have a house in which to unpack, and once again to hang our clothes on hooks and put our food away on shelves. The house had been closed since the previous summer, and it was full of cobwebs and dried leaves and spiders and dust, but it was a joy to sweep and clean it and gather light brush from the surrounding woods for the little cookstove. This activity was doubly welcome after the long strain of sitting in the car. Jimmy and Donald were as delighted as myself. We carried the bedding, cooking utensils, bags and boxes into the house, while Wayne and Henry began some repairs on the car.

Our equipment for housekeeping was so meager—just our camping outfit—and we had become so accustomed to quite a methodical system of living, that we soon had leisure for a drive around the mountain top.

That afternoon we spent visiting with the Turners, and Miss Edith's family in the next house. We were looked upon as company by all on the mountain, and made to feel very welcome. Such chats of old times as I heard. Such questions about our trip and about California, that goal of all their dreams.

We ate home-made cake, drank cold buttermilk, and enjoyed the melons that had been brought up from the gardens in town. It was a pleasure, too, to turn the crank of the well that served all three houses, and draw up the long galvanized tube of wonderful cold mountain water. I was surprised to find that even though this plateau was so high, yet there were deep wells close by most of the houses.

Many of the houses had a part of the porch screened off with sheets to give added sleeping space, and here I saw huge, puffy feather-beds, made up and covered with spreads of unbleached cotton, all very elaborately embroidered in gay colors.

After supper—supper at night, dinner at noon—we walked out to one end of the mountain where one could overlook the town and river. Above was a round golden moon. Below, far, far away, we saw the soft twinkling lights of the little town across the river. The river itself gleamed under the moonlight like a silver band, appearing and disappearing as it wound around the bends far down the valley. Soon there were eight of us, strolling along the road in the moonlight—for folks we met, out walking too, hailed us, talked a bit, and joined us in our walk. We ate a watermelon by the side of the road on a flat rock, and all together sang, "Long, Long Trail A-winding." Although it grew late, yet the air was still warm and balmy, with swarms of fireflies in all the open spaces. The silence was accented by the zee-ing of the crickets, a sound that almost overwhelmed our unaccustomed ears.

Next day I stared in amazement as Mrs. Turner introduced me to the process of clothes washing. Many of the clothes which we had worn on our trip, and washed in Lake Tahoe and in numerous rivers and creeks across the country, were gray with grime, but even so, I feared they would be made worse instead of better after being put into the huge black iron pot that was being made ready, for I was accustomed to boiling clothes in a copper-bottomed wash boiler, and washing in stationary wash tubs. I failed to appreciate at first that the black iron pot, which was so rough and sooty outside, was as clean inside as my own wash boiler.

The wash pot was propped up, each of the three short black legs thrust into a discarded lard bucket, and was filled with the soft well water. A big chunk of home-made lye soap was then shaved into this. Broken branches were used for firewood. Light limbs were always falling from the trees, owing to the sudden heavy winds and rains. These sticks had to be continually pushed beneath the pot as the ends burned off, and the water was stirred as the soap melted. Then into this boiling mass, the dry, soiled white clothes were dropped, and were kept poked down with a discarded broom handle. All during the washing process this work went on, replenishing the wood which burned so quickly, and poking down the heaving clothes. At last the clothes were lifted out on the end of the wash-stick and plopped into a waiting tub of water. From this they were wrung out into another tub of water for rinsing, then into the tub of blueing water, then to the starch, and so to the line.

I was entranced with this strange method of washing and eagerly helped, surprised to see the clothes emerge so snowy white. The same process was repeated for the colored clothes, and then the overalls and socks were worked through. Only a few pieces were hand-rubbed, as the boiling process was usually sufficient. This method of washing was used not only here on the mountain, but, as I later found, down in the town as well. Nowhere did I ever see a wash-boiler in use, and I never saw the clothes boiled indoors on a stove. In all seasons, the women would wait, until, weather permitting, they could make a fire out in the yard under the big black pot.

We found that it was the custom here, in the early evening, for people to gather at Sunset Rock, a huge rocky point that jutted out over the valley to the west where the sun went down. The sunsets were glorious. Away in the heavens would be piled soft turrets of foamy cloud, mound on mound, touched with a beautiful pink, or deep rose, or the sheen of pearl. We sat on fallen logs placed so as to face the wonderful sky, and often, just before the sun finally dropped, leaving the ever-changing afterglow, Miss Edith would step graciously forward. How we enjoyed her singing. There was a quality in her rich round tones like that of a broken-hearted bird.

"Land of the Sky-Blue Water," "Roses of Picardy," and the "Indian Love Song" were often asked for and were always received with smiles and nods from the appreciative audience. This gathering was like a nightly ritual, for the songs she sang were varied, and the glorious sunsets were painted anew each evening, fading out at last into the darker shades of purple and orange clouds as the darkness deepened.

At times, however, the valley below us lay shrouded in fog. We were so far above it that we could look down and see the out-flung blanket of gray and white, dark-hollowed and awe-inspiring, yet the stars above seemed so close to us and so sparkling.

Southern nights under the moon! One evening a group of us gathered on a rocky point across the table-land from Sunset Rock. This place was known as Lover's Leap. A breeze, but the air comfortably warm. The moonlight so bright one could easily read. We lit a small fire of pine knots so we could sit and watch the flames, and Edith, who knew so many songs, sang to us for hours. Sometimes we played cards on a blanket spread on the big flat rock. This rocky crag jutted out over a deep canyon, extending free and away from the main mountain, and giving one the unforgettable sensation of space and height. We were told of an old Indian legend that from this point, in bygone days, had leaped broken-hearted lovers.

Nebo was indeed an ideal place to rest. Everyone lived in a very simple manner, lazying around, day after day. There was one small boarding house. Mail was brought up from town every day by someone; and one family kept some cows on the mountain and supplied milk and butter. The women, who in town were busy with social duties and church affairs, now seemed relieved and free and except for their household tasks seemed to do nothing. A few, I noticed, had a Negro woman with them to do the housework. Some of the women drove down to their homes quite often, but they were always glad to escape from the humid summer weather in the town back again to the breezes on Mount Nebo.

But, oh, the detailed knowledge they all had of everyone's private affairs. It amazed me. They were alert at the mere mention of anyone's name, eager to con over all the family history of that person, to trace their family tree and prove their kinship with someone they knew. Each time I was introduced, I was soon asked what my name had been before I was married. That puzzled me, since I was so completely a stranger to them all, but once I answered back and wanted to know how they could be at all interested since they couldn't possibly have known my people, who incidentally, were from the Highland Grants of Scotland. Maybe I wasn't overly friendly, but I felt their polite questions were entirely too one-sided.

Long, lazy days we spent exploring Buzzard's Cave, a deep ravine among the broken rocks at one side of the mountain, or we would visit the iron and sulfur springs on the "bench," where the huckleberries and blackberries grew rank. Here, years ago, a high-roofed shelter with open sides and rustic seats had been built near the springs, but now this had fallen into decay. Up on the higher land stood an old deserted hotel, which had once been the popular headquarters for Normal School and Institutes, but this, too, was now only an old wreck. We explored this building with interest, finding pieces of old-fashioned furniture and queer old dishes. Mount Nebo had once been the summer gathering place of "Society," and had drawn people from many miles around. Up there, too, at one end of the mountain, still stood a big circular pavilion which had once been a public dance hall. It was roofed, but was open on all sides, and commanded a magnificent view of the valley below.

From here we could trace the course of the Arkansas River down the valley for many miles. Here and there the river was hidden by trees, or disappeared where the high banks hid it from view. The river was given a fresh impetus where it cut through the chain of rock just above Dardanelle, and the force of water was thrown farther down to the opposite side, carving out Holly Bend. Then the swirling eddy, on its outward curve, again crossed the river to the right side and cut deeply into the shore line there. This zig-zag was continued back and forth all down the valley.

The hills of the Petit Jean were toward the south, and the neighboring town was across the river to the north.

Away from the river, and stretching back from the town, the cultivated fields of corn and the green fields of cotton appeared to us up on the mountain in an immense panorama, a variegated checkerboard. Below, in the lesser foothills, one could plainly see the farm houses and barns, and clearly distinguish the cotton rows.

Jimmy and Donald were eager to follow Wayne and me as we walked the mountain paths, exploring the connecting trails and finding the ripe grapes among the rocks. An unexciting game of croquet was the only definite form of amusement. This was played in the late afternoons on a cleared spot under the trees. There could be no picnicking or lounging in the woods here on account of the many insects. The ants, large black fellows, were everywhere. The devil horse, a beastie looking as though he were a miniature saw-horse crudely fashioned of tiny twigs, sat on the tree branches and had to be carefully avoided. This insect was supposed to be deadly poisonous. Each branch and leaf harbored chiggers. They were like minute grains of cayenne pepper, and nothing but liberal applications of kerosene seemed to ease the torture of the itching they caused. I was shown a small green snake that lay along one of the smaller branches over our heads! Sometimes we rested on a bared rock or ventured to sit on a fallen log that had been carefully examined, but usually we returned from our walks tired and hot, to lie on the cots on our porch or to visit among the neighbors, always looking for a shady spot where the breeze would help to make breathing easier.

One afternoon I had my first experience with chills and fever. Before the doctor arrived in the evening, I was out of my head. My fever was very high and my neighbors had given me large doses of quinine and calomel and had applied a poultice of peach leaves to my stomach. I heard one of them say that she thought I had "pernicious malaria." The doctor prescribed quinine, to be taken regularly for several days. He said no doubt the mosquitoes that had bitten me as we camped out in Kansas and Oklahoma had caused this onset of malaria, but that it would be wise to keep a supply of quinine always on hand, and he advised all of us to begin to take regular doses every third day.

Wayne and Henry had gone up to Little Rock to visit their brother; so during the long nights, while my fever lasted, I kept the lamp burning and took the quinine capsules regularly. The light attracted all the daddy-long-legs in the house, ugly creatures on stilt-like legs with round staring eyes that crawled up onto the chair by the lamp, onto the bed, and onto my face. Ugh!

As I began to feel better, I spent most of my time resting out on the porch, but I was overwhelmed by a feeling of lassitude and I seemed to shrink from meeting any of the other people on the mountain. In the early evening, the fireflies came out by the millions, and the air above would be thick with bullbats swooping to catch them. The locusts in the trees kept up their never-ceasing "zee-zee," and the katydids, which actually seemed to say, "She did, she did, Katy did, she did, Katy did," went on and on and on, or came to such a sudden pause that one thought they were surely through for the night, but back again would come in heavier volume, "Katy did, she did, Katy did!"

On the oak leaves we found honey-dew. It glistened in the morning sunlight like tiny rain drops—and was like a small drop of clear, sweet syrup. I tasted it many times, always with amazement. But, there it was— honey-dew. How or why I could not find out.

And with it all I was homesick. Sick for home; for faces that were expressive, for hearts responsive, for understanding of my bewildered, unsettled spirit. I should have been braver. The eager attitude of adventuring with which I had started on our long trek should have carried me through the newness and the strangeness; but it didn't, and my heart was a physical ache within me.

Sometimes the pain of homesickness was a daily heaviness that was so acute that I would look over the out-flung river lands below and deny and deny that it was a glorious view or a wonderful panorama. I even hated the strange weeds and the different grasses and wildflowers. I hated even the way the wind blew.

Oak woods, mockingbirds, fireflies, sunsets, heat lightning—all interesting and beautiful—but try as I would I could not get one speck of liking for any of it. I would lie at night out on the porch where we slept, and picture the ocean, the beach, the hills I knew, the town, the people, the flowers, and the fruit.

The many loved and familiar scenes I had left behind seemed constantly to sweep before my eyes and to fill every waking hour. All the honest-to-goodness beauty of California was intensified in my memory until I felt as though I were walking on air, but it only made my actual surroundings harder to bear. Homesickness is a real and dreadful disease!

My great delight was in the rainstorms. It had been carefully explained to me that it was the lightning that was dangerous and the thunder only noisy; that chickens were often electrocuted when they roosted on the wire fences; and that I should never stand in an open door or at an open window or under a tree during a thunder storm. Yet it was the heavy blasting of the thunder that I found hardest to bear as I braced my nerves against the great crashes that echoed from the clouds. Of course the lightning was terrifying enough, and from one storm to another I found it hard to believe that an amazing spectacle could develop from those first intermittent shimmers of light, the great zig-zags of fire that leaped the clouds.

Yet I grew weary of withdrawing into the house during these summer storms, for the rooms became dreary and dark. There was a terrible fascination for me in the eye-searing flashes of lightning and the surge of the water as the rain swept across the low roofs and dashed on the window panes while the ground outside was inches deep in water, with big raindrops splashing a foot high.

Once I took off my shoes and stockings and went out between showers down the road alone to Lover's Leap rock. I stepped along the tiny trail, my feet pattering flatly on the clean washed earth, around and below the rock to where it ended at a small grotto directly beneath the high out-jutting pile of stone. It was dry here, and large enough for two or three people to sit comfortably. Below, deep in the gorge, and spreading far across to the next mountain, lay the bank of gathering mist, but this quickly changed to gray sheeted rain, as a fresh storm broke above. The trees shivered and swayed; boughs snapped. How scared I was, yet how thrilled! Suddenly the clouds thickened, the light faded and a thick blackness fell like a curtain. I couldn't leave until the rain stopped again, so there I sat alone in the cave, tight up against the rock wall, watching out over the tree tops, half afraid, half exultant. Such a deluge of water I had never seen before. The rain fell in sheets. I felt the earth under me quiver with the volleys of thunder which burst like heavy artillery, and the lightning pierced through the clouds like flaming swords.

But the clouds lifted at last, the thunder rumbled off down the river valley, and the sun came out. So back to the house I went, singing at the top of my voice with the very joy of living, and breathing deeply of the fresh cool air.

Along this road, seldom traveled, were many low places filled with water. The earth was a pinky-brown color and of a clay-like consistency. I waded into one of these puddles, holding my dress above my knees and paddling my feet, working the mud into a soft mass that felt delightful between my toes. A faint memory stirred me of some far-off childhood day when I had waded in a puddle in the rain. I stood there with the mud oozing between my toes, working my feet up and down and trying to out-sing the birds in my exultation.

Hearing voices, I looked around, to see Miss Hallie and a friend of Miss Edith's, a wealthy society woman from Little Rock, coming down the road. They stopped with exclamations of astonishment. I thought they might be patronizing about my predicament, as it now began to appear to me, so I consciously allowed myself to fairly exude my happiness and joy. I talked to them of the freedom of wading in puddles of mud, and told them of my adventure in being caught in the cave during the storm. To my delight they were so pleased with it all that they removed their slippers and stockings, and there we stood, three grown women, with our skirts held high, our legs spattered with the red mud and our feet buried in the soft ooze. Oh, how we laughed—just like Riley's "Fool Young 'Uns"—and how we enjoyed it all! Then a horseman appeared, his big horse splashing the bushes with mud on either side of the road as he pranced along. Miss Hallie whispered to me that this man was the town marshal, and she turned her back and hung her head, hoping he would not recognize her. He rode past us, then turned and looked at us, and throwing back his head, laughed and laughed.

Naturally, being a greenhorn, I had to have the old persimmon joke played on me. Miss Edith, on one of our walks around the mountain, came to a persimmon tree and persuaded me to climb up and get some of the fruit which hung like large green plums. She gave me a boost and I managed to get a toe hold on a lower limb, and picked two persimmons. Upon reaching the ground I was easily induced to try the fruit, which I did, taking a bite and commencing to chew all at once. How she enjoyed the bitter expressions on my face. My mouth was so puckered I could not speak. But I had been initiated. Later I was simpleton enough to chew some bark of the "slippery elm" and I "spit cotton," as they called it, for some time afterward.

In the old one-room schoolhouse, where Sunday school was held every Sunday, was an old piano. Sometimes a crowd would gather there in the afternoons during the week and one of the girls would play, while Miss Edith sang for us.

Miss Edith was about twenty-nine years old, dark, beautiful, and inclined to be stout. Her days were filled, when she was not sleeping, with aimless walks around the mountain or down to the bench, all in an effort to reduce, she said, but as she insisted upon wearing high French-heeled slippers and a tight corset at all times, she and I had many arguments as to the futility of her exercises. However, she had such a sweet disposition and was altogether so charming and cultured and entertaining that she was the center of our little group and popular wherever she went.

One evening, at her suggestion, we planned a ten-mile hike to the higher mountain top that we could see toward the southwest. We were all up early, for we planned to start soon after breakfast so as to be back before dark.

Even that early in the morning, with the locusts whirring in the trees, and the katydids chirping shrilly, there seemed heaviness in the air. The heat arose all around with a blasting force as the hot beams darted through the haze that hung over the dusty yards and roads.

The clouds soon thickened and hung lower. The sky seemed to be composed of layers of gray of varying shades, with broken places, enabling one to look beyond and beyond again, and see deeper blue-gray against the far-off blackness. Then it rained. The trip had to be put off. I was disappointed and watched the rain fall with great impatience. The large drops skipped across the puddles that soon formed in every low place, and the water ran in streams in every direction. However, toward noon the rain lessened, and the sunshine was so strong again that the surface water was gone and the trees and bushes gleamed and sparkled.

Miss Edith and I had made such nuisances of ourselves all morning, bewailing the delay and resenting the postponement of the trip, that Wayne and Henry finally gave in to us, and we all started off, late as it was, planning to spend the night and return the next day. Wayne's brother Frank, visiting us from Little Rock, was delighted to join us and he soon assumed command of the hike. The men strapped up our blankets while Miss Edith and I prepared food to take along. Since we had decided not to go until the next day, when we saw the heavy rainstorm, we had made no further preparations, and there was very little food that was ready to take on an overnight trip. But we wrapped up cold boiled potatoes, biscuits, jelly, milk, bacon, coffee, oatmeal, salt, and sugar. This odd assortment of food we kept from sight, for fear the trip would be put off until a later time when more suitable provisions could be made ready. We were restless just like two children, tired of the daily idleness on the mountain and anxious to start on the trip.

Miss Edith wore a black satin dress and tripped along on her high heels. I had dug out my khaki pants, leggings, and heavy shoes. Jimmy and Donald were left in the care of Grandma, and away we went.

Frank and Henry set a good pace as they realized what a long climb was ahead of us. Miss Edith and I frisked along singing and chattering. We were soon at the edge of our plateau and descended the rough steep path that dropped down past the bench into the gorge. Here we followed a logging road for a while, then began the long climb up the farther side of the gorge to the top of the other mountain, where we were to spend the night.

The sun was hot as we went on and on, following the rocky, winding road. Wild roses, frail and pink and strongly aromatic in the sun, grew on every side. Goldenrod, mullen, milkweed, and tangles of dusty brambles filled all the spaces between the chinkapins and the hazel bushes, and tall sycamores crowded right up to the narrow path. We saw muscadines hanging in dark clusters and eagerly gathered the pungent grapes. Their thick foliage served as background for the blood-red splashes of the sumac berries. I could well believe that these woods were bowers of beauty in the springtime, with the bright blossoms of the red haw, redbud, and crab apple, and the white blooms of the dogwood, wild plum, cherry, and locust that grew on the hillsides.

As evening came on, the road grew fainter and we had a hard time keeping to the trail through the woods. The going was rough and at all times steep. The men forged ahead, but try as we would to keep up, they often had to wait for us, as more and more we found ourselves left behind. Sometimes, we would all walk along in a group, but again and again there Edith and I would be, standing stock still, aching and quivering from the continuous hard exercise to which we were not accustomed, and gazing hopelessly upward at the trail that still twisted ahead.

We realized, however, that we had to get up on top of the mountain before we dared stop, for there was no clearing along the way. Frank told us of a good spring of water and an old house where we could camp. Frank and Henry knew this mountain well, for they had been here often, years ago.

We had no water with us, and so far had passed none. Mile after mile. Poor Edith. She was game, though, and never complained. My legs ached and finally got so numb that down I sat for a while vowing I couldn't go another step. Yet on we went. As it grew dark, we had to walk in single file like Indians, keeping close together on the narrow path through the thick underbrush. Along our way we saw phosphorus glowing from the rotten logs. This added to the sense of vague fear and strangeness. We were in a very quiet world, dark and starless.

Finally Henry, who was ahead, called out that we had arrived at the spring, and we hastened forward. Edith and I were cold, and huddled together on a log while the men investigated the surroundings and cleared away the green scum from the top of the water. It was a small pool between two old tree stumps. They cleared the spring by the light of matches. I was so thirsty that the thought of the green slimy scum didn't bother me at all, and the water was really fine. There was no dry wood handy, but Frank, who had been scouting, said the house was just a little bit further on, so he went ahead to find pine knots and get a fire started. As Edith and I sat there, huddled in the dark, we were so hungry that we reached for the nearest package and dug out the cold boiled potatoes and ate them, skins and all.

Soon we were all gathered around a campfire made of pine knots and broken branches and small logs. We didn't really need a huge fire, but as the wood was so plentiful we indulged our pleasure in a roaring bonfire that lit up the forest for yards around. Here we gathered for our supper. Biscuits and coffee!

We found that the old deserted log house was floorless, and the big thick planks which had boarded the inside walls were nearly all gone. The roof was fallen in and the outsides were rotten. Between the outer logs and the inner boards that still remained were thousands of empty nutshells—from hickory nuts, walnuts, and chinkapins. (The chinkapin was a dwarf chestnut with edible nuts.) This place was used, apparently, as a store house by the squirrels.

Dreading snakes, which were sure to be all around here, we decided to sleep out by the fire, and worked for more than an hour making wonderful pine-bough couches. We were so tired we were a little hysterical. My memory of that night is of gay talk and lots of singing and much coffee drinking and peals of laughter—yet we all managed to get some sleep. I would doze off, only to sit up in a daze and see the three brothers earnestly engaged in prodding the fire and discussing plans for farming, while Edith lay curled up beside me fast asleep. Once, only Edith and I were awake together. We spent the time talking and singing. I tried to think of some song she did not know, and failed, for she seemed familiar with most of the old songs I had been accustomed to at home. Those hours by the firelight, enchanted by her wonderful voice, are a precious memory now.

What the men thought of our culinary efforts I won't attempt to put into words. They never tired of making fun of the food supply we had carried all those miles. For breakfast we fried the bacon, and then cooked the oatmeal in the same small skillet. We had coffee with canned milk and sugar, but the biscuits were pitifully few. At noon we had only a few biscuits with bacon, and coffee without cream, as the can was empty too soon.

Until the middle of the afternoon we wandered over this mountain, which was flat table-land like Nebo, but with many more rocks. Pines and oaks and scraggly chinkapins grew close together, with occasional open clearings where lush grass grew among the fallen and rotten tree trunks. Blue jays scolded in the scaly sycamores and swooped through the clearings, eagles sailed importantly high overhead. Large hawks glided on stately wings above us, and the air seemed full of bees. Frank said that long ago several families had lived on this mountain top. He used to know an old man who kept goats here and who had planted the orchard. Now all that we found were a few gnarled apple trees by the ruins of one old house. We noticed several kinds of oaks here, the overcup and red and white and post-oak, and found also the pig-nut hickory, yellow pines, aspens, maples, and sweet gums. The dogwood leaves were turning crimson above their white undersides and the scarlet berries were so beautiful that we tried to gather some of them, but they seemed to grow in the steepest places.

It had been our interest in the stories we had heard of buried treasure on this mountain that had given an added zest to the trip. We went to a rocky broken slide where the mountain fell away into a deep gorge and saw a big block of stone that had fallen down and wedged itself in what was supposed to be the cave where an iron pot of gold had been buried during Civil War days. In proof of this, Edith and I helped the men pace off the distance between some black walnut trees that we found growing at regular distances apart, forming the shape of a huge diamond, and pointing directly to this cave. These special trees, forming a rough arrow, were part of the legend of the treasure.

We searched a long time to find a certain tree that was said to have directions carved on it, all pointing to the buried gold, but an old grown-over arrow carved in the bark of an oak, and a few vague scars, were all we discovered.

I was quite disappointed when all our efforts brought to light only these indications that something might have been hidden here at one time, and as the afternoon sun was lowering, we had to start back home.

That long trip back! It was long after dark when at last we arrived on the top of Nebo again. I have a never-fading memory of Edith, dozens of times, sitting beside the road refusing to budge, and begging us to go on and leave her as she only wanted to die. We took turns coaxing and encouraging her, though in truth it was now almost impossibility for her to walk, yet we could not really assist her as she was so much heavier than any of us. So she literally staggered on, in her French heels and tight clothes, suffering every step of the way.

Next day, after much sleep and rest, we could laugh at our experiences, but we never told the neighbors of our foolish preparations for that trip.

2. Preview

THE SKY WAS OVERCAST, the heat intense, and the air sticky and close, the day I went with Wayne to visit our farm for the first time. Jimmy and Donald were very eager, also, to see this much-talked-of place. They were weary of the quiet sameness and mild activity of our mountain days and this trip seemed like an adventure to them.

We were soon off the mountain. We stopped the car near the blacksmith shop on the main street of town, beneath a huge cottonwood tree where one of the cross-streets cut over the river bank. Here lay a river boat, a big stern-wheeler which was used to carry passengers across the river only when the water was so high that the pontoon bridge had to be taken out. This meant that the floating pontoons were disconnected and half the bridge swung down against either shore. This pontoon bridge was more than two thousand feet long, I learned. It had seventy-two pontoons and had been here for nearly thirty years.

While Wayne was talking to an old friend in the doorway of a high-pillared white house, the man's wife picked me a bouquet of fragrant yellow jasmine and honeysuckle that climbed luxuriantly over the porch and fence. I saw a limp sack that looked to me like an old jelly-bag hanging by a rope from the lower branch of a small tree on the lawn and I inquired why she had hung her jelly out-of-doors under a tree to drip. At her surprised look, I pointed toward the tree. "Why that's a 'pappy-dad,'" she said. It was a canvas sack that had once been half-filled with straw, upon which the children had swung, seating themselves astride the bundle. The stuffing in the sack was now nothing but a small wad, for the children had worn it out.

Jimmy and Donald thought that was the most amazing swing they had ever seen.

There were lawns before nearly all of the white painted houses. They were of Bermuda grass, a stiff, hardy grass that stayed green until the heavy frost killed it in the fall. Many of the homes were made beautiful by the protecting Virginia creeper. Huge cottonwood trees and sycamores, and magnolias with their glossy green leaves and milk-white blossoms, were planted along the streets and towered above the houses, while on nearly every porch was a swing-seat. The crape-myrtle bushes with their rosy blooms, and the mulberry, haw, and chinaberry trees attracted my attention also. There seemed to be so many of these along the side streets and in every yard.

After the cool breezes of the mountains, the heat in town seemed unbearable. My clothes clung, wet, to my arms and shoulders and I could feel the prickly heat popping out on my back. The leather upholstery of the car was hot in the sun. My head was damp all over and the loose hair around my face clung, while the perspiration trickled beneath my hat over my forehead. The breeze made by the moving car was a relief.

The fields we passed were mostly of cotton, long, even rows of beautiful green, but I saw some vegetable patches and some fine plantings of pole-beans and dark-leaved stalks of corn growing at the ends of the cotton rows.

The river on our left flowed muddy and deep below banks twenty feet high. Sometimes the road went along the very edge of this bank, where the soil was loose and sandy. Again it curved inland, cutting directly through the cotton fields, where the soil was black and shiny from the recent rains.

The cotton plants, with leaves like those of the thimble-berry, were of an even deep green, stretching away for miles toward the foothills, the fields broken by rough roads, the ruts filled with water.

The car sped along, Wayne negotiating a muddy stretch where the two ruts were submerged, out driving high and carefully to keep in the two ruts that looked the best and most traveled.

There were no more real houses—only Negro cabins and unpainted shacks; yet these people paid big rents, and made no demands for improvements, for they knew they wouldn't get them if they did. Some of these little houses had the protection of a cottonwood tree for shade, but most of them were bare and ugly, with a brick or stone chimney trying hard to cling to the warped wall of the house.

A few of the houses had apparently once been white-washed. I saw stove-pipes sticking out of windows, and pillows stuck in window-frames where glass had never been. Nearly all of these shacks were two or three feet off the ground, supported at the corners on rocks of various sizes. Often single tall stones would serve as support at the corners, but generally three or four stones were placed one upon the other. Beneath the houses thus lifted from the bare earth chickens and pigs were lying in the heat of the day. Rags and papers and bottles and cans had been thrown under, too, and the ground was generally white with chicken feathers that never got all blown away.

Only in a few instances were the houses fenced away from the fields or the barn lots. Generally the two-room shack sat right beside the road with cotton growing up to the house on three sides. A castor-bean tree, a few gay zinnias, or a morning glory vine were usually the extent of any flower garden. Gourd-vines grew luxuriantly here and there, and the fences were often hidden with the reddening sumac bushes.

As we got farther down the river, we drove between high hedges of Osage orange, and we were relieved from the sickening heat at times by the cooling shade of overlapping cottonwoods whose branches were thickened with trailing possum-grape vines. The Osage orange thickets, Wayne told us—or bois d'arc—were a persistent growth, with roots that seemed to sprout eternally. They had heavy thorns and made a good strong fence, so they were often found planted along roadways or sometimes formed hardy hedges in conjunction with the old fences to keep stock from breaking through. Large yellowish, orange-like fruit grew on the bodark trees—they were commonly called bodark—and the hard yellow wood was used sometimes for railroad ties or fence posts. The Osage Indians had made fine bows and arrows from this tough, pliant wood.

At last we turned into our farm, the car following the deep ruts in the sandy road. The big gate was sagging from one of the gate-posts, and I began to look closely at the buildings. We drove past a huge, old, weathered barn, the roof obviously in need of shingles, past two small shanties to the left and the right, each perched up on their high pile of rocks, and then past a big hay shed, that stood aslant to the roadway. Then we came to a small house that was built on a sandy bank near a row of pecan trees facing the meadows and the river.

No trees, no vines, no shrubs, not a flower here—only cotton. Cotton crowding to the very sides of the house; and flat, wide-spreading weeds, burr-laden, covering the ground all around except where a beaten path led to the back porch and a narrow sandy strip went around to the front porch.

It was a box house, made of foot-wide boards running up and down from the eaves to the ground. At the bottom, the irregular fringe of board ends left ample room, so that roaming pigs and chickens could crowd underneath, and I saw several pigs asleep there, stretched out in the cool sand.

Long battens covered the cracks between the boards, or were supposed to, for many of them had been broken off halfway up the side of the house.

The house was composed of two main rooms under the peaked roof, with a porch extending all the way across the front. A lean-to had been built on behind. This small lean-to, I soon saw, was divided by a rough partition into a kitchen and a dining-room.

We were given a royal welcome by Mr. and Mrs. West and her two sons with their wives. They were preparing to eat their dinner. I stood back wondering just how to get up onto the back porch, as there were no steps, just the half of an old log. One of the tall boys stood watching my hesitation. "Best climb right up," he said. "We're allus aimin' to fix some steps there, but never get 'round to it."

These people, Mr. and Mrs. Eric West, had worked for my husband's parents years ago, Mrs. West having "dandled Wayne on her knee many's a time," she said. She was a big woman, her hair gray, her eyes crinkled from the sun and good nature. A snuff-stick protruded from one side of her mouth. Enveloped in a clean calico apron, Mrs. West was just about to put dinner on the table. "A short horse is soon curried," she remarked, dishing up 'Tucky Wonder beans and hot cornbread and pouring coffee. One of the young women was sent hurrying out to pull some more green onions and get cucumbers and tomatoes—all three of the very finest quality.

"Make out your dinner," we were urged, hospitably, and we were soon eating with relish, despite the terrific heat under the low ceiling.

Mr. West, who wore a sweat-stained blue chambray shirt, clean faded overalls, and heavy shoes, was tall and thin and white-haired. His eyes were blue, peering from under bushy eyebrows, and he was always blinking them. When he spoke, he often stuttered, especially if he grew very much in earnest or became excited.

The table, put together with long boards, was covered with brown oil-cloth and laid with old china plates, tin cups and bone-handled knives and forks. On one side of the table was a long, rough, weathered wash-bench, but there were several cane-bottomed chairs, too. The young men and their wives were living here, too, to work in the crop, so there were ten of us to enjoy the excellent meal.

After dinner, the younger people having gone back to their work, we were invited to sit out on the front porch, in the cool.

As I passed through the front room, I noticed the many nails driven into the walls on all sides. From each one, hanging by a string, were bottles of pills, bottles of chill tonic and Black Draught, and snuff sticks or brushes belonging to the different members of the family. Nails were easier come by than boards for shelves.

The walls of the rooms were covered with strips of tan building paper, nailed on with large tacks put through circular disks of tin. This paper was very soiled, and there was much evidence of tobacco juice having run down the walls. But the floors were scrubbed clean and the two beds that I saw were evidently composed of puffy feather mattresses. The un-bleached muslin spreads and cases were starched and ironed. There was a high pile of neatly folded home-made quilts on the floor in one corner, and I wondered about them.

While the men talked of crops and rentings and conditions in general, I was busy admiring the skill with which Mrs. West managed her snuff dipping. Her snuff tin, a small round can about two inches high, she carried in her apron pocket.

"You jest get you a good stick offen some tree like the elm, that chews up good to a brush, that's all," Mrs. West advised me. "Didn't ya never dip none?"

The snuff stick was well wet with saliva, then worked around in the powdered tobacco until a ball of the snuff was sticking to the end of it. This she put carefully into one cheek, the rest of the time spent, of course, in spitting, while the brown stain ran down the sides of her mouth.

As we sat on the porch, we looked across toward the Bermuda mead-ow, thick with cottonwood trees, but the river could not be seen. It was about a mile away across the meadow and beyond the sand-bar. The heavy growth of willows and sycamores and cottonwoods and the "bodark" thickets along the fences, with their festoons of wild grapes, cut off the view. As the afternoon wore away, the plaint of the mourning doves and the cheery, brisk call of the mockingbird came to us from the meadowland. Never did I weary of hearing that soothing, spaced, "Coo-coo-a-coo" of the mourning doves. The sound simply enchanted me.

The sandy bank before the house slanted down about ten feet to a narrow strip of land planted with cotton. Along the top of the bank grew huge pecan trees. These were the ones planted by his grandmother, Wayne reminded us. Between these trees and the house ran the narrow road that wound along through the farm from house to house. This narrow road was not used for public traffic. The main county road was sixty yards away down beyond the alfalfa field. Then beyond the county road, the meadowland stretched over toward the river. In the meadow, further to the right, was a large pond where the cattle and mules drank. The pond was in a natural depression and seldom dried up, for the water was usually on a level with the river, a mile away.

I had climbed onto the back porch to enter the house, for there were no steps, only part of a rough log, and now at the front porch I saw there was no way at all to the ground except by jumping. There had once been steps here, Mrs. West told me, but the last folks who lived here had cut them up for kindling.

We helped to roll out huge watermelons from beneath the porch where they had been buried in the cool sand. These were cut lengthwise and the crimson pulp eaten with spoons. Someone brought out the salt gourd, and never have I tasted such melons. When Mrs. West got up to go and get water and the broom to sweep off the porch where the melon juice had dropped, I asked if I might pump the water. I enjoyed priming the old pump which stood on the well platform and was reached from the back porch by walking across two planks. I worked the pump handle up and down twenty times to get a lard bucket full of water.

Oh, that pump! How often was I to stand there, in anger, in despair, in high hope, working that old pump handle and gazing far off to the low-lying hills in the west, where the sun sank and the gorgeous sunsets blazed. This was my preview had I but known it.

The water was sluiced over the melon juice and was swept neatly off, to keep the flies from gathering. The window screens that I saw were old and broken and worthless for protection.

I knew I was an object of curiosity to Mrs. West, just as she was to me, but her kindness and generosity and wholehearted interest in my comfort, together with her clean poverty, quite touched me, and I felt greatly at ease with her, but mentally up in arms to think that she should be forced, through circumstances, to live here and work hard and do without even the barest necessities.

I was astonished that this farm, that I had heard the family discuss so often back in California, should be in such a deplorable condition. It was very obvious that there were many improvements to be made, and I surmised that further examination would surprise us.

My disturbed mental attitude I carried back with me to the mountain. That evening when the group gathered to see the sun go down and to listen to the singing, I kept looking around and wondering in just what way these people differed from those I had met on the farm. I thought of Mrs. West, in her clean clothes, with her good cooking, her fine garden, her generous hospitality and good nature. She had made no complaints. I had always heard the words "poor bottom folks" and "sharecroppers" spoken in a most slighting manner. Poor they might be, yes, but only in material wealth. They were surely rich in good qualities.

I got an opportunity to talk to one of the businessmen from town and asked him why, if the farm lands were as rich and productive as everyone said, did no one paint the houses, and why were the farm houses all such little shacks? I was making no accusations; I was honestly puzzled. I hoped to hear from him something more than evasive answers or polite admonitions meant to calm my troubled mind. He listened to my questions and my praise of the farm people with a look of tolerant amusement. When I waited for a reply, all I got was, "Well, if you don't like it here, why don't you go back where you came from? We're satisfied with things as they are."

I did not know until years afterward, as I came more and more in touch with the problems of farming, that these merchants themselves were caught in the meshes of an economic condition that had been evolving since long before the Civil War.

3. An Exile from Erin

THE NIGHTS WERE beginning to be chilly and the days much cooler as we prepared to move down into town.

Edith and her family had gone back to Little Rock. The schoolteacher had left the mountain some time ago and was already back in her home, ready for the fall session of school. Grandma and Henry would continue to board with their friends, who were now moving down to their house in town. Our case had not yet come up in court, and the present farm leases would not be up until the first of the year, so no new arrangements could be made until then. We would just have to wait. There would be no difficulty in getting tenants or share croppers, for the farm lay in the very best cotton-producing area along the river.

Wayne rented the vacant house his friends had advised, out toward the hills at one end of town, and we got our furniture from the railroad freight office.

Those months spent in town were strange ones to me. I kept hoping my "Exile from Erin" feeling would pass, yet I couldn't seem to take any real interest in settling down, nor could I find in my mind any contentment. I felt the need of something definite on which to build.

To me the town seemed so impossible. I missed the things I had been accustomed to. There was no library—no lodge. Never before had I been so conscious of being a complete stranger, and it was becoming apparent to me that if I were going to live here for any length of time, I would have to settle down into the close quarters of this small community, and reconcile myself to the gossip and activities of the church groups and the school. Wayne and Henry told me of a doctor and a merchant indulging in hand-to-hand combat on the main street, and there was a shooting affair between two prominent citizens because they got into an argument over politics. I was amused to find that as I expected there were several men who were socially known as Colonels. I had been cautioned not to mention politics or the Civil War for fear of offending someone, and I had also been told that in the South no lady voted or bothered with political affairs. Those interests were not mine especially, but I resented the narrow-mindedness and the dictatorial attitude behind these cautions.

A keen sense of helplessness and defeat oppressed me, and I wished that I were back home where I belonged—where life seemed to hold untold possibilities, with a sense of bigness and a scope of living that was an active incentive to daily effort.

Somehow I dreaded unpacking and only dug out of the big boxes those things we most needed. How I longed to have the time pass so the lawsuit that had caused us to come to this sorry land would be settled and we could go back to California!

I got out of bed every morning wondering what there was to live for, and kept myself going only by taking an intense interest in making my own bread and keeping house. To me it was a most heartbreaking country, and I was thrown entirely upon my own resources. The daily sight of the old black wash pot and another "pappy-dad," that poor swing, hanging like a lifeless body from the bent limb of the cottonwood tree right in our own yard, put me constantly off my bearings; they epitomized so much.

I washed all the windows one day, and a neighbor walking by paused to remark, "Looks like you-all are just a-hankering for work. Seeing as you don't aim to stay but a while seems kinda like a waste of time to do extra cleaning. I wouldn't if I was you!" I stared at her in amazement, unable to think of what to say in my own defense. Now I knew why they had laughed, people we had passed on the highway in Colorado. The sign that Henry had hung on the back of our car had read, "California to Arkansas."

Wayne and Henry and Grandma didn't say much, but I knew that they, too, were disappointed with the conditions they saw everywhere.

My two large panorama pictures of the ocean, with the breakers rolling in on the rocks, placed above the fireplace on the mantel, only added to my homesickness, my longing for a sight or sound of the ocean. I thought of the many times we had ridden through the foothills to the sea, in fall, in winter, or in fruit-blossom time. Day or night, whenever I thought of California, the tears would start in my eyes. All the states away from the ocean seemed to me like the back rooms of a house, either dark and cold and dreary, or stifling hot. Nothing like being able to sit on the front porch, as it were, near the Pacific Ocean, in the breeze. Pictures kept flashing through my mind, and in fancy I rode along the Coast Highway and through the low foothills. I saw again the green, rolling hillsides where splotches of deep blue lupines accented the rounding curves. I gazed on the fields of lush green grass where each spring the silver-leafed poppy plants, with their golden silk petals, caused us to exclaim anew with pleasure at their unbelievable beauty.

I seemed to see the sea gulls that paraded on the wet sands and the little sandpipers that scurried back and forth where the white combers rolled with their everlasting regularity, and the coast road that curved and dipped along past the little blue-roofed Italian farmhouses. Here were the windbreaks and hedges of dark cypress, bent shoreward from the ocean winds, and the little streams that ended in tiny beach coves where the salt water rose with the tides to push inland and form swampy lagoons.

Here were the rushes and here the sand dunes. Following the stream beds toward the hills were tall, silvery growths of eucalyptus trees and the dark towering redwoods beyond the chalk-rock cliffs. How we had delighted in finding the new growth of redwoods that made fairy rings around the ancient stumps. Toward evening how the fog would drift in from the ocean and sink into all the soft hollows of the coastal hills.

My mind was often filled with the sound of the meadowlarks singing in the hills at home, and I would picture the look of the purple haze over the canyon near Loma Prieta. I would drop off to sleep at night lulled by the memory of the regular swish and roar of the big breakers, or filled with thoughts of the times, in the fall, when the waves were small and stole quietly to the shore with something of vain regret, a sadness, in the repeated soft surge and retreat, surge and retreat of the green water.

Jimmy and Donald stayed close with me. They were kept from starting school when it opened because they had caught a skin disease that had been epidemic in town all year. It settled on their scalps and faces, and no amount of ordinary medication seemed to help it. I do not know if it was ringworm, for even the doctors were puzzled as to its cause or cure. The old home remedy of lard and gunpowder, that my mother wrote me about and advised me to use, finally dried up the sores, but it took weeks to make a complete cure.

While Jimmy's and Donald's heads were clipped in places and bandaged to keep the thick salve on the bare spots, I began a series of daily lessons, making up games and methods to hold their interest and speed them along into their first primers.

I was greatly disturbed about my boys. In California they had always been kept close at home, going with Wayne and me on rides and on visits to their two grandmothers. They had never been left with other people, nor indeed had they been allowed to visit or play with the children of friends or neighbors unless all the children were under careful supervision. All the mothers I knew agreed with me that little children needed their mothers' constant care, and it was not a common practice to leave our children unattended or to hire strangers to stay with them.

Now just before time for the school to open for the fall term, I went one day with a neighbor for a walk around town. On our way home we passed the school and sauntered in to look around the grounds. The main building was closed, of course, so we examined the smaller building back of the schoolhouse. This was divided by a high wall into two divisions, and when I saw that the walls were covered with obscenities, both in pictures and words, I expressed my amazement and righteous indignation to my neighbor in good set terms. She listened very calmly to all my expressions of disgust and surprise.

If this disgraceful condition was the work of unknown trespassers during the school vacation, I told her, then I felt that the premises should be painted before the children returned. If it were a permanent condition, then I wondered who was responsible and why there could be such apparent indifference on the part of the teachers, or the townspeople. My neighbor, whose two young daughters attended school here, agreed with me that something should be done about it, but suggested, in a rather defensive tone, that perhaps the conditions here were no worse than anywhere else.

I could not help but make the comparison between this school and the ones I had known at home where all sanitary conditions were under careful supervision and where the playgrounds had a teacher in constant attendance.

So now I was sick at heart at the thought of sending two inquisitive little boys to a school where their daily surroundings and their companions would surely supply more evil than I might successfully counteract at home. I began to think of what I could teach them that would be a defense and a preparation for the shocking vulgarities that seemed to await them at school. I started several times to have an earnest discussion with them, but had not proceeded very far before I began to hesitate, appalled at the evils that I seemed about to bring to their attention. It angered me to think that I should have to face such a situation, and so I was really glad of the excuse I had to keep my boys at home, and glad that it was already too late for them to enter for the term.

I made a set of cards with the letters of the alphabet in both printed and written form. Very soon we were playing a game where each of us pretended to be a letter and stepped forward in rotation, saying his new name, the letter, according to the new phonetic method. Ka for C; then ah, for A; then t; and then in unison ka-ah-t, cat. My scheme worked well, and the boys were soon reading the silly little stories in the first reader. Then I began teaching them to count, using knives and forks and chairs and articles around the house. The effort made in carrying these things back and forth to put one chair with another or add a fork to those left in the drawer, seemed to make addition and subtraction easy to understand.

Anyway, Jimmy and Donald eagerly spent hours at their work, and learned quickly. Colors and shades were distinguished in leaves and flowers and in our clothes. Corn bread and pies and apples were divided into halves and quarters. We occupied ourselves with lessons each day until the boys' attention would begin to wander, but these lessons soon became a regular part of our day's routine that we looked forward to with great pleasure. Often, when the housework was done and the lessons over, we would ramble over the nearby hills, picking up the long, narrow pecans that covered the ground and gathering armfuls of the red-lacquered leaves, which seemed to glow when the sun shone through them.

The days were calm and still, with a dreamy haze over all the land. From the hillside we could look back over the town and the river and see the fields of dry yellow corn and the big orange pumpkins scattered through them.

The smoky blue atmosphere was background for the mellow gold of the hickory trees, the flaming gums, the brown oaks, and the scarlet sumac bushes. Faded stalks of goldenrod bordered the paths and fluffy heaps of freshly fallen leaves still glittered in the sun. These lay lightly on top of the brown leaves that had only a few days ago been so flaunting and so gay. As we wandered along we interrupted squirrels gathering nuts, or found an occasional cluster of refreshing muscadines and 'possum grapes.

To our surprise we found the ripe Southern persimmons far different from the large glossy orange Japanese persimmons we had known at home. The trees were usually about the size of apricot trees, although some grew much bigger and taller. The fruit was a dark orange, at first, as it ripened, small and quite puckery to eat from the tree, but after the frost came and the fruit fell to the ground, the skins turned brown and thin, like that of dates, and the flesh was soft and very sweet and good. We were always glad, in our wanderings over the hills, to find another persimmon tree where we could gather up the dried fruit to take home. The persimmons that were left on the trees were eaten by the birds and squirrels and opossums.

We watched the cotton ripening in the fields nearby, the tight green bolls slowly turning brown and bursting open, and the cotton, gleaming white, beginning to hang loose against the dry leaves. A strong weed odor filled the air. The sun beat hotly down in the middle of the day, but a cool breeze sprang up as the shadows lengthened.

One afternoon I watched some Negro women picking cotton, in a

field near our house. I heard them singing, so I went over to the fence
to watch and listen. I was thrilled to be in contact at last with an actual
field of white cotton that looked just as I had pictured it in the old songs,
"Gwine Back to Dixie," and "Old Black Joe."

I finally climbed over the fence, little realizing how it must seem to a
Southern Negro to see a white lady climbing a fence to get into the field
with them. But I did just that, and eagerly gathered up my apron and
asked one of the women if the cotton was picked one piece at a time.
She grinned and gravely and wonderingly explained to me that the divisions were called locks and that with a certain twist of the fingers the
four divisions of loosening cotton could be picked out all at one time,
but cautioned me to be careful, for the sharp ends of the dry bolls stuck
the fingers. Much of the fluffy white cotton was dangling from the bolls,
although the seeds in the bolls still held tightly, for the cotton was pulled
out gradually by its own weight to be blown by the winds and drenched
by all the rains.

How happy I felt to be there among those Negro women, actually
picking snowy cotton into my apron. They all stared at me, and got little
picking done while I was there, but yet they seemed glad to talk to me,
seemed to know who I was, and spoke so kindly and courteously that I felt
my heart expanding to their friendliness. They each dragged a long cotton
sack of lightweight canvas that hung from their shoulders by a strap. After
a bit the women went on with their picking, although they kept watching
me and talking together. But I had great pleasure coming. They began to
sing. I couldn't follow all the words, although I listened with rapt attention. They sang in short bursts of melody, one starting, the others joining in—then silence. Then again a song with sweet mysterious melody,
part chant, part a humming cadence. Then I heard "Beulah Land," and
"Nobody Knows the Trouble I've Seen," and another hymn, that seemed
almost familiar. Suddenly one young woman lifted her voice in a high
sweet call like a bird that caroled downward in luscious tones. Her face
was calm and almost illumined with an inner joy. This song was about
the Savior's loving care and the lost sheep returning to lay their hearts at
Jesus' feet. I will never forget that scene, while the sunset was beginning
to glow redly in the west and the air to chill, for the simple Negro strains
I heard had such purity and spontaneity in them that I left the field much
shaken. No wonder their songs have lived and moved so many people in
so many lands. The daughters of music!

4. The Profit of the Earth Is for All

ONE DAY WHEN Wayne returned from the lower farm he brought Mrs. Rollins back with him to stay a couple of days, while he figured out some way to help her. She was "in a tight," she said. She was a widow with three boys who had been on our lower farm for three years. Her boys were able to do the heavy field work, but she herself worked at hoeing and picking, too.

Mrs. Rollins had rented the lower farm of one hundred and thirty acres for fifteen hundred dollars cash, due at a certain date. The bank lent her the money with which to buy three teams of mules and pay for feed and wages for hired help. The banks had always lent money at a high rate of interest to those who could get a lease on land and who had a crew to work out a crop. Wages then for cotton picking were a dollar and a half a hundred pounds. She also got credit for her living expenses, managed through a store in town, giving as security a mortgage on her tools and teams and crop. The landlord's lien came first so the bank took the second mortgage.

She raised an excellent crop of cotton down there below the hill along the Petit Jean River where the soil was black and rich. Although the farm was close to the woods, the boll weevil had not yet molested the cotton there. She raised three-fourths of a bale to the acre. There would have been seventy-five bales if all were picked. That was the last good crop on that farm, however, for the boll weevil arrived the next year.

During all the time that the crop was being planted and worked, up until it was "laid by," with cotton at forty cents a pound, the bank, and Mrs. Rollins and we, the landlords, expected good money and all the debts cleared.

When the time came for cotton picking, the price began to drop. First to thirty cents. Mrs. Rollins refused to sell at that price and when the cotton was ginned had it piled by the gin, feeling that the price would surely rise. By the time about twenty or more bales were piled by the gin, cotton was down to twenty cents. Still Mrs. Rollins didn't sell, for during her three years on this place she had seen the price go from twenty-five cents to fifty-five, so fluctuations were not unusual. By the time it had dropped to ten cents, the notes held by the bank all came due. And the rent was due.

The picking was to be paid for. Two of her boys got married, adding two more to the family group. "We ain't bought any clothes, nor had much to eat, and we've all worked hard every day from morning till night." So she sold all the picked cotton, paid something to the bank, and paid for the picking. Her boys, who were nearly in rags, bought new overalls, blue chambray shirts, and shoes and heavy underwear for the winter. In other words, the landlord's money went to pay others. The bank demanded the rest of the money due, and the landlord wanted his rent.

Wayne and Henry had made many trips down to the bottom to over-see the work and to beg Mrs. Rollins to sell her cotton while the price was high. The courageous woman refused, living in hope, but she promised faithfully to pay her rent. Then, without Wayne's knowledge, she sold the rest of the cotton at the low price, leaving Wayne out. Again she paid something to the bank, and paid all those who had helped to pick the cotton.

"The rest is in the field now," Mrs. Rollins told us. "But the boys jest ain't aiming ta pick no more with nothing to eat. Don't know as I blame 'em either. Ain't nothin' fear them anyway, and they have stuck by the work right willin'. Well, what air you a-goin' to do? I guess you can do as you like. Go to jail, I guess, like the bank said, is what I see fer me, but I jest cain't see that I could manage no other way. I knowed in reason something bad would come of it all. Ain't never seed nothin' anyway but sickness and hard work, and here I am."

Wayne looked across at me, where I sat sewing by the fireplace. He didn't need to try to get my opinion nor I his. We were both completely engrossed with the details of this story, and were already on her side, no matter what she had done, even though we were the losers. As Wayne said afterward, "I don't blame her one bit." I kept trying to remember where I had heard the words, "The profit of the earth is for all."

There were about twenty-five bales still unpicked, with the price at the lowest. This was in the black land, the gumbo, and rainy weather had set in. Lots of the cotton had fallen out of the bolls into the mud, and that still in the bolls was badly damaged. The tips of the cotton would turn blue and down by the seed it would be yellow, as a result of weathering. Wayne had one bale picked and had to give the seed for the ginning. Seed had sold for forty or fifty dollars a ton. Now it took five dollars to gin a bale. Later it got down to two and a half dollars a bale for ginning. Some share-croppers simply went off and left their crops because they couldn't pay for the picking, and the cotton rotted in the field.

Thinking that all this condition was only temporary, Wayne said, "You go ahead and pick. Store it, and make another crop, if the bank will renew your note. I'll take one-fourth of the cotton, instead of cash, for the rent next year.

But the bank refused a renewal unless the landlord signed, which Wayne wouldn't do, so they took the mules and wagons and tools and tried to get Wayne to go ahead with them and take action to have Mrs. Rollins put in jail.

Even if she had sold the cotton to pay the rent, it still wouldn't have paid that debt in full, and though Wayne could have prosecuted her and demanded his money from the bank, and got it, too, that would have meant jail for her. He couldn't do that, so he lost the rent and the bank got only a part of its money. I tried to figure it all out, but got the figures so confused that I felt there was about four hundred dollars that couldn't be accounted for.

Meanwhile, joined by some relatives of her daughters-in-law, the last of the crop had been worked out, and the bank was charged with the cotton picking and finally lent the money to pay it. True, the Rollinses had had the use of this cash, but what else could they do to get any money to live on? They would be expected to pay ten percent to the bank for what was lent on their store bill, and, as I later found out, the arrangements for over-charging credit customers were intricate and very, very profitable. A monthly borrowing at the bank, giving a note for the same, was repeatedly renewed, the compound interest piling up, so a year's transactions in this customary manner was indeed excellent business for any bank.

I sat there in the lamplight looking at this poor woman, aged and stooped with hard work, years before her time. I saw the dumb resignation on her face, and listened to the long complicated account of a widow's struggle to make an honest living against every odd. I marveled that she had had the courage to try at all. I couldn't figure out the intricate bank dealings, try as I might, and Mrs. Rollins could neither read nor write. One of her boys had been through grammar school. The others—well, I don't know what to say, for it was the one who had had some schooling who told me later, when he was working for us, that he did not believe that the world was round. "I can see for myself," he said, as he stopped by the well for a drink one day. "Don't need anything to go by but my own two eyes."

"But didn't they teach you that the world was round when you went to school?" I asked. "Oh, yes, I read all about it in the geography book, but you know you can read anything in a book, I've found out. Got my own two eyes to see with, ain't I? I can sure tell if the world's flat or round. Looks flat, all I ever see. 'Cepting the hills, of course, but I figger them in."

What interested me personally was Mrs. Rollins's wonderful ability to make up biscuits with sour milk and soda, and I was an eager pupil during the few days she stayed with me as she helped me in preparing the meals. In a bowl of flour, an unmeasured amount, a hollow was made, and into this went about two cups of buttermilk or clabbered milk, some salt, soda and shortening. The amount of soda depended upon the age of the milk, freshly soured clabber taking less that if it were older and more acid. The mixture was carefully stirred with a spoon, the flour being gradually brought from the sides of the bowl to form a mass that was quite workable, not too sticky, nor yet too dry. This dough was then kneaded by hand until smooth and elastic. Meanwhile the bake-pan had been set on the stove and a generous amount of shortening melted in it.

As a bit of dough was broken from the mass and shaped into a biscuit, it was laid in the melted fat, then turned over and placed in its proper spot in the pan. Thus all the biscuits were well greased on both sides and when baked had a delicious brown crust. This method, although calling for so much shortening, was commonly used because it insured good biscuits even with wet wood and a slow oven.

The gravy, too, that Mrs. Rollins made was delicious. She used just the right amount of grease and flour combined, cooked thoroughly to the right stage, and gradually diluted with milk as it was stirred until it was of the right consistency, with pepper and salt added to taste. Pork grease or drippings when chops or ham were fried made the best gravy of all. Chicken gravy was made in the same way. Those who had no meat bought lard or cooking fat at the stores. This white milk gravy, I found, made up a major part of every meal, and, with biscuits and corn bread, the latter often made with water and grease and no milk at all, formed a large part of the diet of Southern people the whole year round. Salads and fresh green vegetables I was missing already, although I was beginning to like the general use of turnip greens.

When it came time for Mrs. Rollins to go to bed, my offer of a night-gown was pleasantly but firmly refused, which puzzled me greatly, as she had brought no package with her and had only her snuff stick and her box of snuff in the pocket of her dress. Later, however, I learned that her gray flannelette gown served equally well for a robe at night and a petticoat by day, so the mystery was solved. I remember telling Frank about this later and seeing him take a piece of chewing tobacco from his pocket and eye it speculatively as he remarked sagely, "You'll never get ahead of these people. They are always prepared for whatever comes. I doubt if you will ever understand them."

Mrs. Rollins was anxious to have me come down to the lower farm to visit her and kept asking Wayne to bring me. So one day after she left, Jimmy and Donald and I went along with Wayne in the Ford and had dinner with the Rollins family.

We went into the farm by the road through the cotton lands that took us to the little ferry across the Petit Jean. This ferry was smaller than the one we crossed in as we came through Oklahoma, just a flat scow with an apron that could be raised or lowered at the ends, and was pulled across to the other side when a passenger stood up and pulled hand over hand on the steel cable overhead.

There were nearly forty acres of virgin timber circling the high rise of land that formed part of the foothills of Petit Jean Mountain. This tongue of rough land jutted out like a spur from the high hill. The timber had been left untouched here to prevent the land from washing away and the trees were tall and beautiful. There were big scaly-bark hickories, oaks, walnuts, pecans, ash, and sweet gum trees. It seemed quite like a park, for with the absence of underbrush one could admire the fine big tree trunks.

The house was built just at the place where the ground fell away steeply from the rocky hillside to the woods and fields below. It was a two-room shack, the front room built on a level place on the rocky ground, the back room extending out into the atmosphere, upheld by heavy timber and big tree trunks that left a space where a team could be sheltered and where wood was occasionally piled whenever the men got around to cutting an extra supply for the cookstove. Built close up to the hillside nearby was a little one-room house where one of the boys lived with his wife.

Mrs. Rollins's boys were all home, and I met Avis, the wife of the oldest boy, Ben. The other young women were away for the day. Avis was a tall, well-built girl, with flaxen hair and beautiful blue eyes and pink blushing cheeks. I liked her instantly.

We entered by the only outside door and stepped through the little kitchen to the front room. Here there was barely space for chairs for us all, for three beds took up most of the room. The three boys were sitting by the heater trying to compose a letter to Sears Roebuck to order a part for their rifle, which had been broken. They had a torn piece of wrapping paper and a very short stub of a pencil. I watched and listened, at first thinking that I should not let them know that I noticed their difficulty, but it began to seem heartless on my part to sit there so indifferently, so I offered to take the order home when they had figured out what they wanted from the catalogue and mail it for them. They had no envelope and no stamp, so I got the number of the part they wanted and that night I made out the order and mailed it.

As I sat there while the boys were struggling to write out their order, I counted the cardboard cartons that were stacked in the corners of the room and beneath the beds. Mrs. Rollins was ironing in the kitchen and the freshly ironed overalls and shirts were brought in and hung on a wire stretched high across the front room. Later she folded the clothes and packed them neatly in the boxes. There were no shelves or closets in the house except the tiny shelf behind the kitchen stove and the two where the dishes were stacked. Various pots and pans hung on nails around the stove; but everything was extremely neat and orderly and clean.

In the afternoon, Avis and Jimmy and Donald and I took a walk on the side hill. We found patches of yellow violets in the moist rich earth and we picked big bouquets to take back home with us.

Behind a stump, under a small bush, I discovered an empty gallon fruit jar, with a tightly-fitting cap. Avis explained that it was hidden there to be exchanged for a full jar of "white mule," for someone in the district was a moonshiner. She said that this was the common way of delivering the whiskey. She knew where one still was located, but had promised not to tell anybody.

We followed a narrow, winding road around the side of the hill and came to a small house where a friend of Avis's lived. Avis wanted me to meet this woman and see her beautiful phlox and marigolds, so we went into the yard. We called out and found that no one was home. Still, the door was wide open and all the windows had been taken out. I saw chickens perching on the porch, in the house on the chairs and table, and even on the beds. Jimmy and Donald were amazed by this and they began to shoo them out, so we chased them all into the yard and shut the doors securely. I found four eggs on one bed, and two on the other, for the hens had evidently been in the habit of laying there often.

There was a huge cellar dug in the side of the mountain near the Rollins house. This was used for the storage of various things, but Avis told us of the times when they were all glad to take refuge in it when the storms were bad. Yes, it was dark and they had no lantern, she said, but at any rate no one thought of anything like that, they would be too excited, too scared. We could hardly imagine what it must be like to be shut away into the mountain with that heavy door closed against the rain and wind, to stand inside in the darkness waiting for the storm to cease. Many times, Avis told us, they had stayed in there all night. They took in the chairs and wash bench to sit on.

In the hard rocky ground near the kitchen door was the well. It was the pulley and bucket type, and we pulled up the water with such pleasure. Cold, good-tasting mountain water. Avis said that down below the house and nearer the creek they had a small garden patch and grew some sweet corn and beans. "It does all right unless the creek overflows. Then we lose everything. But that is new land down there and raises extra good cotton, only the boll weevil is beginning to work there now. It didn't used to."

As we drove home late in the afternoon, we talked of Mrs. Rollins and her problems and wondered just how she had managed to make a crop and feed the whole family. Of course, Wayne reminded me, they didn't eat the same food we did. They ate beans and corn bread and fat meat, mostly, and whatever they could raise in the garden patch down by the Petit Jean.

5. The Story of Cotton

DAY AFTER DAY I kept asking questions until at last I understood the full cycle of cotton, the man and mule crop. I was told that it was planted in the spring, mortgaged in the summer, and left to rot in the fields in the winter.

But in all the telling of the cotton story I noticed that the cotton and its chances under the changing weather conditions were always stressed, but never the personal welfare of the owner or of the share-cropper. Their inconveniences and hardships were taken for granted.

Weather permitting, the cotton was usually planted in April or May, in soil that had been well prepared. Those preparations began the previous year sometimes, if, weather permitting, the last crop was out of the fields in time. Then the stalk cutters were driven up and down the rows and the dry stalks cut and broken into short lengths which could be plowed under, thus adding eventually to the enrichment of the soil.

Some farmers always burned the cotton stalks to get rid of them entirely, for when the cotton had been tall and rank, the stalks were more often mashed than cut up by the stalk cutter, and the broken pieces would be sticking out in all directions through the freshly plowed field.

To put the ground in better condition for a crop, the land often was flat-broke first: They carefully plowed the old stalks under, and then went through again with a middle-buster to form new rows. If this could be done early enough, the ground would settle and the stalks were more liable to rot and not be turned up again to tear loose the delicate young plants during hoeing or cultivating.

In sandy soil, if the furrows had not had time to settle, some farmers used a culti-packer to pack down the soil. If the soil had already settled, the next step would be to harrow the beds down to a proper height on which to plant.

In heavy soil, or black gumbo land, if the soil had already settled after the first plowing—and this was generally the case, for rains were sure to come at this time of year—a harrow was run across the tops of the high ridges, or a stalk cutter might be used to break the top, or the top would be dragged with the harrow weighed down and a stout one-by-three wedged between its teeth.

These were the methods when a one-row walking planter was in use. For riding-planters, the usual method was to use the middle-buster early and plant later, for the plow of the planter cut down the row as desired at the same time that the seed was planted.

The planter was a wonderful thing to see, yet so simple. It dug a small furrow, the hopper dropped the seed at regular intervals through the gears in the bottom of the planter, and the two tiny plow points set at either side behind the hopper covered the seed while a small roller ran along behind, pressing down the furrow where the seed lay covered. So there the ridge was, smoothed down and ready for the rains to start the quick growth.

What was most to be desired in raising a cotton crop was to have a long-staple, drought-resistant cotton, and a boll with few seeds and lots of lint. Then if the enemies of the cotton field, the boll weevil and the army worm, did not descend on the poor farmer, all was well.

All was well indeed if the winter had been cold enough to kill all the hibernating insects, and then enough rain fell in the proper season to give the right moisture to the soil. Then spring, too, should arrive early, with rains. The summer should be dry, with hot, growing nights, for in cool nights the cotton stands still. The fall should arrive dry and sunny, with late frosts, for excellent picking conditions.

However, everything was variable and unpredictable. The farmer planted in hope, worked from daylight until dark, and never knew at any time, until the cotton was at last sold, what he would have. Neither seed, nor soil, nor weather, nor health, nor selling price could be depended upon.

It was amazing the amount of gumption and tenacity these cotton farmers displayed. They had a dogged quiet determination to carry on. They and their mules alike, tough and wiry and stubborn, suffered from the awful conditions under which they labored—poor shelter, poor food, and little hope of reward.

As soon as the young plants popped out of the earth, and the long rows could be clearly distinguished, side-plows scraped the earth along beside the young fast-growing plants. If this was done early enough, all the small weeds and grass would be covered, leaving the clean-looking rows of tiny pale green plants down the center of the ridges.

Then the field hands thinned the plants so that there were from one to three in a place, and hoed out the grass between them.

Some years, by careful work, all the weeds would be taken care of in this way, by plowing and hoeing at the proper time, weather permitting. If it continued to rain and the grass and weeds got too big, the scrapers had to be used on the cultivators. The scrapers were set at a slant on either side of the ridge on which the cotton grew, and they cut the grass and weeds loose and turned them over into the furrows between the cotton rows. This buried the weeds, and one time through was generally enough for this job. Later, the cotton on the ridges was hoed again to keep the grass out—while the plants were still small. It all depended upon the amount of rain just how many times the hoeing had to be repeated. The plowing was usually repeated every two weeks until the crop was "laid by," by which time the plants shaded the ground enough to keep the grass down. Sometimes the plowing had to be done five or six times, all depending on the weather. If blossoms came by early July, the crop was doing well. The blossoms were first white, then pink, then they fell off and the small bolls appeared.

If the weevil punctured the young boll, the boll might never develop, or it would be partly developed and partly dried up. The worm, which grew from the weevil's egg, ate inside the boll until it, too, emerged at last as a new weevil.

If arsenic of lead could be applied in a manner and at such a time that it would be on the plants when they were attacked by the weevil, many of the weevils would be destroyed. But it generally rained at the very times when the boll weevil was rampant.

The arsenic of lead was generally applied by filling thin cotton bags with a mixture of lime and the poison and then tying the bags onto both ends of poles which were laid across the saddle of a steady mule. Up and down the rows the mules were ridden, the jolting causing the dust to shake out and settle on the plants. This job was always done when the dew was on the cotton, in the early morning, so the powder would have a chance to stick to the plants. There were regular dusting machines, too, which could be bought for this purpose.

The field hands, men, women and children, all chopped the grass and weeds from the rows of cotton. Even if the rows were half a mile long, one man could easily chop six to eight rows in a day, if the crop was not "in the grass" and the work had gone along without too much rain; but if the grass got a start it was a good day's work to get two rows chopped. By the middle of July, cultivation was generally ended and the crop was "laid by." Then the time was spent attending camp-meetings, singings, dances, visiting friends, or just lying around the farmhouses, idle, sick, chilling.

Early in August the first bolls began to open. The pickers worked from "can't to can't"—"can't see in the morning" to "can't see at night." They picked from two hundred to three hundred pounds a day. The fields were dew-soaked in the early morning, and feet and legs and dresses and overalls became soaking wet, as did the long canvas sack that hung from a strap over one shoulder and dragged along behind the picker through the wet cotton. Men and women and children of all ages were in the fields now.

Few were left behind to care for even the youngest children, and tiny babies often lay on pallets in the shade of the cotton plants or were left beneath the cotton wagon where the sacks were emptied. Little children played or whined and cried all day long in the fields, or trudged back and forth in the heat, carrying drinking water from the nearest pump. But the children seemed to learn early that the cotton field was the most important part of their lives and they were seldom rebellious.

When the sacks became too heavy they were carried to a waiting cotton wagon, and there weighed and emptied. A pair of balance-scales hung from the wagon tongue, which was generally propped up with a neck yoke. Later on I learned that many of the pickers tried to get their cotton picked early when it was wet with dew, for then it would weigh more. I heard also that sometimes a picker would manage to throw a pail of water into the cotton sack to increase the weight, and sometimes the scale itself was tampered with, for a blob of lead would be soldered beneath on the lower side of some of the weights to deceive the owner. So, a lot depended on having an honest observing weigh-boss. The man who acted as weigh-boss entered the weights in a book.

Wages for picking ranged through the years from fifty cents to two dollars a hundred pounds. Cotton eighteen to twenty cents a pound usually paid a dollar and a quarter a hundred.

From early September until Christmas and even into the next year the cotton wagons rumbled along the roads to the gins. High side-boards were added above the regular wagon bed, and the loose cotton was heaped high. Bunches fell off along the road or caught in the bushes and soon lost their snowy whiteness.

At the cotton gin the wagons waited their turn to drive in beneath the huge suction pipe that could empty the load in half an hour. An average load was about a five-hundred-pound bale—or about sixteen thousand pounds of loose cotton. The circular saws tore at the cotton, and loosened it from the seeds. The cotton lint was then carried through a suction pipe to the big baler and the seeds blown through large pipes to the seed house adjoining. The big bales were bound with metal bands, each bearing a number. From here the bales could be taken to the compress and reduced in size. These were easier to handle and took up less space in the warehouses.

A renter usually paid one-third of the corn and one-quarter of the cotton he raised to the landlord and had his own teams and tools. A share-cropper paid half of the cotton he raised and the renter furnished him with teams and tools. His clothes, food, and medicine were bought through the year—using the renter's credit at the nearby corner store, or in town at some larger store.

Some of the landowners ran their own stores and the poor share-croppers were compelled to trade there.

There were also renters who paid cash rent. They fared better at times, but if crops failed they were utterly lost. This then was the story of the cotton.

6. You-all Come!

I GRADUALLY BECAME acquainted with my neighbors, there in town, for they often came a-visiting in the afternoons. When I visited them we usually sat out in the porch swings, long high-backed seats fastened to the porch roofs with heavy chains; sat and sewed and talked gently and agreeably. All the conversations I ever had with any women in the town were very like that—sweet and low-voiced, with no arguments, no issues raised, no subjects discussed that might cause the least difference of opinion.

When I had dared, at first, to remark on the conditions that were strange or unjust to me, I was met with only a sweet bewildered attitude, a seeming indifference or ignorance, I couldn't guess which, and a clever maneuvering of the conversation to smoother channels.

One afternoon when I was visiting my nearest neighbor and we were sitting out in her swing, a fat old Negro woman ambled down the street, dragging a child's wagon heaped high with ironed clothes. "Just look at her!" said my neighbor, calmly, lifting her eye from her embroidery. "The triflin' old vixen! I just know she's got my pink silk teddies, probably wearing them right now. I counted every piece of last week's wash, but she said she never saw my teddies. You can't trust them," she ended complacently.

My first introductions to the neighborly art of borrowing began when a little black-eyed girl appeared at my back door early one day to ask if her mother could borrow some "baking powders." Again the same child came as a messenger to ask if I were going to wash, for they had seen me bringing in pails of water from the faucet in the yard. Her mother wanted to get the soapy water when I was through with it to scrub the kitchen and the porch at home. Later the child carried the warm soapy water back across the road in pails, where the floors were scrubbed with a broom and rinsed with water from their well. Thus they had not only the good warm suds but were saved the work of drawing water up from a forty-foot well, for even though they had a pulley arrangement and didn't have to work a pump handle, hauling up the water was hard work.

I admired the quick obedience of these same children and their constant training in manners and in little courtesies. This early training of children was what produced so many of the young men and women famous for their good manners and genteel ways. Most Southern men, I noticed, were very much at ease inside a home, and gave a lot of attention to what were usually thought to be only women's interests. They seemed to have a marvelous degree of understanding, especially of women, hence the renowned fame of the charming Southern gentleman.

Whenever I was a visitor in their home, these children upon entering the room where I sat would speak to me pleasantly, and they always came to say goodbye when I left, their mother hanging on their every word and forming the proper phrases with her lips and correcting the least mistake they made. I used to try to tease them when they left my house with their pleasant "You-all come," by saying, "All right, only all of us can't come—won't I do?" but I soon gave up, as I found that "You-all come," upon leaving, was the Southerner's way of expressing approval of you or a desire to continue your acquaintance. Someone told me that if ever my visitors took their leave, or I left their home, without those magic words being spoken, I would know that I had been cut off from further social favor.

Even later, down on the farm, I was often in the midst of the weekly washing when a neighbor would appear, to chat or to borrow something. Upon leaving there was always a pause, and then the friendly, "Well, you-all come and go home with me." Even little children, sent in on an errand, would edge away mumbling, "You-all come," or "Come and go home with me." Once I said in reply to a near neighbor, "All right, just wait till I put on my bonnet and I'll go." It was about nine-thirty in the morning and I was busy frying sausage, sealing it up in quart bottles. She looked surprised and bewildered and did not understand. Smooth, meaningless, all-embracing, hospitable, friendly, was that expression—rather clever after all.

In the fall, the white men and boys as well as the Negroes went 'possum and coon hunting in the nearby hills.

Often, when the nights were clear and frosty, the stars shining and the moon out in all her glory, I would awake startled by the unearthly, blood-curdling sounds I heard. It was only the hound dogs out with the hunters, baying in the hills near town. The baying of the dogs was deep, penetrating, and mournful, and gave me a fresh appreciation of all the old stories I had read of runaway slaves, and I remembered, too, the baying of hounds across the wild moors in Arthur Conan Doyle's *Hound of the Baskervilles*.

Once, in the middle of the night, we heard a neighbor's chickens squawking wildly. They were roosting in a rudely constructed shelter— just open roosts covered with branches, near an empty lot. Wayne dressed and went out to investigate, for the chickens belonged to a poor old widow. After a long time, he came back carrying a 'possum by the tail. It had become wedged between the fence and a large bough as it tried to get in at the hens. I wouldn't touch the ugly looking beast, but we locked it up that night and next day another neighbor looked at it and offered to take it home and cook it. He said he would bring us half of it. I refused any share in the "greasy doings," as he termed it, but later I did taste a bit of the meat and found it much like very soft, fat pork.

The colored men who lived in town worked at odd jobs, or hired out as field hands to the farmers. Many of the women were hired to cook or wash, or they went out working by the day doing housework, but the most of them worked with the men in the cotton fields. Many of these Negroes were the descendants of old slaves, and were very law-abiding and seldom in any trouble.

When I was up on Nebo, Miss Hallie had told about a young Negro who killed a white woman. The young man knocked at the door, late one night, where a widow and her daughter lived alone, and when she opened the door he felled her with an axe. The daughter escaped through a window and got away.

The woman had been known to keep several hundred dollars in the house and robbery was presumed to be the motive. The Negro boy was later judged insane and put away.

The thought of this ghastly affair used to sicken me, especially at night if I couldn't sleep. Then I accidentally found out that the house we were living in was the very one where the murder was committed, and the same family of Negroes still lived across a vacant lot from us. I could see the little black-eyed children playing around their cabin whenever I looked out my kitchen door.

What havoc one's imagination can work! One day Wayne and Henry and Frank hired a team to take them down to the lower place. It had been raining and the roads were quite impassable, so they left the car down at the mule lot in town. They "aimed" to be back by nightfall. I was busy all day. I had made "starter-bread" and was waiting for it to rise, but the day was cool and the flour rather poor, so the dough was slow in rising. I covered it carefully after lunch and went with Jimmy and Donald up to the nearby hills for our usual walk. This "starter" yeast was given to me by Mrs. Turner. It was a yeast that was kept by adding sugar and flour and water each time the yeast was used for bread making. The neighboring women seemed to keep this yeast going year after year, those who neglected it borrowing from others and adding the sugar and flour and water until it bubbled in the fruit jar and was ready for use. Some one told me that the original "starter" had been brought from Missouri.

We gathered leaves from the red gum trees, leaves of clear yellow and of beautiful shades of red verging to black. They resembled so much the gorgeous Canadian maple leaves that I used to gather long ago. We returned with our arms full of branches from the trees and filled our little house with their beauty.

When the lamps were lit and our supper over, I put the bread into the pans, but I already knew that it would be a failure.

It was the first time we had been left alone. We sat in the kitchen near the stove. I sat and rocked and sang to the boys and answered their countless questions, keeping them amused with stories, and hoping every minute to hear Wayne's step on the porch.

The hours went by. The fire was built up again and again, and the bread—white, "sorry" looking loaves—was at last put in to bake. Either the yeast was too old or the flour was poor, for the dough did not rise properly. I was glad that the bread-baking kept me occupied, and I kept the children up with me, for I dreaded going to bed.

I was thinking of that family of Negroes that lived so near, and realized that this kitchen was the very room where the woman had fallen when the boy struck her down with the axe.

As it grew late, I could feel my spine stiffening and my voice sounded queer and far away when I answered the boys' questions. I drew my rocker back so that I could watch the doors and the window, too. The boys got drowsy at last and wanted to go to bed, but I would stir them up with another story and a promise to go to bed in just a few minutes.

At last it was obvious that the bread was as well baked as it would ever be. I was disgusted and disappointed at the failure of my baking so I opened the back door and carried the heavy loaves out on the porch,

hoping some stray dog would carry them off. I was out of breath with nervousness as I hurried back into the house. By this time I began to realize
that something had detained Wayne and I decided that he would not be
home until the next day. The children were soon fast asleep in their bed,
but I was afraid to undress.

I got the butcher knife and the axe from the storeroom and hid them
near me under the spread on my bed. Frank had left me his gun before
they left, a thirty-two Colt automatic, saying in a half-joking, half-serious
manner that I might need it, and although I was almost afraid to pick it
up, yet I hid it under the pillow. The lamp I left burning, and got into my
bed, removing only my shoes. I lay there reading, but listening tensely to
every sound. Sometimes I would doze off, only to jerk awake in as real
and acute fear as I have ever known. It was about daylight when suddenly
I heard a loud rapping on the back door. I put on my slippers and took the
gun from beneath the pillow. Frank had told me how to fire it, and I had
my finger on the trigger, although I had never fired off a gun in my life.

I remember standing, wondering what to do—whether to go to the
door or not. I knew it wasn't Wayne for he would have come to the front
door and called my name. Still the rapping continued.

I took the lamp and set it on the kitchen table and slowly unlocked the
door. A feeling, half-curiosity, half an undefined fearful urge, drew me. I
felt a strong conviction that I must open that door, must answer that sum-
mons. With one hand held behind me, gripping the gun, I stood peering
out through the screen. It was just that time before the real dawn when
the world is in half-darkness. There on the porch stood a Negro, about
four feet away from the door. Yes, my heart stood still, just as they say
one's heart does. I remember thinking, "I must shoot him—I must shoot
him," but I was so scared I nearly fell down.

The man had on a dark overcoat and a cap pulled over his eyes.
"Ma'am," he said, "has you all got a shotgun? Us wanted to go hunting!"
I don't know how I managed to answer. I heard myself saying, "No, we
haven't got that kind of a gun," all the time thinking, "I must shoot him—I
must shoot him." I tightened my finger on the gun, and was just bringing
it around to my side, when the Negro turned away and went down the
steps without another word. I watched him leave the yard, then calmly
locked the door and went back to bed with the tears running down my
face. Yes, I was sure enough scared.

When the sun rose and the new day began, my courage came back.

The children went out and looked to see if the bread was gone, and it was. Some stray dog had taken it. I spent the day helping the boys with their lessons, doing my housework and sewing.

Toward evening I walked across the road to my neighbor's home. She sat sewing by an open window.

"Well, Mr. Wayne isn't back yet, I see. I reckon they found the roads pretty bad in the bottom." I assured her that they would be back any time now, and started talking of our walks up on the hills.

I had thought to tell her of my experience of the night before, but somehow I couldn't—I don't know why. I suddenly realized that Wayne was not back yet and though I was still hopeful, I was not sure that he would get back that night, either. I felt the tears start in my eyes and turned my face away. I know now that I should have boldly asked permission to stay at her house till my husband returned, but somehow I had a sense of being a stranger to the town and to everyone in it. My lack of understanding of the true kindness and hospitality that would have been so freely and gladly given held me from speech.

I walked across the Bermuda grass slowly, my head hanging, my feet heavy, and my mind in an agony of dread. I just couldn't stay in that house another night.

Suddenly I heard a car coming and was hailed by a cheery voice. They were back! I hurried forward and we all entered the house together. How glad I was; so relieved. Wayne gave me a searching look and said, "You were scared." "No, I wasn't," I said gayly and nervously. "No, I wasn't. I had the gun." It wasn't till months later that I told how I had spent that dreadful night.

Somehow in all the years afterward, I never felt the least fear, although I stayed alone many a night with nothing but screens on the doors and windows. Many Negro families lived back by the bayou down near our farm, and passed along the little road by our door when they used the farm for a short cut, but I felt I could never be more frightened than I had been, and as time passed I became accustomed to many things.

Our days were really uneventful, yet each day brought a new interest. In a box of books which I had shipped along with our furniture was a big volume on medicine, an old-fashioned German doctor-book, profusely illustrated in bright colors. Pictures of children with measles and smallpox and all kinds and degrees of fever, and many with faces half-eaten with dread diseases, were all between its covers.

One afternoon an old woman who lived down the street came to call on me, and seeing the big book, asked if she might look at it.

She sat turning the pages, looking with interest at the pictures. Finally she said, "I sure wish I could read. Like nothin' better than to follow through some disease to see if them doctors done what we usta do for the same things. Would it be asking too much if you was to read me a little? I'll call out the ones I used to nurse, and maybe you could read me the medicines, or maybe it says what you do fer them diseases. I allus hankered to know them things."

So I read the routine care to be given a patient, or the medicines that were listed, while she sat there nodding to herself, following each word eagerly. She named the cases of illness she had known in her own family or among her neighbors. She had served as a midwife and nurse for years, out in the hills, although she had never been to school in her life.

As the cold weather began, I noticed in all of the houses where I visited that the family life seemed to drift to the fireplace. The rest of the rooms were unheated, so after the meals were over and the fire out in the cookstoves, everyone naturally gravitated to the main living-room around the open hearth.

In summer the front porches were generally used as outdoor sitting rooms. The heat and the idle days gave a needed leisure but the people generally seemed to be without avocations or hobbies. That old joke of the ignorant Southern farmer who just sat and thought, or sometimes just sat, was equally applicable, I found, to many here who had had more advantages. With the exception of a few of the elderly men, who were referred to as "readers" because they had a library of books in their homes, the usual thing was to see all the men idle. Just empty-handed, empty-minded, and idle. Resting. Many of the women drove aimlessly about town in their automobiles in the afternoons and evenings, visiting from house to house, or taking their children for drives out along the country roads, so as to keep cool. There was a popular place for swimming in the river over near the Rock and we often saw the cars driving past filled with young people going to swim. Once we walked over there too, and watched the fun. The water was deep and warm, but it seemed to me there were more well-dressed young men and ladies watching the swimmers than there were those in the water. It was a Southern social gathering, I thought.

Saturday afternoon in town was the general shopping time for the colored people as well as the farmers, for the town women did their shopping mostly during the week. The field hands didn't want to work on Saturday afternoon, and so it had gradually become a settled custom among the farmers to finish up odd jobs Saturday morning and then go into town to buy the week's supply of groceries, or to loiter along the sidewalks and in the stores, talking with friends.

The drug stores would be busy serving Coca-Cola and ice cream, and a lively trade went on with the occupants of cars parked along the street who had their ices and cool drinks served to them on trays which were attached to the car doors.

In the evening the tiny motion picture theatre would be crowded, although during the hottest weather pictures were shown in a small open-air theatre, and even though a sudden storm or a heavy rain might bring the performance to an abrupt end, this was preferable to enduring the intense heat indoors, where the electric fans only seemed to make the heat more insufferable by contrast.

Early evening was the time, the country people having gone home, when the short business street appeared quite gay for a while under the street lights, as the town people in light-colored clothes shopped or loitered with friends beside the parked cars, laughing and talking, speaking to everyone. No use here for even the lightest jacket or wrap, I found, and the light trousers and white shirts of the men and the thin bright muslins of the women gave the street an atmosphere that was sadly lacking in daylight.

In my early ignorance of Southern customs I went down the Main Street one Saturday afternoon and ventured into a general merchandise store. While I was busy looking at a fashion book, the store filled up with customers all milling about in the aisles and laughing and talking excitedly. To my amazement I saw that they were all Negroes, and I was struggling to work my way toward the front door when Wayne discovered and rescued me. The druggist had told him of seeing me enter the store. I was actually conscious of being rescued, too.

On the street in the fall it was a common sight on Saturday to see country boys sauntering about eating bananas, or squatting along the sidewalk munching crackers and cheese. They had come in to get their cotton ginned, to meet their friends and amble up and down the street seeking excitement and companionship, or to wander in and out of the stores to purchase new shirts or shoes to wear to the usual Saturday night dance.

Once I watched with amazement a cotton wagon, heaped high with loose cotton, being driven down the street to the gin. An old bewhiskered man sat up on the high seat, hunched over the reins. The sway-backed mules stepped along, indifferent to everything. Behind the wagon, walking in single file, were a woman and four half-grown boys. The woman looked tired and old and dirty. Her dress was torn and faded, and the sunbonnet that she carried in one hand was a gray wilted rag. The boys, in rolled-up overalls and torn faded shirts, plodded along behind their mother. None of them looked around at the people on the sidewalks. They just plodded along, their heads down, the dust rising in a heavy cloud around them. No one seemed surprised but me.

Around the corner from the drug store were several cotton wagons piled high with baled cotton, waiting for the cotton buyer. Each bale was encased in coarse brown cotton webbing and bound with metal strips. The gin number of each bale was stamped on the metal. I watched the buyer climb up over the wheel and slit a gash in one of the bales with his knife and pull out a handful of loose cotton. Then he stood by the wagon and picked the cotton from hand to hand to straighten it out so he could judge the length of the fiber and estimate its staple and worth. The handful of cotton was then thrown carelessly down on the street and the rest of the bales tested in the same way. A Negro who worked around the drug store sauntered out and pocketed the discarded cotton. He saved enough in this way through the fall, I was told, to make considerable money, and this was one of the reasons, too, why so much dirty lint was blown about the streets.

Yet, what surprised me and pleased me most was the courtesy and gentleness of everyone in town. Whether rich or poor, white or black, whenever I passed any man or woman, as I strolled around the streets admiring the flower gardens and the huge magnolia trees, a peculiar emotion swept over me. The women all had a warm friendly smile and a "Howdy," and the men removed their hats and nodded, or drawled "Howdy" in their deep sweet voices. They were all so very appealing with their low, slow speech, for they seemed to express a sense of real kindness that gave one a feeling of being recognized and considered personally.

This was, indeed, an old Southern custom, and I liked it and unconsciously adopted the habit myself. It was as if there was a real recognition of the bond of the great human family, and no one was ignored. This was the feeling I found, that fostered the deep interest that was taken in every least detail in the life of their own kinfolks as well as those of their friends and neighbors.

I learned to nod and speak to all I met whether I knew them or not. I saw that everyone, upon meeting a friend or neighbor, would stand right there and begin a long conversation, for time meant so little here, and friendship so much.

As time went by, I was somehow pleased and satisfied to see so fully exemplified, in the life around me, all the little verses and mental pictures that I had held through the years as representing the sweet serenity and unruffled life of the South.

I used to hum the words:

> *It's an old Southern custom,*
> *When you're walkin' down the street*
> *To bid everyone "Good mornin',"*
> *Don't you think that's kinda sweet?*

7. The Bottomless Pit of Misery

OUT FROM TOWN near the foothills were many small fields of cotton, the stalks gray and leafless now, but heavily spotted with snowy cotton still unpicked. Across the browning fields the houses that were nearer town became more and more exposed as the trees that had sheltered and shielded them from the hot summer sun lost their leaves and became branched outlines against the autumn sky.

On our walks Jimmy and Donald and I often climbed up on a big out-cropping of gray rock, where, through the openings in the pine trees that grew here, we could see the wide channel of the river, brown against the gray banks. We found a narrow twisting road, too, that ran through the woods close to the river, where tall elms and hickories and slippery elm and paw-paws grew. One day we gathered yellow goldenrod and fringed purple asters. The blue jays and woodpeckers darted through the woods, and here and there in the gray fields we saw the brilliant red birds making a scarlet spot on the limbs of some leafless trees.

Everywhere, as the afternoons waned, the crickets' shrill pulsating chirp filled the air and great flocks of blackbirds gathered in the trees and called with the throaty, liquid sweetness that stirred in me a sad, half-forgotten memory.

All the Southern landscape now reminded me of pictures I had seen on old-fashioned calendars where there was just such a glory of autumn sky and blazing woods. The afternoons now were sunny and clear like those in the fall days in California, yet there was a soft pearly-white mistiness in the air, accenting the blue, blue sky, and holding a definite feeling of the cold whiteness that lay behind. At last there came a real cold snap with deep frost, and we found ice on our water pail in the kitchen one morning. The oak and gum trees that had been so beautiful with their gorgeous colors now dropped most of their leaves and even the last blades of grass were dead and brown.

One day we noticed a peculiar whiteness in the air. It seemed as though it might come from the very snow that we felt was so soon to fall. In the morning we awoke and the whole world was white. It was the first snow that I had seen since I was a little girl and I looked out in ecstasy on a land of enchantment, strange and beautiful and unreal.

In the days that followed we built a large snowman on the porch, and took long walks up on the hillside where we had wandered in the sunshine such a short time before. We never tired of admiring each separate laden twig or noticing the beautiful shadows made on the snow by the overhanging branches that bent beneath their soft weight of white.

We wandered up through the woods and climbed again to the high rock above the river, going single file and stepping in each others' tracks to keep our feet and legs dry. Here we took snapshots of ourselves, perched on the fallen logs in the bright sunshine that made a glittering paradise about us. Such clean, crisp breaths we drew, with reddened cheeks and sparkling eyes, but, becoming chilled at last, we would hurry home and heap wood on our fire and gather around the fireplace, feeling doubly snug and warm after our trip through the winter wonderland.

From the neighbor children we learned the trick of making smoked snow-balls. Snow was packed tightly in a ball and held on the end of a stick in the heavy smoke of the fireplace where hickory wood was burning. The outside soon turned gray and became saturated with the smoke flavor as it began to melt a little, but it really did taste good. We learned also to make snow-cream by adding sugar and cream and vanilla to a dish of clean snow.

Wayne and his brother were busy those days. They were undecided yet whether to put a new tenant farmer in charge of the upper farm or move onto it ourselves. It would be a long distance for Wayne and Henry to go each day from town down to the farm to work, and unless the farm was put in charge of one man who would have the authority to hire share-croppers, we could foresee difficulties arising, even though we were such close landlords.

Although we had come to no open decision as to our future plans, still I saw that mules and pigs and cows were being bought and taken down to the farm, and some farming equipment and good tools were bought at a bargain from farmers out in the country. Nearly every evening Wayne and Henry sat by the fire talking endlessly about crops, prices, and mules, and the people they had met that day, and making plans for the coming year's work. Too sleepy to listen any longer, I would finally go to bed to awaken hours later and hear their voices, as they laughed and talked, enjoying their own society immensely.

I was surprised to find that I was beginning to feel eager, even willing, to put my shoulder to the wheel in helping Wayne to have a good trial at his heart's desire, farming here in Dixie, but I was beginning to wonder whether I would be of much help, a rebellious, homesick woman, my eyes seeing little good in anything around me, my heart always aching for the dear familiar life we had left so far behind.

It was while some snow was still on the ground that my neighbor called me to the window of her sitting-room, one afternoon when we had been sitting by her fire sewing. "Look out there and you will see something that will make your heart ache," she said. I looked and saw that cotton pickers were picking the cotton in the field close to her house. Men, women and children were there, with their pinched faces and ragged clothes. One very young woman dragged along a little soap box, with heavy disks of wood cut for wheels. In this box was a little boy, just old enough to sit up. The child wore nothing but a short bit of a cotton dress and a tiny ragged cotton sweater. The poor mother wore a thin cotton dress, and had no wrap whatever. There were tears in my eyes. We were standing by a roaring, fire, and I knew how bitterly the wind was blowing outside. But my neighbor assured me that her daughter was hunting up some of her children's old clothes to take out to the girl, and that I needn't cry, for if I did I would be always in tears. "There are lots who are worse off than those you see there. You'll get used to it if you stay in the South." But would I? It was my first inkling of the bottomless pit of misery.

That same evening, as darkness came on, Wayne, who had been down to the lower farm again, came rushing in and asked me to hurry and get some old clothes, as a team would be along any minute from the upper farm, with cotton pickers who had been sent off because of a fight between some of the men. "I wouldn't give those men a thing," Wayne said. "I saw them as we came through there a while ago and talked with them. They are just no good at all, and that is saying a lot for here; but there is a woman and her daughter, a young girl with a little baby. They sure need some warm clothes.'" I soon found an old coat, a heavy skirt, a sweater and two knit caps. Wayne took these, and went out to meet the wagons. He told me about it later. After the first wagon passed by, with the two men, chewing tobacco, perched on the high seat, he stopped the other wagon and passed the clothes up to the two women who sat huddled on the old worn quilt. The girl, for she was only that, was trying to keep her baby warm by hugging it to her with her thin unprotected arms. She reached for the bundle and the tears ran down her face, so that she turned her head away. The old woman tried to thank Wayne, but she too broke down and cried openly. They were both shivering with cold, the baby was sick, and Wayne said it seemed to him that the thin dresses they wore were all the clothes they had. Both were barefooted. They were headed up the river road to camp out on the hill outside of town, where we ourselves had spent the night when we first arrived.

Grandma, who was visiting us, was so worked up about it all, that she immediately started off for town, and succeeded in demanding, for this family, food and clothing from some of the merchants, and medicine from the drug store. She had always been known as the "Old Blue Hen's chicken," when she had lived here years ago, and woe betide anyone who stood in her way when on charity bent. Missionary work was her delight. That night, after seeing that these poor people had been properly attended to, she talked to me for hours about the sad cases she had taken care of all her life. I seemed to notice, however, through all her stories, a calm acceptance of such conditions, taking them all as a matter of course, evoking much sympathy but little surprise.

As we sat there by the fireplace talking, Wayne and Henry began to tell of their experiences that day in the lower bottom. I never saw them so disturbed and so angry. Grandma darted her sharp eyes at me several times during their story. I felt my heart contracting and I became cold and nervous as I listened. They had parked the car and were passing the one-room shack on the bank of the Petit Jean near the ferry crossing. The door of the house was swinging on a broken hinge and they saw that in the old broken-down bed lay an old man, so they stopped. When they heard him groaning, they went inside. He was suffering from a large carbuncle down low on the back of his neck. Beyond him, in the same bed, lay his daughter, a girl about twelve years old, sick with malaria fever. They were both covered with a pile of old ragged quilts.

The house had only an open hole for a window, for there were not even the usual boards that should have been fixed to cover it, boards nailed together and hung on strap hinges. There was a little iron heater, not over three feet high, with one lid on top, but no wood was to be seen, and the old safe held only a few broken dishes—no food. And, oh, it was such a cold winter day. The man lay moaning and groaning, saying he wished, like Job, he could curse God and die.

His brother had been coming in to see him, he told Wayne and Henry, bringing some vittles, but had now gone to see about renting a place for the new year. His daughter had tried to carry on with the cotton picking, but she had fallen sick. He was a widower. He had a young son off somewhere staying with kinfolks. The doctor had sent down some medicine by his brother, but knowing that it was poison, the sick man had put it away somewhere for safety and couldn't remember where. And, anyway, it had to be put in hot water and poured on the carbuncle. He turned his face away and again groaned that he wished he was dead.

The nearest house was over a mile away, but Wayne and Henry got a man and his wife to promise to go and attend the sick man and the girl. As they came through town, they arranged at the drug store to have fresh medicine sent down with a man who was going to make a trip that way the next day.

That little shack was built right on the bank of the muddy river. The bank was high and steep there and the house perched right on the rim.

Cotton pickers who lived in this shack always had to get their water by following a trail down to the water's edge and dipping it up. This was the only house available for pickers while gathering the cotton in the nearby fields. Other pickers had either to camp out or make trips to and from other sections. The owner did not live in this neighborhood, and was interested only in the amount of cotton picked. He seldom saw his tenants and cared nothing for their problems.

The river was generally muddy from rains, but in summer, during a dry spell, when the creek was low, a green scum must be cleared away before the water could be dipped. Then it had to settle before it was fit for use. Of course it made them sick.

The old man had mumbled, "It's a right unhealthy place. The skeeters is sure bad, but there's lots of carp and cat. But I wished I was dead."

— · —

I remember that Thanksgiving Day was clear and cold, with bright sunshine during the afternoon. Grandma and Henry had dinner with us that day. It seemed that we were more conscious of expectations than of thankfulness, for all our thoughts were on the future as it lay ahead, full of uncertainty and change.

But that first Christmas in the South was a delight. For the first time in our lives we had the opportunity of choosing our Christmas tree right where it grew, a beautiful young cedar out among the young trees on the hills. Wayne cut the tree and we rolled it down the slope to the car. We pasted strips of colored crape paper into loops and made long chains of red and blue and pink. Little wax candles of delicate lavender, green, yellow, and red, were set securely on the out-flung branches. Popcorn and cranberries were strung and hung in festoons, and we made popcorn balls and roasted peanuts for peanut-brittle candy. High at the top of the tree, beneath the twinkling tinsel star, we placed our beautiful Christmas paper angel that had hung on the tops of our Christmas trees for years, all silver and pink and blue. This was "Mary" Christmas, Santa Claus's lovely daughter.

The surprising use of firecrackers at Christmas time was a great pleasure to Jimmy and Donald. I never became reconciled to the idea of fireworks being used in the season that rejoiced in the birth of the Prince of Peace, but all over town firecrackers were set off and everybody seemed very happy.

When the firelight danced on the walls and we pulled down the shades in the evening, we would light the little candles on our tree with great care and enjoy the wonderful perfume of the wax candles and the cedar boughs in the warm room, admiring our own handiwork that had helped to create such Christmas beauty.

PART II

That Heap o' Livin'

8. Poor Little House

IT WAS A DAY late in January when we finally got moved down onto the farm. As the big packing cases and pieces of furniture were being unloaded onto the front porch, I climbed up onto the log at the back porch of the little share-cropper cabin by the pecan trees and walked into the tiny kitchen, my mind in a kind of daze. Wayne had thoughtfully told me the advantages we would gain by choosing the middle house where the Wests had lived. It was away from the other houses, and we would have more privacy. Grandma and Jim, it seemed, were happy to have the house nearer the barn where the mules and cows were kept.

Conscious of the fact that everything was definitely settled at last, and that we were actually moved down into the bottom lands, still I was far from happy, accepting it all dully, inwardly rebellious. We wouldn't be here very long; oh, no, we couldn't stay here. Not for more than a year at the most. The lawsuit would soon be settled, the farms would be put into good condition and left with a good tenant farmer, and then we would all go back together again to California. Surely.

Part of me was benumbed and dead as I looked at the dirty empty house. It was a definite shock to realize that this was really going to be the place where we would live. I couldn't cry, but I felt helplessly ill and beaten as I wandered from room to room and looked out the dirty windows across the fields of dry cotton stalks. Cotton stalks that were bent and brown and crowded up against the house on two sides, their bareness now exposing for yards around the accumulation of gray weathered rags and old tin cans and bottles that had been thrown out from the house into the fields.

The packed earth near the back porch was covered with a dried white soap film from the countless tubs of wash water that had been overturned there. Greasy dishwater too, I saw, had been tossed out through the broken screen of the tiny kitchen window—and remnants of food and caked grease still were smeared along the boards where the window slid open.

Because the ceiling was so low the window had been built so that it could be pushed along sideways in a wooden slot. All the house screens, I now saw, were very rusty and almost entirely gone.

The men had to stoop to pass through the little dining room, and I noticed now that the ceiling of tongue-and-groove boards had broken loose from its moorings above and sagged low overhead.

However, my rescue from sick depression lay within my own powers. I was soon directing the work, seeing to it that our things were left outside on the front porch until the big black pot, which Mrs. West had left for me, and which lay on its side by the pile of ashes behind the house, was filled with water and a brisk fire of driftwood kindled underneath. Mrs. West had also left us the tiny kitchen stove, for it belonged with the house, and there was a stove in the house to which she was moving.

Boiling water, with lye added, was used to clean the floors and walls. First, we took a zealous pleasure in tearing the old paper from all the walls. It was about four layers thick, some of it put on twenty years before, Wayne told me, for he knew when his father had had these houses built. The paper was fastened with large tacks and "shiners," as Donald called them—round, tin, concave disks about the size of a fifty-cent piece. We found each layer of paper stained and spotted.

The baseboards in the living room were streaked with brown snuff stains, too, so we pried them off and carried them outside. These were scrubbed with old brooms, using the hot lye water and soap. In the sitting room, too, on the north wall, we removed ten layers of paper. Underneath, stuck fast to the bare walls, were empty mud-daubers' nests. I counted two hundred in that room, small gray mounds of clay. These we scraped off with hoes. Dirt and dust and caked mud were plentiful between all the loose paper and the wide boards of the house walls.

When the paper was all burned in the yard and the walls sluiced down, we walked around the house and discovered that many of the narrow laths that had been nailed on to cover the cracks between the boards were gone. They, like the shingles of the roof, had gradually become loosened through weathering, and we were told that many had been deliberately snapped off and used for kindling by the various tenants.

By nightfall the house was dry. We spread our two carpets over the bare floors of the front rooms and carried in the furniture. We had been ill-advised in shipping any furniture at all, and some of the chairs were broken, but the huge lounge, the sewing machine, the dining set and the rockers and dresser were in good condition. We made up the beds, and filled the lamps—and so began the years of the "heap o' livin'" that makes a house a home.

Those were busy days. In the fields the stalk cutters were soon at work clearing the fields for the plowing. As soon as the stalks were cut near our house, Wayne had the ground plowed all around, plowing deep and turning up the furrows to expose the polluted soil to the sun and wind. Where we would walk around the north side of the house and near the pump and the woodpile the sandy soil would soon pack down again, but it would be clean. On the south side of the house we planned to make a flower garden.

Then the excitement we had, sending Jimmy and Donald crawling in under the house, armed with rakes, to drag from the farthest corner the deep accumulation of chicken feathers. The ceiling of the dining-room, where the narrow tongue-and-groove boards held together like a platform, had to be raised, and was lifted carefully with a two-by-four while one man crawled up above and spiked it back in place. The little kitchen was not ceiled and the upright boards were missing over the open doorway into the dining room. This left a large hole through which refuse had been thrown into the tiny attic over the dining room. Broken dishes, old cans, rags and papers were dragged from there with rakes, until a wagon was heaped high with the trash. Parse, who drove the wagon away, remarked to me, with a friendly smile, "You-all shore aim to be clean. Goin' to keep you right busy." He was the one who explained to me the common custom of tearing off the loose battens for kindling.

The old roof had many holes, and the storms of years had filled every crack with dust, and dust lay in a thick layer on top of the boards of the dining room ceiling. It was impossible to get that cubby-hole as clean as I wanted it unless I could have climbed up and done the sweeping myself. Later on, when the winds swept in gales across the sandy fields, the dust sifted into the house from the outdoors and sifted down, too, through the ceiling onto the paper we tacked there. When it rained the muddy water ran down through the ceiling from the leaking roof above and broke through the ceiling paper, too.

Yes, poor little house! It had been built originally along this "sandy front" for the housing of cotton pickers. It had been put together with unseasoned lumber and with a roof of hand-hewn split-oak shingles, large, rough and irregular in size. Neither paint nor white wash had ever been on it, apparently, so now it looked like the hundreds of others in the bottoms, a clear silvery gray, weathered with the sun and rains of years.

Through the years many of the shingles had warped and loosened and been blown off, as they became old and rotten. Parse told me that they had often tried to patch the holes in the roof but I would find that the worst leaks were still around the chimney and over the windows. He said that wherever an attempt was made to patch the holes with new shingles the only result was more leaks in fresh places, for the old shingles were loosened more and more.

I found that this was very true the first time we had a big rain storm. The dull thumping of the raindrops on the roof gradually increased to a steady beating that drowned out all sense of everything else, and soon the "drip-drip-drip" and gush of water pouring down onto my bed and dresser and onto the table and stove caused real alarms.

Nearly everything had to be moved or shifted. Every bucket and pan was put into service. The big wash tubs were also used, one on the kitchen stove, and another on the bed, where a regular stream of water was coming through the ceiling.

We all thought it was exciting at first, and interesting, to lie awake in the darkness and listen to the light-toned "ping" or the heavier "plunk" of the drops as they landed in the shallow wash-basin and the deeper water bucket, but we knew that a new splash meant that the water had seeped through another spot.

Yes, all exciting and interesting, but soggy and stained curtains and wet bedding and the inability to make a fire in the cookstove because the water ran down the chimney and filled the stove, soon became more than an annoyance.

One night we had to get up twice and move the beds to different places, but even so the bedding became damp. The wind blew so wildly that we were afraid to make a fire, and we endured the cold wet hours until daylight.

One night as dusk came on I noticed a peculiar damp grayness to the atmosphere, and a sort of hovering warmth. It was as if a hush had fallen on the earth. Even the animal sounds were quieter. The pigs went hurrying past the house toward the barn, and the chickens gathered early in the hay shed, and the cows came to the pasture gate and stood and waited to be driven up to the big barn.

When we awoke in the morning we noticed long thin lines of white snow across the floor where the laths were missing from the outside of the house, and we knew we must hurry and start putting on the new wall paper.

It was a beautiful world! It snowed nearly all that day and the next night and part of the next day. Three feet of snow! How we enjoyed it! I let Jimmy and Donald take some of our mats for sleds and slide down the bank, and we poured water on the slide at night to insure a frozen surface next day. A snowman now stood on the porch and one out in the yard, and with the addition of pipes and hats and faces of charcoal, they gave the boys much pleasure. Several of the children of our neighbors came up and the boys made big snow forts and threw snowballs. I made sorghum molasses candy and poured it on the hard-crusted snow in the shade to cool for pulling, and the children felt that we were having a winter party.

Down in the meadow across the road toward the river a large patch of old corn stalks still stood. What fun it was to go trooping in a crowd, all bundled up until we could hardly make our way through the deep snow, to trace a rabbit track right to the foot of a corn stalk, where we could see an air hole, and throw ourselves flat on the snow, and run our arms down to the ground and pull up a rabbit by the ears. Fried rabbit for supper!

Every evening Wayne and I worked at tacking the tan-colored builders' paper on the walls. We began in the sitting room, perching on a long ladder under the high ceiling, and sweating from the heat of the big tin air-tight heater that was kept almost red-hot against the cold winds. We held the long strips of heavy paper against the walls, and the large tacks were inserted in the tin "shiners" and hammered in securely. We took turns tacking and holding the paper, for both tasks were very tiring. The high ceiling here did not require paper.

When we had tacked on two layers for extra warmth on the north wall of the living room, and before the third layer was put on, I took a crayon and made a big notice for future readers—giving our names and the date and the record of the two hundred mud-daubers' nests knocked down, and ending with these inspired lines:

> *The snow is deep and wood-pile low;*
> *The sun shines beautiful on the snow.*
> *Red birds are hopping all around—*
> *The snow is deep upon the ground.*

We set up one of our beds in the sitting room. The winds that blew against the north side of the bedroom made that room very cold, but the boys would be tucked into their blankets with a hot water bottle at their feet and would be soon asleep. I always referred to the bedroom in winter as the North Pole. Each morning I would put on cotton gloves to keep my hands warm, go into the bedroom and put everything in order, and then fasten the door securely against the wind. We stayed in the sitting room nearly all of the time, for even with a fire in the kitchen, when the cold wind blew, the draft lifted the rug on the floor, while it blew against the walls as if to force them in. The paper on the walls billowed and strained between the "shiners" and we had to watch it carefully to keep it from coming loose.

The door into the bedroom and the one between the dining room and sitting room were each fastened with a home-made latch. The doors themselves were home-made, of planed tongue-and-groove boards, and the bar that was used to fasten them was a strip of wood whittled smooth and fastened to the door with a wooden peg. This bar dropped into a slot made from a piece of hand-worked hickory. To the bar was tied a piece of stout string, which ran through a hole in the door above it and hung down on the other side. Anyone wanting in had only to pull on the latch string and so lift the bar and open the door. The back and front doors had broken locks, but with regular door knobs. Since the house had warped and shifted so many times during the years, these doors did not fit properly and so could not be tightly closed. We always took a certain pleasure in using the latch string on our inner doors and decided to keep this old-fashioned arrangement.

Soon after we moved in, I had gone back to bed one morning after breakfast, as I felt rather ill. Everyone was out of the house, gone up to the barn or to Grandma's, and I was enjoying the warmth and quiet while I read the last San Francisco paper from home. It was freezing cold outdoors, with a strong wind blowing. All at once a loud stamping on the front porch alarmed me, the door opened, and in walked two men, strangers, who seated themselves comfortably by the stove after a casual "Morning" to me. They had just stopped in, they said, to warm, as they were on their way to town. They saw by the smoke that we must have a good fire going.

I was so surprised and embarrassed that I had nothing to say, but that didn't matter, for I was completely ignored. The men seemed quite at ease and talked and laughed with each other, chewing their tobacco, and spitting on the sides of my heater. Finally they left abruptly with a mumbled "Thank ya kindly."

I lay and thought that over and made up my mind to get the outer doors fixed and put on good locks and to keeps the doors locked.

Mr. and Mrs. West, with Parse and Vena, had moved from my house into the little two-roomed place which stood across from Grandma's, on the bank where our little roadway made a turn near the big barn. The other brother, Claude, and his wife, Vera, went back to the lower bottom. Parse was staying on awhile to work on the farm. They settled down into their accustomed routine with none of our excited housecleaning, although it seemed to me that there was no end to the repairs that should be made in that cabin, too. Their chimney leaned so far away from the house wall that all sorts of rags and pieces of boards had been used to fill the gap around the fireplace, and the wind blew in and chilled the room even when there was a roaring fire. Wayne got all the men on the place together and propped the house back to an even keel on the piles of rock on which it sat, and pushed the wall up against the chimney again.

Strange as it seems, we could not persuade them to tack on the wire netting we bought for all the windows. Since they believed it was the night air and not the mosquitoes that caused malaria, they decided that the flies that came in could be endured, for the screens, they said, cut off the breeze in summer when the heat became unbearable under the low ceilings, and in winter they were not needed anyway.

There were four small windows in the house but most of the panes were broken and the holes stuffed with old clothes. The front porch was full of large knot holes, and here, too, the steps were gone.

I helped Wayne and Henry clean Grandma's house for her, finding there the same dreadful conditions we had found at Mrs. West's house. This house, too, looked as though it had never been thoroughly cleaned for twenty years, and with the house being so close to the big barn, the pigs and chickens had long been accustomed to sleeping beneath the house. Until the bottom of the house could be enclosed with hog-tight wire or boarded up, the pigs went right on burrowing back of the chimney out of the wind.

It gave me an awful start, sitting quietly sewing by the fireplace with Grandma, to feel the floor beneath my feet lift and sway, and hear the sounds of grunting and squealing coming up through the floor, and sense the rough, grating pulsations of the house walls as some pig down below scratched and rubbed his back. A broken hole in the kitchen floor made new flooring there an immediate necessity, so new steps were built then, too, and placed at the front and back porch, and steps made for all the other houses.

The weather was so cold. The sky was gray with threatening clouds. Two men were plowing in the new ground and the rest were hauling loads of wood, using up the logs and huge fallen branches from the trees in the meadow.

We enjoyed the big heater. Wood was plentiful and we burned it all day and far into the night. Great piles of driftwood and logs were gathered from the river banks, to be used until a slack time came in the work, when the men would go to the meadow to cut down the cottonwoods that had been killed by lightning or were old and decaying. Behind the house they had piled a huge crow's nest of bleached branches and chunks. A similar pile was hauled for the other three houses.

Mrs. West expressed her thankfulness and surprise that her need of wood should be taken care of in this way, even while she recognized the saving in time for the men to work together with the teams so as to get back to the regular farm work. For years the tenants had been accustomed to making trips to the river for driftwood in their spare time, and often the houses were left unsupplied even with wood enough for cooking. Cottonwood, when dried and split fine, burned well, but left no coals. The stoves were always full of heaps of fluffy white ashes. Later, good oak and hickory wood would be brought from the lower farm down the river.

"Sure makes a body feel good to be treated like they was somebody, again," said Mrs. West as she stood grinning in her doorway watching the men throwing off her load of wash wood. "Reckon I'm feeling younger already," she added, working her snuff stick carefully around in the snuff can as she prepared to take a fresh dip.

The changes we made were watched with great interest by our tenant families. A long stretch of barbed wire, which sagged from the corner of my little house out to the hayshed, and which had served as a clothes line, was one of the first things I had taken down. We had a real clothes line, with proper posts, put up. Clothes pins were a novelty, I found, as was my dustpan, for "A body don't need all them newfangled notions," I was told by Mrs. West. She had stopped by on her way down to the mailbox to see what I was doing.

"But, didn't the barbed wire tear your clothes?" I asked.

"Well, yes, maybe some, but a body can watch out and take them in if the wind blows right hard; and I been a-sweepin' the dirt right outside all these years. Didn't need no pan as I can see. Well, live and learn. You're right full o' notions I can see, ain't ya?"

Our other family of share-croppers lived past my house down at the corner of the farm where the two county roads met. There was a mail box there nailed to a post, and the mailman left in it the mail for everyone on the farm. He drove a horse and buggy and arrived at this box sometime in the forenoon.

This family consisted of a man and wife and six children. The older boys were fifteen and sixteen years old. The only girl was twelve. The youngest was a baby of four who toddled everywhere after the others. In between were two young lads of eight and ten.

Matt Wood belonged to that certain type of Southern men we were meeting here and there who were really good men—good in every sense. There were many like him, I found, and as time went on I kept my faith in human nature by clinging to the thought of such men as Matt. These Southerners were simpler, unaffected, real, going their own way, aping nobody. Each had definite peculiarities, but each, in his drama of life, with all its problems, struggles, hopes and despairs, met his days successively, year in and year out, with a peculiar calm philosophy, even in the face of almost fated failure. Each seemed to take the same indispensable place in the social scheme of men here on the cotton farms that the common earthworm does in the plan of creation.

I once listed the good qualities we found in this sharecropper of ours. Matt was tolerant, kind, low-voiced, calm, hard-working, clean, and God-fearing, with a sweet gentle smile and friendly brown eyes.

He was a small man, thin and wiry. His jaw was not heavy but was very firmly set. Matt could often be seen helping his wife with the big weekly washings, pumping water, hanging out clothes in cold weather, and emptying the big wash tubs. He worked steadily in his crop and in his garden, and never rested a day, unless he was having a chill, until the crops were "laid by."

His children had all been taught to respect their parents as well as other people. They stayed close at home unless invited to the homes of their friends. When cotton-picking time came in the fall, all, except the youngest child, worked daily in the fields. All the children were very clean even in their work clothes. Their little house, too, was beautifully clean. They gathered wash wood and fireplace chunks from the river banks regularly, and as we allowed them to use the cottonwood from the meadow, they had plenty of stove wood, which was always split and drying in advance of the daily need.

The use of clean sand from the river, with lye and hot water each week when the washing was done, kept their bare floors smooth and spotless. Old brooms were used for scrubbing, and a regular deluge of water flooded all the floors regularly. Fresh newspapers were kept pasted on the kitchen walls, and fresh builders' paper was often retacked on the walls of the other rooms.

They had only a little furniture—four beds and a dresser, a dining table and six cane-bottom chairs; the kitchen stove and the kitchen safe. This safe was a tall cupboard, with doors of wire netting. Here the food and the dishes were kept away from the flies. Extra pots and pans hung on nails near the stove. The beds were spotless, with pillow cases and spreads all embroidered and starched. Their personal clothes were kept clean and ironed and neatly mended.

Each Sunday saw all this family off to Sunday-school or to a singing. Often they brought friends back home with them to spend the evening. Evening in the South, I soon learned, was what we had always known as the afternoon, the hours between twelve noon and six o'clock.

Mrs. Wood told me that one Sunday, years ago, they had left their home to go to church, and returned just in time to see fire consuming the last of everything they owned. It was pitiful, she said, to stand there with the children around her and see their table, which had been ready for their return, burning up, and their dinner eaten by flames, while her children clung to her and cried with hunger and fear.

Mrs. Wood's zeal amazed me for she was so thin and bony. She had such a tiny waist that I felt that even my two hands could encircle it. Always she sang at her work—church songs. Her religion was her life, and she kept as firm a grip on her own temper and emotions as she did on the lives of her husband and children. When she was telling me how much she approved of my efforts to bring Jimmy and Donald up in the way they should go, she explained some of the evils and temptations with which she had to contend in protecting her older boys. She didn't "hold" with dancing. "One Saturday," she told me, "some big boys from along down the road a piece came by to get my boys to go out to the hills where they was having a dance. I sent them on their way right smart. "Call my boys to church—call them to Sunday-school, call them to a singing, but don't you dast to come by a-calling my boys to sinful livin'.'"

I never saw Mrs. Wood idle. She was either working in the fields, hoeing or chopping the cotton, or else she was gardening, washing, cooking, or sewing. And she was always making quilts. Some were piled on the floor, carefully covered with sheets to protect them from the dust, for there were no closets or shelves in the rooms. These quilts were instantly available to make an extra pallet for company—a bed being quickly made up on the floor in some corner of the house or out on the porch in the summertime. She told me that bedding was a constant source of trouble during the summers and fall because the damp hot weather caused mildew to form on the quilts just as it did on shoes and clothes, and they had to be sunned and aired frequently. That was why, on fine summer days, I saw so many quilts hanging out on the porch railings and fences all through the-bottom land.

9. Pussy Wants a Corner

THROUGHOUT THE PASSING years the changing families of sharecroppers in the houses on our farm always reminded me of the game of "Pussy Wants a Corner."

In Mrs. West's little house it was very crowded now, for besides Parse and Vena, another son of Mrs. West's, with his wife and two small children, had come from the hills to visit. Now Grandma and Wayne and Henry and the Wests were busy preparing for hog killing, for it looked like we might have settled cold weather. Among the hogs that had been bought when we were living in town were two that were now selected for butchering.

The butcher knives were sharpened in advance, clean cloths collected to wipe and cover the meat, and the big hogsheads rolled out and set at a slant in the ground to hold the scalding water from the big iron pots.

Lots of oak chunks were piled ready by the crow's nest of driftwood, and the fires were started early so as to have the water hot.

One of the men shot the hogs, and then they were stabbed in the throat so that they would bleed well. The carcasses were dipped in the scalding water, then laid out on rough planks upon saw horses, where the hair was scraped off with knives until the skin was clean and white and slick.

The hogs were then hung up beneath the walnut tree and cut open, the entrails all removed and the carcasses well sluiced out and rinsed off and left to drip and chill all night. The next day the meat was cut up. Each piece was wiped and rubbed well with a mixture of pepper, salt, saltpeter, and brown sugar. This meat was then put into wooden boxes between layers of coarse salt and left until the slabs had shrunken and the meat was somewhat permeated with the sugar and pepper and salt.

The fat that was to be rendered was taken from around the intestines and from the flanks where the hams and shoulders were squared off.

Hogshead cheese was made from the heads and feet. It was a tedious task to wash and scrub these parts, scraping and digging until they were immaculate. After boiling, the water was well skimmed to remove excess fat, the skin and bones were taken out, and the meat and the seasoned liquid put into crocks. As it cooled it jelled into a firm mass which could be sliced easily.

The backbones and spare ribs had to be used soon if they were not to be salted or smoked. Generally they were divided up among the neighbors, as were the liver and brains. If any sausage was desired it was made from the loin or meat from the backbones or shoulders.

Now up by Grandma's house preparations were under way to "try" out the fat from the hogs. Grandma was bossing the job with Mrs. West and Parse and Vena helping. Grandma was as agile as anyone around the black pot, and was certainly enjoying herself.

Grandma was tall and thin, with very black sharp eyes, and hair very black with only a few gray hairs. She was now having an opportunity to repeat many of the farm chores she had done as a girl when she lived in the mountains, and here on the farm she had lots of help and advised and assisted in everything, from making lye soap and hogshead cheese to hominy, which she boiled in a black iron pot in her fireplace.

All the fat was cut into small pieces and slowly melted down in two of the large black iron wash pots, which were set up as usual on old, blackened, empty lard buckets. When the grease had been rendered out it was poured off into the many waiting empty buckets that had been collected from far and near. The yellow crisp remnants, the cracklings, were saved to add to corn bread in place of shortening; this made "shortnin'" bread.

Much of this tasty brittle pork was given away to all the neighbors as a treat, too.

The little grandchildren were running about, trying to help, too; a boy of five and his sister, aged three.

Mrs. West told me afterwards that she had kept telling the two children to go away from the big black pots and stay in the house, for they were always trying to poke wood under the pots so as to watch the flames and smoke.

She said that they would go into the house, but the little boy continued to make many cautious trips out to his mother and generally went back each time with bits of crackling she gave him. These he was sharing with his little sister. He had climbed up and opened the cupboard and both children were sitting under the kitchen table eating corn bread. Finally the little boy made one trip too many, and was harshly told by his mother to go to the house and not come out again. Mrs. West was watching. He had in his hand a large chunk of cracklings and went again under the table with another piece of corn bread. His little sister, seeing that this time she was to go without any cracklings, set up a howl. Stuffing his mouth, little Bud looked over at her and said in a lordly manner, "Oh, you eat your corn bread and quit studyin' about the cracklins."

The perfect panacea for worry!

The black pots of hot grease made me realize almost with nausea the amount of grease that was used in the daily diet of these people. I thought of the story Mrs. West told me about the little boy who opened the kitchen safe and examined carefully the several bowls of cold white gravy he found there, looking to see if they were mildewed. He held one up at last and called out, "Ma, when was this sop fried?"

Mrs. West said she didn't aim to keep any old gravy or any other leftover food; and in this I saw she was only following the common customs for I noticed that biscuits were generally baked fresh and hot for every meal, and big pans of gravy were regularly stirred up. Often during the week, and always on Sundays, cakes were baked, and fried-pies, which absorbed so much grease, were a common part of the diet throughout the year. These fried-pies were made from circular pieces of dough that were usually piled with dried-apple sauce, then folded once across and the edges pressed together. These were fried in a skillet of grease on the top of the stove, each side being well browned as they were turned over in the smoking fat. Dried apples and dried peaches made the best fried-pies.

Even turnip greens were invariably cooked with lots of grease, and fresh lettuce would be wilted with hot bacon drippings and served with vinegar.

Henry used to argue with Grandma about the lard the Wests used, and declared he would not buy them any more of it. At last, as an experiment, he took a little pig and fenced it up in a small pen by Mrs. West's house, and told her he knew he could raise the pig and fatten it on what she threw out. Mrs. West laughed tolerantly at Henry's earnestness, and agreed to save all her waste food. Henry put a big bucket on the old stump by her back door and each day he took the swill, added some bran, and fed the little pig himself. The pig grew very rapidly, and no wonder, for it lived on the best of the groceries that our money bought. True, indeed, the old saying "A woman can throw out more with a spoon than a man can bring in with a shovel."

Out in the smoke house, which was a small shed with a mud floor, behind Grandma's house, the salted slabs of bacon and the hams and shoulders were hung by wires from the rafters, for smoking. Hickory wood and corn cobs were brought for the fire, which was built on the floor but not encouraged to blaze. The fire smoldered away day after day until the meat turned dark and dry on the outside and took on the aroma of good ham and bacon.

Sometimes a preparation called "Liquid Smoke" was bought and swabbed on the meat after it was well cured. In case a warm spell came and the meat did not take the salt, and so was not ready for smoking, a pickle of salt with saltpeter and brown sugar was made and the meat kept in this until it was cured and ready for smoking. This meat was often painted, too, with the "Liquid Smoke," but in either case we always had a smoke house full of hams and bacon, and a good meal could be made ready with little trouble.

I canned dozens of quarts of sausage. Now we could have pork and apple sauce. We had the sausage seasoned and ground at the butcher shop in town. The sausage was shaped into patties and fried, and then packed into sterilized quart jars. The hot grease was poured in on the meat and the tops put on. Then the jars were turned upside down, and the hardening grease served as an added seal. When the jars were opened we heated the sausage in the grease from each jar, and used some of the grease for gravy, to be served with hot biscuits and there you had a real Southern meal.

A short time after this, Avis and Ben arrived from our lower farm to live with Mr. and Mrs. West, and Parse and Vena moved out to a place in the hills soon afterwards. I was glad to have Avis now so near to us. I was finding it hard to be a "Twelve O'clock Feller in a Nine O'clock Town," as the song has it, so one evening Wayne and the children and I went up to visit with the Wests. Mr. and Mrs. West were over at Grandma's. So Avis played her Blue Amberol Cylinder Edison records for us.

That was the high peak of the visit for Avis to ceremoniously play one cylinder after another, each time waiting for our expressed appreciation with as much happiness as though she herself were the musician.

Living day by day, as I was, with no amusement or entertainment and with no music except my own singing, I found that "My Pretty Snow Deer," "Waikiki" and "Down Hawaii Way" made a deep impression on me. While the music played there were no distractions of any kind and so each note and vocal tone could be heard distinctly and relished to the full. We played "Down Hawaii Way" dozens of times, to learn the words, and a feeling of jubilant intoxication swept over me, making me feel excited and exhilarated inside, though there I sat on the straight-backed chair, smiling, and realizing how out of all proportion my elation was. I had no one to explain my feelings to, for I knew they wouldn't understand me if I did.

10. My Baby, Sure Enough

WE HEARD THAT our neighbors' young boy was very sick. He was only twelve years old, strongly built and pink-cheeked, two uncommon characteristics here, for the faces of the boys in the bottom lands were usually of a peculiar yellow-tan, and their bodies were lank and bony. But malaria, and inadequate doctoring and nursing, had proved too much even for Buddy, and now they told me he was dead. That afternoon Avis and I went down to the little house that stood just across the county road on the next farm, to visit Mr. and Mrs. Randall, as was expected of us.

The day before, as I went for the mail, I had called in to see how Buddy was, and found him having a convulsion. It was all his father and I could do to hold him on the bed during the dreadful spasms. His mother was completely worn out, for the boy needed constant attention. His skin was the exact color of a lemon. I made the comparison with some lemons that were on a table by the bed.

Now when we entered the front room we were given a seat near the stove. About a dozen neighbor women were present. The poor mother was in the back room sobbing by herself. The father was out in the back yard with a couple of men.

The beds had been removed from the front room and the long coffin set up on two chairs. In it lay the body, packed in ice and covered with a sheet. They were trying to keep the body until the brother and sister, who were away from home, could arrive. Nevertheless, in the little iron heater a roaring fire was burning, as the day was quite cold. The women had arranged all their chairs around the stove.

I sat down and slowly looked around. None of the women looked at me nor spoke to me. I was a stranger to them and this was no time for visiting. They were all using snuff. Some had their snuff-brush sticking in one side of their mouth. Every so often, one of the women would rise slowly, push her chair back behind her as she rose, go to the stove, swing the cover around, spit, drag the cover back again, and shuffle to her chair, scraping it a little on the floor as she got it back in position and seating herself without even so much as a glance to right or left. In a few minutes another woman would rise up and go through the same performance and sit down. Then another, and another. Each time the chairs scraped on the floor, the stovetop clanged, and the heavy shoes dragged and scuffed on the rough boards. Finally one of the women leaned across me and whispered to her neighbor that she knew this was coming; this, by her jerked-back thumb, referring to the coffin, because only the night before she had seen the death light dancing over the house and had come over and told the mother that she might as well give up hope, for her boy wouldn't get well. This whispered information was received in silence, and the women went on dipping snuff.

By the time about seven had gone through this solemn ritual of advancing and retreating to the stove, I could stand no more, so I went out onto the porch, where a large box made of fresh lumber waited by the window. On it was perched a little girl, the daughter of one of the women inside. She was thoughtlessly swinging her legs and banging her feet against the side of the empty box. Then along came a couple of boys, who dashed past me into the front room and out into the kitchen, calling in a loud voice to the crying mother, "Can we have Buddy's ball and bat?"

I heard Mrs. Randall hushing the boys and putting them out the back door. While I stood there hesitating in the front yard, she came around the house and sat down on the end of the porch near me. She thanked me for coming, and I tried my best to comfort her a little, being all the while overwhelmed with a sense of utter helplessness, seeming to hear my every word strike on the empty air and echo, as though it were impossible for mere words to even approach the mother in her stark grief. Her voice was calm but heartbreaking as she talked to me. Her seeming acceptance of her boy's death with such complete resignation seemed to me to be quite in keeping with the way she had lived—placid, submissive, enduring. Her last words to me still ring in my ears. She said, "I lost a little baby several years ago, and now this was my baby, sure enough."

But life went on as usual. Avis and Ben suggested that we might like to go down to see the school play, so Wayne took Grandma and we all went. We were eager to go for we were hungry for some kind of entertainment. Besides, I wanted to see the place where my boys might attend school.

We drove up in the darkness and parked. There were a few other cars and dozens of farm wagons already there in the school yard. When we made our way inside with the rest of the people who were eagerly crowding in, I found that the seats were all plain benches, without backs, and the floor was of hard-packed earth. No attempt at flooring at all. The roof had a hole in it through which I could see the stars.

Previously when we had driven down past the schoolhouse, I had carefully scrutinized the place and remarked on the apparent absence of any sanitary arrangements. Avis now casually informed me that the boys and girls just went over the bank as plenty of bushes grew there.

Avis also whispered to me, seeing me gazing intently around, "They aim to build a new schoolhouse some day. Heard talk of it for quite a spell now." Grandma seemed to enjoy it all but she made no comments. I felt that she was doing a lot of thinking, though.

Calico curtains were strung on a sagging wire that hung so low across the platform that we could see behind it and watch the preparations being made to send the younger children out as fairies and butterflies with wings made of paper-covered wire. These wings, though they drooped and sagged and flapped loosely, did help amazingly to create a stage setting different at least from the usual school atmosphere.

A few coal oil lamps were set in brackets along the sides of the walls, and in the poor light it was hard to see just what equipment was there for the teacher's use in her daily school work. My whispered questions to Avis brought out the facts, however. There was nothing else at all. There was a small blackboard, but there were no maps, no globes, nothing except the seats, but near the low platform, where a table and chair served for the teacher's desk, stood an old organ.

Right then I made up my mind anew to continue teaching Jimmy and
Donald. A California high school pupil should really be able to continue
on with the work through the grades at least. Here was one teacher with
about forty pupils, all classes from the first to the eighth grade. It meant
a two-mile walk twice daily, a cold lunch, wet feet, sickness, my little boys
gone from me daily from eight to four-thirty—to this place. I sat and pon-
dered. I knew I was capable of teaching my boys, and I wondered about
the teacher's ability.

Someone had hinted that I would be compelled to send my boys to
school, just as all the others did. Not that the truant laws were enforced—
far from it, for when there was work to be done in the fields it was hard
to get many of the parents to send their children to school, as they were
needed in helping to make a living, yet no one was supposed to come out
in open defiance of accepted ways in the bottom, and my keeping the
boys at home had aroused some antagonism.

Since Buddy's death I had found myself more and more nervous and
apprehensive, and my eyes seemed to be opened to find what lay behind
many things I had been accepting. I felt critical and alert and on the defensive where Jimmy and Donald were concerned.

By the middle of February the farm work was well under way. The
cotton and corn stalks were cut, the oats planted, the ground disked and
plowed, and quite a large vegetable garden planted in the strip of black
soil down below the bank. A large patch of sweet-corn was planted there,
too, and away off in a corner by itself a row of popcorn.

Back of our house and extending over to the hay shed the loam was
very rich and quickly made ready. Where I walked along, dropping seed
corn in the hills, the corn could be quickly covered with the hoe, or even a
dexterous pat of the earth with my foot was sufficient to cover the grains,
so loose, so rich was this ground. This stretch of ground was going to be
my very own garden, and here I put in seed for "roasting ears" and peas
and cucumbers.

This garden spot was truly wonderful, for the soil was just what one
dreams of when thinking of an ideal seedbed; dark, soft sandy loam. The
hay shed had once been used for mules and cows, for there was a row of
stalls down one side, so here was some of the richest soil on the farm. I
was so enchanted with the planting of the tiny globular onion sets that,
in my enthusiasm, I planted several gallons of them, in long evenly measured rows. Under the daily showers, and in the warm sunshine, they
sprouted a quick green. In the evenings I used to walk up and down be-

tween the rows, pulling up the little weeds with their long clinging roots that were so quickly grown and so easily released from the loose soil, or gayly hoeing a bit here and there, with infinitely more pleasure than Eve, poor dear, ever lived to enjoy in her garden.

Wayne and Henry sent specimens of the soil on the farm across the river to the Agricultural College at Russellville, and got advice from the farm experts. There were places where the soil was too sandy and several spots that were low and water-soaked. So now they set the plows deep as they turned the furrows, and lime and potash were added where needed, and peas planted along the sandy strips to be turned under later. Many of the farmers living near us declared it wasn't wise to plow so deep. They were afraid we were ruining the land, yet how astonished they were to see the good crops that resulted.

11. Outa Heart

"YOU'RE JEST OUTA HEART, ain't ya?" I looked up, hearing the sympathetic drawl to which I was becoming accustomed, to see Mrs. West, sunbonnet on head, snuff-stick in mouth and an empty lard bucket on her arm, eyeing me with an amused smile, where I sat, down on my knees with a hammer, thumping the caked dirt out of the worn cracks between the old tongue-and-groove boards of the kitchen floor. As I hammered away, the dried dirt would break up and pop out. I was red faced from my exertions, and about ready to cry with exasperation. "Tain't no use," she went on. "A body better let the dirt settle. Jest fill up again, seems like." I told her that I planned to fill the cracks with clean white river sand until I could get a floor covering, but that idea only seemed to amuse her. "Jest a-goin' to keep yourself wore out, ain't ya, ya poor thing. You know I feel right sorry for ya, ya got so much to learn."

She soon went on her way, down to a neighbor's to get some turnip greens, and I sat there thinking. I realized that I was like a little girl playing in a play-house. Everything in the house was washed and ironed. The stockings were mended and ready. Every picture and cushion and ornament was out in use. From my scrap-bag and sewing-box, I had used up all my embroidery cotton and pieces of cloth to make extra sofa cushions. I had replaced every button and worked over all our clothes carefully. Everything I owned was in its best possible order.

Although I was tired to death of hearing the lawsuit discussed, yet I knew that all our available cash was going to pay for the cows and pigs and mules and farm tools, and to the lawyers went large checks for their expenses.

The cost of necessities, for our share-cropper families as well as ourselves each week, was heavy too. I knew that the money we had brought with us from California would not be enough to see us through; already we had borrowed at the bank. I felt that I should do my share toward saving by helping with the garden and the chickens, and be content with just the necessities for our daily needs.

My intense interest in cleaning and making what small changes I could about the house had quite filled my mind, and as all the talk was still only about putting the farms in suitable condition to rent again, after the lawsuit was settled, I was not concerned with making any very big changes. To make a really good house of the little shanty was impossible, and useless, too, as we were making no plans for staying in the South permanently. Yet I had been studying that partition that divided the kitchen from the dining room and decided that I would knock the boards down myself if I could not get some of the men at the job.

It almost made me weep to see the real work and effort Wayne was spending to try to make me comfortable and satisfied. I kept myself worried and nervous all the time because I could not be content, when Wayne came in from his business dealings with the "dear Southerners" I was really pleased to hear him give his opinion of them and their business principles and ways. Even in his quieter moments I overheard him damning the country, people and all. Wages for common farm labor were from a dollar to a dollar and a half a day.

Poor white trash! A surprising thing to me about these people was that they were really human beings after all. They got sick and got well, married and had children, ate three meals a day—and there the resemblance ceased. Not "contented wi' little," they were "contented wi' nothing." They didn't indulge in ideas—so the weather was the common topic of conversation. They thought we were queer, because we read books and papers.

Each day the routine of necessary tasks about the house took several hours, yet I was putting in lots of time on the school work of Jimmy and Donald, and encouraged them to write many letters to my mother in California. We always enjoyed watching the heat lightning that so often shot up low from the horizon in a burst of glory, and the sunsets with their soft pastel colors. Jimmy and Donald were always running to get us to come and watch the beautiful colors or to admire some particularly beautiful cloud effect where the heaped up billows of white seemed to stand out like glacier peaks against a soft rose or yellow sky.

One evening we learned the words to "Tinkle on the Shingle," and sang them over and over. That song came to have so much more meaning, now that we were really experiencing such humid shadows and lay at night listening to the rainy tears that tinkled on the roof above.

Sometimes we sat out on the platform of the well, in the evening, and looked across the cotton fields, seeing a few scattered trees, three or four unpainted shanties, and in the distance the low undulating foothills. There we watched the sun as it sank behind the low line of blue mountains far in the west. Night after night we watched the moon change from crescent to half, enlarging until it became a huge glistening orange ball that floated high in the clear sky. We watched the distance lessen or increase between the crescent moon and her neighboring stars, and carefully marked the changing positions of the constellations. Orion, the Pleiades, the Big Dipper, and the North Star were all familiar friends to be looked for night after night.

We seemed to live with the stars, the moon, and the sky.

The children, I hoped, would remember this and appreciate it all as they grew older.

In the space that we had left for our flowers we set out dozens of little zinnias that Mrs. Wood gave us. There was a small peach tree already growing there. Later on it had a few small branches of beautiful pink blooms. On the sandy bank down beyond the pecan trees were three more of these small trees of Indian peaches. Grandma said the peaches would be small but they were red in color and the flesh would be red and would make fine peach pickles.

But the little peach tree there by the house was always our treasure. Jimmy and Donald and I spent much of our time together out around the house, raking, sweeping, hoeing weeds and carrying water. We encircled our tree with wire netting from the barn, digging holes for the three posts, with the heavy post-hole digger. We spaded the round around it, brought fertilizer to it and watered it regularly, but every time a cow or a mule got loose on the farm it would wander around and finally make straight for our precious little peach tree. How often, on looking out, we would see one of the mules stretching his neck to bite off the new growth. We moved the posts and the wire several times but were always amazed at the reach of a mule's or a cow's neck as they placidly mouthed the tender leaves. Yet stunted and ill-shaped as the little tree was, and growing so slowly, it always remained our cherished pet.

Around the front of our little house we set out dozens of young castor bean bushes for immediate shade for the porch. The castor beans were tiny spindling shoots at first with delicate baby leaves. We had carried them home from a neighbor's yard and set them out in the warm damp sandy soil and each one grew quickly and sturdily. When the tall clean stalks developed they reached to the edge of the porch roof. All summer their large spreading leaves gave us welcome shade. Some of the leaves were dark maroon, some had a purple cast; others were a bronze-green, each growing on a clean red stem. The rain would run down the red and green stalks or drip heavily from the leaves and fall on the soft sand beneath the trees, making round, smooth depressions. In autumn when the burrs dried on the metallic-green and scarlet spikes they split open and the large black seeds covered the sand, and so the young shoots would come up again each year with the first warm days of spring.

We brought some young willows up from the meadow and transplanted them behind the house near the pump, hoping that tall willow trees would some day be growing there, spreading out and supplying us with much needed shade.

We planted a bulb that someone had given to us and we watched eagerly to see what would come up. It was a lavender hyacinth. It grew out in the corner of our little garden spot, with all its wonderful waxen bells strong with perfume. The spring winds blew, the sand drifted day after day, the bleary sunlight came and went, but it held its head up firm and strong as though in utter indifference to the wearisome, cheerless days. We shielded it carefully and each day went out and looked on it with love and admiration, but at last it drooped and died, and we felt very sad as we looked at it.

Jimmy and Donald and I often went down in the pasture and picked up gallons of pecans. We noticed some beneath the bark of one of the fence posts, so we began to search every post, finding dozens of pecans in some of them. It seemed a shame to steal from the birds, but it was such fun. Hunting Easter eggs wasn't to compare with it. We were so excited.

On Valentine's Day I had made a little party for the boys, with a white cake and candles and lots of games. We made peanut brittle and sorghum-molasses candy. Everyone on the place came in and we really had a good time.

We would be glad when summer came for even though it would be hot, we knew it would be easier to keep well in the warm weather.

Now from the table where I wrote by the window I could see gray bare trees, gray fence posts, and off to one side a stretch of yellow plowed land and the dried yellow grass of the meadow.

March! Weather very windy, very cloudy, but as yet no rain. Plenty of frost and plenty of cold days. We had ten little baby chicks, three new kittens, some little baby pigs and a new calf. A Negro woman who lived by the bayou went past one day and asked me if she could have one of the kittens when they were older. "All of us colored folks is might fond of cats," she said, "and you can give them all away if you is a mind to."

There were four mules plowing and harrowing out in the fields and Wayne planted the alfalfa down below the bank. The oats were four inches high, and the radishes and lettuce and green onions were up, though hardly large enough to eat yet. Late one afternoon we planted carrots and beans and tomatoes and three kinds of cabbage. We examined the packages of vegetable seeds with excitement, and felt as though we wanted to plant some of every kind.

Often after supper Wayne and I and the boys would lie out on the front porch on an old cot with a quilt and cushions on it, and watch the splendid electric storms. The sky would be full of thunderheads with an intermittent shimmering and flashing of light, bright and dazzling, low on the horizon.

Later on in the night a wind would come up with a few spattering drops of rain; a soft wind, that at first only puffed gently, but that soon increased in violence and at last would almost blow the house over. But in the morning the sun would come out and a mockingbird would be singing on the roof. In the fields the mules would soon be going 'round and 'round.

I was using a large-size coal-oil stove for cooking, now, a three-burner, and we were up at six-thirty each morning. Breakfast was soon over, dinner "on," the dishes washed, and the irons heating. My! Soon we were enjoying new potatoes and green peas from our own garden. I felt much better when the sun was shining, though the crows could be heard squawking in the trees—a sure sign of rain.

Again big clouds in the sky presaged another rain soon, though the days had been so cold that our two fires were going all day and we had put the extra quilts back on the beds. Those sunless days were clammy and damp. It would be much better to have the cold weather all at once and not have the blooming peaches and apples all over the state killed by a late frost.

The wind was blowing steadily, but the sun was shining. This kind of day was called a weather-breeder.

Then a day of clouds again; another rain coming. I liked the gentle rains, but the heavy storms were too strenuous. We always prepared for a rain in the attitude of mind of a captain undertaking another voyage to sea, or as Noah must have felt when God began fastening the doors.

We brought in wood, pumped the buckets full of drinking water, filled the tea-kettle, and put the washtubs out under the eaves to catch the rainwater for washing. Everything that was loose we brought in from the porches, for the wind would carry even the wash-pan and milk-pails away.

The sugar box from beneath the dining room window, where the sugar sack was kept, and the flour barrel in which stood a sack of flour, both had to be moved. We cleared off the shelf behind the cookstove, and pulled out one of the beds, and put folded mats beneath two of the windows. One of us would run out to see that the big hay-shed doors were fastened securely so they wouldn't be broken off their hinges. Lard buckets were set on the heater to catch the drips there, and others on the dresser in the bedroom. We made certain that the lamps were filled and cleaned, for it got dark about four o'clock in a storm, and then we braced ourselves for the first big crash of thunder.

"Why don't you shingle your house, old man?" How often I thought of that! Because if the roof was an old one, you had better let well enough alone until a new roof could be put on. Everyone's mind was on getting the year's work started. There had been no time yet to think of such a thing as shingling the roofs. Our drips came just where it was most convenient to put the bed, and the sugar-box, and the flour barrel, and fresh leaks broke through in different places with each rain; and that was life in Arkansas.

I had to keep my thoughts of California locked away in a separate apartment in my head or I wouldn't have been able to get by at all. When I pictured to myself that dear town by the sea my heart would come up into my throat, as I would seem to hear the ocean, and smell the roses, and see the people I knew.

I had been idling about the house for several days. The air so delightful, the sky so blue, and wind so softly warm. Yes, a weather-breeder, they said, this kind of weather in May, but they said that in March and in April, too. But to myself I kept repeating, so that the few lines, half-forgotten, might perhaps bring back all of the verses of the old poem learned long ago:

> *Oh, to be out, to be out and away*
> *Through a roseate morning in garlanded May:*
> *With every vein throbbing in raptured acclaim*
> *To the music of odor and hue on the brain.*

My feet so light across the floor, my spirits so exhilarated that I seemed to be actually in contact with the blue freshness and sweep of the ocean breeze, the salt spray on my face, or the comforting half-gloom of pine woods and hidden redwood trails!

I had made a small bulletin board to hang on the wall where I stood when I was washing the dishes. On it I pinned bits of poetry that I wanted to memorize, or favorite poems that I enjoyed repeating to myself.

I glanced up now at the lines:

> *And what is so rare as a day in June?*
> *Then, if ever, come perfect days.*

From these lines I went on and on, repeating the poem to myself aloud and enjoying every carefully placed word of it, until it seemed that I could actually hear the very rooster of the poem crowing, just as I had heard one when I was a child in Canada, crowing out behind the big barn in the springtime, with the snow lying in small batches here and there, and the green grass springing so lush and delicate, the sun so brilliantly bright, the heavens so high above, and with it all a miraculous sense of uplift and lightness and unreality. Ever since I was a young child I had carried with me the memory of a certain wonderful morning, and through all the years it had strengthened me, and been a source of mental ease and relaxation. It was the kind of memory saved for contemplation on a sleepless night when the "inward eye" supplies "the bliss of solitude."

Such a deep reality I had in remembering the stinging sweetness of that spring day, with the air, though sun-warmed, still carrying the crispness of ice and snow. I saw again in memory the thousands of pink and white daisies scattered over the common, that corner lot with the narrow path cutting through, where all the little girls walked sedately after Sunday School, and where the boys played baseball on weekdays. I saw again the rows of blowing daffodils, the hedges of prickly English holly, and the tight green bursting rosebuds that so soon would blossom under the cherry trees.

And now it was on just such a day in spring, here, that Jimmy and Donald and I went for a walk in our meadow. Wayne and Henry and Grandma had gone into town early in the morning. Closing our front door we walked through our little house and I noticed with pleasure how well the rooms looked, for everything was orderly, and neat and clean. The table, laid for supper, was covered with a large white table cloth that hung down low on all sides. Each corner was weighed down by a little rock tied in the cloth, so that the windows could be left open for air, and the wind, should it come up, would not blow the cloth away, exposing the food to the dust. This was a trick I had learned from my neighbors. A lovely lemon pie, that had been my special work that morning, was carefully placed in the center of the table with tall glasses and bowls grouped around it to hold up the cloth.

As we closed the back door we felt like the three bears that went out for a walk leaving their little house so inviting and so in readiness for their return. Down to the Bermuda meadow and over toward the river we went, among the mourning doves, mockingbirds, budding trees, and willows green and feathery. The sky was cloudy, so it was pleasant walking with the sun out of sight. The recent rains had been heavy. Here and there in the depressions were large pools of water which we skirted or leaped across with much laughter. We cut willow for whistles and proudly succeeded in getting one to blow well. We lay on our backs in soft grassy spots gazing up at the sky, picking out animal heads and faces in the undulating white clouds above us. The clouds rolled over and over or drifted slowly across the heavens, exposing edges of gray or black, or again they formed deep caverns that contrasted with the white foamy billows against the blue sky.

Suddenly, glancing back toward home, we noticed a brown cloud of blowing sand up above the bank near the hayshed. We gazed at this with apprehension as we began to walk back toward the road, going faster and faster, for we realized that the sand blowing so heavily up there meant that quite a wind was rising even though as yet it had not reached us in the meadow below the bank. The sky was darkening quickly, and the white thunder heads had faded away while we had been engrossed with our willow whistles.

Soon we were running at top speed, for loud peals of thunder and distant flashes of lightning put us in a panic. We finally reached the hayshed and took refuge there for a while. Straw from the open runway filled the air, dust whirled and eddied, sand came in heavy gusts that blinded us, and the rain, driven along by the shrieking wind, began to pour down in torrents, as though the heavens were opened.

At last we made a wild dash for the back door of our house, got inside and stood there aghast.

The rain was coming in streams through the tongue-and-groove ceiling, bursting through the building paper that we had tacked over the boards, and spouting down on the table cloth. The sand that had been blowing so hard had already sifted in over everything in the house, and now water seemed dripping everywhere. The table was a wreck, the pie unrecognizable, the floor flooded. Drips—here, there, and everywhere. We rushed madly about, moving the phonograph, shifting the lounge, rolling the sewing machine and putting in the windows where the white curtains flapped madly or hung like wet soaking rags. The water was coming down over the side window casing, and was running down over the stove pipe, too.

The windows were all held in place by large nails that were pushed through holes bored in the window sashes and fitted into holes bored in the window frames. Since we had numbered all the windows, we fell to like sailors manning a ship, and quickly put them into their proper places. Soon all was done that we could do.

We gathered in the sitting room, keeping together, in fear of the flashes of lightning and the deafening thunder. It was a terrible sensation, to be shut in, helpless, with the wind fairly lifting the small box house from its foundations. The walls seemed to be pushed in a foot with each blast of the gale. The partition wallboards ran from the floor up between the high ceilings of the two front rooms with nothing to brace them between. We held the door to the bedroom tightly so that it would not be sucked open when the great gusts hit the house, for the latch did not seem strong enough to hold the door against the wind. The wall swayed and any minute we expected the whole building to go over like a house of cards. We even wondered whether we should stand by the partition wall or go over to the outer wall, trying to remember something we had heard as to the safest place to be should the house collapse on us.

It was only about half past three, but it was so dark that nothing could be seen outside beyond the little peach tree.

We cowered and shrank from each burst of lightning that flashed, our nerves jumping each time the rolls of dull thunder would culminate in a splitting crash. This was repeated over and over. It was not safe to have a fire or a light, and the day was grown so suddenly cold and full of fear.

But it was lived through, the rain let up, the thunder gradually lessened, and the sun, now low, shot long golden shafts of light from beneath the broken clouds. Everything soon changed from gray to gold, in the glorious reflected light. The light sparkled on every leaf and danced on the beautiful cloud reflections in every rain pool. The meadow trees stood out as if they were all part of a huge painting, for the level rays cut through the woods and across the grass in gleaming bands.

We opened the doors, took out the windows, mopped and wiped and swept and straightened and put everything in order. Outside the birds began to sing again and the water around the house gradually seeped into the sand. Before the sun had set we had almost forgotten the fears and discomforts of the storm in the peace and quiet and beauty that pervaded the whole outside world.

12. Poke and Sassafras

ALL WINTER LONG many of the farmers and sharecroppers had gone without green vegetables, eating biscuit, sow-belly and beans, and sauerkraut when they had any of that. Even lettuce or green onions were prohibitive in price. Few of the land owners would give up any garden space for their tenants' use. Those who had raised some turnips had long ago eaten them up, so among them all there really was an acute bodily hunger for fresh greens by the time spring came. We, too, found that canned fruit or tomatoes or sauerkraut or buttermilk never seemed to quite satisfy that longing.

We had plenty of eggs and milk and butter and meat and canned fruit and preserves, but how well I remember, as the winter wore away, eagerly digging, too, for the freshly sprouted turnips in the buried pile behind the house, so as to cook the pale green sprouts for salad greens, and how the very first taste of the long new leaves satisfied the terrible craving for something green and fresh. Our sharecroppers and many of our neighbors would come carrying empty lard buckets and ask permission to "grabble" in our big turnip pile out back of the house by the cotton field.

How we all watched for the first sprouts of the pokeberry bushes, for these young shoots were as delicate as asparagus. Along the sandy bank under the pecan trees they grew, and down in the meadow around the fallen logs and in all the fence corners. In the first warm sultry days of spring, when the bees were humming in the creamy locust blooms, and the call of the mockingbirds rang out so persistently and cheerfully, we would take a basket and wander around the farm to hunt "poke."

The green poke stalks, from six inches to two feet high, were crisp and leafy and pungent with a peculiar pleasing weed odor, and though these stalks had to be parboiled before they were fried, they still retained their tantalizing flavor when cooked, and made a delicious treat.

Along the sandy bank, too, were sassafras bushes, and many buried roots were still in the ground where they had been left when the weeds were hoed out.

During the long spring afternoons, the housework all done, Avis and I and Jimmy and Donald would go hunting sassafras roots for tea. I wore a gingham dress and checked gingham sunbonnet and old cotton gloves. We each carried a hoe and a big knife, and would laugh and chase each other, and dig frantically in likely places. The roots were reddish-pink, lying close to the surface and easy to dig up in the sandy loam. These roots we laid away to dry and when we wanted sassafras tea we took pieces of root, washed them carefully and simmered them until a reddish tea was made. It tasted like—sassafras. Rather sweet, rather insipid, a lot like half-forgotten candy or medicine of childhood days. But it, too, made an interesting item on our menu, and was said to be good to "thin" the blood that had supposedly "thickened up" during the winter months.

I had heard several of the neighbor's children expressing the hope that Easter would be a "pretty" day. One of them, a bit bolder than the others, had been urged to ask me if I would join in an egg rolling, as they would like to have it in our meadow because it was the largest strip of open meadow land around the neighborhood.

It was fenced off into large areas, one part for cotton and corn, another to enable the mules and cows to be herded down to the pond to drink, and other fences to keep the stock from straying over to the river.

I was also asked if I would bring eggs, as only a few of the children who would like to come had any eggs to bring.

What fun it was to boil and color dozens and dozens of eggs. At home we had always colored eggs for Easter but never in such quantities, and our preparations had never been surrounded with such a stir and air of festivity.

Easter Sunday came, a beautiful spring day, sunny and warm with a soft breeze blowing. Mockingbirds sang and the wood doves called from the thickets of bodark. Soon after lunch we took our baskets of eggs, large split-hickory baskets, and went to the meadow.

Before long there were about twenty-five children and a few parents gathered there, all hanging back bashfully waiting to be invited to join in the egg hunt. Avis helped me count the eggs, and then one of the girls counted out to see who would go "it" to hide them. A limit was set to the area in which we were going to play, and the children all hid their eyes while the eggs were being carefully hidden beneath protruding tree roots, in soft grassy hollows, by rotten stumps or half-buried at the base of tiny clumps of weeds and bushes.

As the afternoon wore away, the egg supply grew smaller. Many complaints were heard from the girls. We discovered that the larger boys were keeping the eggs they found and eating them in advance, so now all the eggs were gathered up, and pepper and salt passed around. Soon every egg had disappeared.

We grown-ups lounged on the grass and took care of the smaller children. Avis seemed especially happy. Her cheeks were so pink and her eyes so sparkling blue. The sky was full of fleecy clouds. The warm breeze lifted and fell with a caressing touch. The soft calls and answers of the birds filled the air.

I had always been impressed, when looking at large beautifully colored pictures where ladies in silken gowns and men in velvets and satins sat outdoors in an atmosphere that seemed to be nothing more than a continuation of their own drawing rooms, by the fact that they were able to dress in that manner, and take their ease, without any apparent thought of protection either from the weather or insects or the rough ground. Here was my first enjoyable contact with this outdoors that was really "Southern."

The breeze that blew was not a breeze, it was just a languid motion of warm air that caressed, that was just enough above body heat so that one was conscious of being fanned as though by some huge unseen fan. The atmosphere had that glistening, newly washed appearance that comes after a rain.

The new green of the cottonwoods and the bright green of the Bermuda grass, surrounding us on every side, made vistas that arched toward the feathery willows along the river banks. The higher open spaces through the branches of the trees showed rolling mounds of snow-white clouds. Beneath us, the grass, though so new, was warm on the warm earth, and was soft and stringy and clean, and in the open sunshine we had no fear of chiggers, insects, or snakes. We could lie down and stretch out in the sunshine and just "be." Nothing has ever reminded me so keenly of those dear remembered words:

> *Now the heart is so full that a drop overfills it;*
> *We are happy, now, because God wills it.*

13. Emma Lee

ANOTHER "PRETTY" DAY. Just enough breeze. Great mountains of foamy white thunderheads to the northwest. The mockingbird on the hayshed singing his heart out. The bees humming in the locust and the elderberry bushes.

"I see you-all air a-fixin' to wash, hain't ya? Well, ya might as well, I reckon, while this spell of weather lasts. Pretty to be out here, ain't it?" Old Mrs. West set her empty bucket on the porch and seated herself on the edge of the well, where I was pumping water for the wash pot. "Geneva," she went on, pushing her gray sunbonnet back from her hot face, "I was wondering if I could ask ya for some peanuts. Done went to your Grandma's so many times, I get kinda ashamed.

"You-all ain't been over to the meadow lately, have ya?" she went on. "Thought the boys might have seen how the dewberries was a-cumin'. Sure aim to get me some of them pretty berries. What say you and me goes over to the meadow this afternoon. You'll be through washing likely, by then." Then with a deep sigh, "Folks up at the house is sure a-chillin'. Maybe you could let a body have ary bit of quinine. Aim to get some come Saturday."

I knew better than that. It took cash to buy quinine and capsules. The few I knew who would consent to take quinine had always taken it in a teaspoon, after a chill had set in and they were burning up with fever. In no way could they be convinced of the connection between the disease in themselves and the tormenting mosquitoes that infested their houses, though some did believe that quinine broke up the fever. They thought I was marvelous to be able to get the powder into the capsules which they had learned to swallow. So I was constantly supplying capsules of quinine besides lending doses of chill tonic and castor oil.

It was very curious, the ideas of medicine some of these people held, and of course they resented any implication that they were ignorant or superstitious. I found it difficult to keep my face from betraying me when at times I was earnestly assured that black walnut juice would cure ringworm. And I never forgot how I had endured the peach leaf poultice applied to my stomach when I had lain sick up on Mount Nebo, and the neighbors decided that I had "pernicious malaria."

Now when I began to have chills and fever I was advised to tie one red woolen string around my waist, and another one around a young peach tree and the fever would go into the tree. And I had heard that some of the men carried a potato around in their pockets to ward off rheumatism. Many people refused to eat the late fall melons because they believed they were the cause of the malaria chills.

Malaria was truly a strange disease, preceded by violent shaking, and a feeling as though the blood were suddenly turning to ice in the veins, when one wanted all the blankets in the house piled on the bed, and heat applied to the shaking limbs. This shivering lasted sometimes two or three hours. Then the body would be burning up with fever and sometimes delirium would set in. Besides this the bones of the body ached, an ache that nothing seemed to relieve. This might last all day or all night. Then the body would break out in perspiration, and the patient was left weak and spent. The quinine and iron tonic and purgative and lemonade which were recommended by the doctors resulted in a ravenous appetite, a wan face and a grim determination to be more faithful in taking quinine regularly in the future.

As time went on I began to realize with a deeper understanding the real condition of the poor bottom folks, always "chillin'." We had all left California in good health and since then had been able to have a doctor whenever we needed one, and to buy medicine and have proper food and whatever was necessary for our comfort. We had been careful, too, to screen all our windows and doors. Yet malaria had already settled in our systems, and already we seemed to have much less vitality and strength. I often considered the conditions where screens were unknown, and quinine ignored for lack of money and lack of knowledge. The malaria mosquito, the Anopheles, was hatched out everywhere, at the edges of the ponds and in the bayou, as well as in the rain barrels and in old tin cans and dishes and pails left scattered about the farms. I did not blame the children for hating to take quinine, for their parents did not know how to buy it in quantity and fill the capsules themselves, and a teaspoonful of the powder, taken with a drink of water, was a dreaded dose.

The share-croppers often took the bitter concoction called "blackdraught." This was a vile mixture supposed to work on the liver, having some quinine in it, but costing far more than quinine. The druggists also recommended a bottle of medicine called "606" and this was bought at a high price in town or at the cross-roads store—bought and seldom used. It too was nauseating to take, with a bitter taste that lingered on for hours. So—chills and fever!

"Reckon you heard about Emma Lee goin' to have a baby, ain't ya?" began Mrs. West that afternoon as we walked across to the meadow to pick the dewberries.

"Funny thing is, she keeps right on a-sayin' that she don't know a thing about how it happened. Her ma's just sick about it, but Emma Lee ain't sayin' nothin'. I heard tell that Mr. Duke was a-gain' to have Jack Dawson arrested, but Emma Lee insists he ain't been around her. She says she cain't remember nothin', she was asleep; reckons it jest happened to her naturally. She ain't namin' no one, and for all I can find out they ain't no one ever been seen near her, either.

"Well, them Dukes is great ones to keep to theirselves. Old Mrs. Duke ain't hardly spoke more'n a civil word to ary woman in the bottoms, and as for him, the men don't pay him no mind. He spends his time fishin'; you've et some of the carp and drums he's caught. He's all the time gone off to the bayou. Does a heap of good fishin' though.

"Well anyways, gettin' back to Emma Lee. They've tried to make her talk by sayin' they won't get her no doctor, but nothin' ain't moved her yet."

Yes, I'd heard of Emma Lee. From Avis, from Grandma, from Wayne. I had never seen her. They were all puzzled. Her story strongly reminded me of an old song of my mother's, about a young girl in a similar condition who had taken her own life:

Because that I found that love was a lie,
And that it is harder to live than to die,
I have leagued with Death to bear me away;
God take no account of the deed, I pray.

Who knew of Emma Lee's private suffering, and what it was that gave
her the strength to hold her ground?

We walked on over to the river, climbing over the fallen trees and finding the dewberries thick where the old logs had lain for years, covered
heavily with matted vines. We each carried a heavy stick, and before we
looked for berries we stirred and poked in the vines to scare away the
snakes. Several long black snakes slid out of sight under the logs, and we
wandered back and forth, stooping low over the vines that trailed on the
uneven ground.

"Mrs. West, Grandma tells me that you have been married several
times. Will you tell me something? What do you think of real love? Do
you believe in it? Do you think you know what I mean when I speak of
romance? I hope you don't mind my asking, but you do seem to know
such a lot of things and to have thought about life." I had been waiting for
a chance to ask Mrs. West this for some time, and her mention of Emma
Lee's predicament seemed to pave the way for a natural question.

"Well, I don't mind a-tellin' you, seems like. I don't rightly know what
folks means by romance. I've heard of it. What love I've saw ain't much.
'Pears what me and Eric has got, now that we're old, is as close as ya can
get to it, even if I didn't allus think so. It ain't somethin' you can talk about
to your own folks, though. Guess I'd better tell you how things has bin
with me.

"All my marriages was hope marriages. You are kinda what they call
sentimental yourself, I notice. Well, I was too, when was young, and it
dies hard in a body. It's a thing you don't hardly get rid of.

"Anyway, I was right happy when a young gal. We was pretty well
fixed at home on a hill farm. We allus had enough, and we all worked and
helped along together. Ma was allus workin'. No time for nothin' else.
Work and babies was her life, same as mine.

"I was kinda pretty then. Had nice hair, thick and long. Lots of freckles, too, I remember. I was a big gangling thing, though, but strong as a
horse. I could ride horseback and plow, and dance, too, you bet.

"Anyways, nobody done told me I was pretty till Walt come along.

"I was pretty to him. He said so, and I sure wanted to be. He told me my long braids was real nice. He liked to see me comb my hair, and many's the time he pulled my hair all out of braid jest to stand and admire it, and run it through his fingers. I looked forward to being in my own house. I used to think how nice it would be, jest us two alone, day and night, too. You know how young gals is in their own minds." Mrs. West gave a short laugh and hesitated, casting a sidelong glance at me, but I wouldn't meet her eye, for I didn't want her to think about me. She had been looking back to her own past, and I wanted her to keep on.

"I remember once, standin' in the back yard, out by myself in the moonlight. I took down my hair, there alone, and figgered to myself jest how I could be a-standin' before and him a-liftin' my hair to his face, and admirin' me, too. It was sure wonderful.

"Anyways, we married right off. Had a house and everything, off over in Greasy Hollow, back of where Ma lived." Mrs. West wrinkled her eyes up at me and laughed, with the soft laughter she had, a laugh that seemed only a smile that had grown so big it must be apologized for.

"Shame on me to be a-diggin' this all up. It's bin buried a long time.

"Well, things was not what I thought, nor marriage neither. Kinda put me in mind of at a square dance, in a set, where the feller bows and scrapes, sashaying back and forth on the corners, till he's got ya for a pardner. Then you're set to one side. You're his pardner, right enough, but the fun goes right on across the opposite corners as before. He only knows you're there when he needs ya. Swing your pardners, yes. He knows you're there. No need to look; no need to smile at you. Takes your hand 'cause he can reach and get it, and knows you'll be right in your own place when it's time to grab and swing. But, la, ain't never a look or a smile really at you. You're too close, I guess. You ain't never off across where he can get a good look at ya any more. You're by his side, and he's by you, but his face is turned so as to see what's goin' on in the set. Do ya see what I mean? It allus seemed to me to be like that.

"A woman's allus there. There ain't no other place for her, at first. We only had one bed and that's where I had to be. Weren't no other place to go at night. Several times I tried to sit up late by the fire, hopin' Walt'd go to sleep, but as he was still gettin' in all the wood for the fire, 'cause we was so fresh married, he got mad at me burnin' it up at night without no reason. That didn't work.

"Later on, after two babies come, we had to get another bed. I was sure glad when they got the colic and cried, fer Walt got so tired o' lis-tenin' and waitin' fer me that he'd fall asleep and forget me.

"You know it weren't long till I was sure it weren't the Lord sendin' all them young uns, and the best I could do was to kinda sidestep. Worked all right till he figgered what I was up to and then the fireworks began.

"You jest cain't know what it means when you've bin down a-sufferin' a day and a night to have ya another baby, and no doctor to help, and then have your man walk in and look at the little mite and say to ya, sneerin', 'It's God's blessing it looks like me.'

"I got used to that. I had to get up all the wood by then, even helpin' to cut down the trees, for we was clearin' land.

"Funny thing though, I can remember at first, many's the time, a-standin' by the side of the bed in my white nightgown I'd starched and frilled so careful, takin' down my hair and combin' it out, slow and easy, standin' so the moonlight was a-shinin' on it through the winder, jest a-hopin' he'd maybe reach over and touch my hair, or say, like he used ta before we was married, 'My you're pretty!' If'n he'd only done that once, I coulda felt so grand. You see, Ma never had no time to praise me or admire me none, and the words Walt had used ta say I kept on rememberin'. I could think o' them when I was out hoein' corn in the hot sun, or doin' them big washings down by the branch."

Mrs. West was really gone now, I knew. Gone entirely back into the girlhood she remembered. She didn't look at me anymore. She had forgotten me and just kept on talking aloud. I turned away from her so that if she looked up she couldn't see the marks of tears on my face.

"Even the birds that sung in the trees over my head used to come and chirp on the lower branches where the wash bench stood and sing 'Pretty! You're pretty!' like Walt used ta, or so I liked to think.

"Them days I never knew what tired was. I felt so good and happy, and hoped each day that maybe come night I could get ta hear them words.

"Vittles didn't get no nice words fer me. I allus cooked good, but he jest et it and said nothin'.

"No, nothin' ever happened. All I ever got fer my pains was, 'Fer God's sake, what ya standin' there fer? Pin up your hair and come to bed out of the cold. You'll be sick, and I'll have you on my hands.'

"I know he meant well, but all the years I was a-married to him till he died, I was jest a-livin' in hope."

By now some of our pails were full to the top with the large black dewberries, so we set them up on a stump and wandered along near the shore where the sand was damp and firm. It was cool here. This was the first time I had been right down to the water. The main current seemed to be a long way out—for near us were many broken sand banks and piles of gnarled roots and branches where the river was building up fresh "made land." The sediment content of the water was so very great that from year to year the changes in the current were taken advantage of by farmers who owned land to the water's edge. They built barriers of wire and poles to catch floating debris. This slowed the current and so hastened the deposit of the rich soil that otherwise was carried down the river.

When we came to a fine clean log in a shady place, we sat down side by side. "What did you do then, tell me," I urged Mrs. West. "What did you do then? How did you have the courage to keep on?"

"That's why I married agin. I knew all about children by now—I had me six.

"I got a-hankerin' to have someone jest look at me once agin like they could see into me, myself, and know what I was really like. I got to goin' to meetings agin, and once I went with a crowd of young uns over the hill to a dance. I admit I only had one idea. I was a-lookin' out fer some lone man I could see that could be a-marryin', and I was a-hopin' to jest keep up the courtin' time long enough so as ta get my fill o' sweet words. I was jest hungry to hear myself praised and admired, yes I was.

"Well, it served me right. This time word had got around that I was a good, hard-workin' woman, even if I did have six children. Tom wouldn't hear tell o' nothin' but we go off to Centerville and come back married. He'd had his eye on me all the months I'd bin a widow, he said.

"Lovin'? Say, now, don't you go to laugh at me none. He had some sorta notion that I was too good to be insulted that way, as he used ta call it. Didn't want no free woman, and didn't take truck with no loose lovin' talk. He aimed to build up a good farm, he did, and be somebody.

"I remember once, I was kinda driv to it I guess, I threw my arms around his neck and kissed him. This was all before we was married. He bent his eyes on me full of sorrow and a kind of reproach and took my hands down and said, 'Now for God's sake, don't start actin' up like no common _____. I thought you was a decent respectable woman. I don't blame ya fer wantin' to marry, so do I, but let's do it right. We'll git married right off and go to housekeepin'.'

"Hope agin. I figgered to myself that maybe it was better that way. Could I get him to lovin' me after we was all moved in, maybe we'd kinda have all our courtin' and marryin' together and he'd kinda keep in the habit, and things would be like I dreamed when I was a girl.

"I guess I'm an old fool to set store by all these things I'm a-sayin' to ya. Nothin' come of it. It was the same as with Walt, only now I learned to have me a headache often when night come on.

"I sometime would like to know does a man have to give out a lot of lovin' words to them paid women, and so sorta gets ta thinkin' he's got ta make a difference somehow with a real wife.

"I'm kinda shamed to go on. Ain't a soul ever knowed this. Through all them years of mine, what with work and eight more babies and a good house and fixings and teams and all—well—I guess it's a curse to be a good woman. As fer children—if young folks only knowed it, they is the ones that keeps their parents decent, maybe; I don't know.

"Many's the time I cast my eyes around, driven most desperate by them same old words in my ears, 'Honey, you air sweet,' or somethin' like that; but it never done me no good. Not ever did a man look my way. I was married. My house run over with children.

"I worked hard them days; done all my own work and field work, too, to help. I sewed and took care of all the sick folks fer miles around. All the women liked me, I knew, for they didn't know that I had looked at all their husbands in turn and give 'em all up. Weren't no use. Not a one ever saw me, in that way I mean. They knew I was a good woman.

"I used ta wish I could have some way o' showin' that I'd a bin a bad woman iffen I'd only had a good chance, but them days no one dast smear paint on, nor powder. Oh, I know that ain't why the girls use it now, it's jest style ailin' them. They still got lotsa hope, and lotsa life ahead.

"Well, I lost a good home after Tom died. Some of my boys turned out kinda wild and to help 'em out o' scrapes and gamblin' and family debts and things, I mortgaged the big farm Tom left me and soon lost it.

"Then I married Eric. Well, somehow, none of that other didn't crop up so often. I guess all hope died. Eric was a just widower himself, with five children, all ages. We just sorta joined forces together. Now we've had us three young uns of our own.

"The years hain't bin too bad. I've seen a heap of work and sickness mostly. Lost me five of my children. Bin a good mother all I knew to them all. Seventeen times I bin through it. It don't get no easier each time havin' a baby neither. Don't let anyone tell ya that."

I ventured a glance at Mrs. West, and was startled at the smile on her face. The saddest, sweetest smile I have ever seen.

"I'm an old woman now, I reckon, but jest the same, I ain't never forgot how it was through all my life, to look up at the moon a-shinin' like it done summer nights and know it weren't fer me. Jest a-wasted, far as I was concerned.

"I've bin ashamed many a time to be a-envyin' my own girls their beaus. But I done all I could ta make 'em hard ta git and ta find all kinds excuses to keep puttin' off a weddin' they was a-countin' on, jest so I could see they had plenty o' time ta get in enough o' courtin', and could get to hear enough sweet lovin' words ta treasure up against the empty years was sure ta come.

"It's the truth as I know it, my body ain't never got in my way none. Cain't seem to recollect ary bother a-cravin' of a man. Nothin' they had to offer ever was ary attraction to me, except jest to be held tight and warm in a man's arms, jest so as to feel I was really wanted. That's the secret of it all I guess. I coulda gone without everythin' and done any amount of work. I guess that's funny, now ain't it, fer I sure did go without lots o' things, and I've allus done all the work that come along.

"I ain't a-braggin' none, but goodness knows I ain't never bin one to slump in my work, nor nag a body none. I bin kinda pleased with livin' when I could ferget myself. It was only that terrible longin' I used ta get at times, and I'd hev give up my life so glad, my body anyone coulda had, I used ta think, could I only hev traded it fer the joy o' bein' made over fer myself, not because I was needed fer a wife, or to help work for a farm, but just wanted once because I was a woman.

"Well, you can see now why I ain't holdin' no blame on poor Emma Lee. I don't know what happened, nor who the man was, but I give ya my word, it was a man, and no boy. A boy'd have done different or else showed up and took his blame now. No, it was a man what was cute enough to hev made a study o' woman and found out what I bin a-tellin' ya.

"I know that the men folks allus laugh at any woman that gives in to 'em and don't make a good bargain, but that's allus after they hev learned to trick 'em with the honey lyin' words they've found means so much to a girl.

"They's a kinda magic in some things a man can say, and I begin to see if us women could only stand together, and there weren't none o' the dry cattle to allus be offerin' free what we're a-tradin' with, maybe we'd be happier. I don't rightly know.

"Emma Lee ain't foolin' me none. Silly of her to expect folks to believe she don't know nothin' that happened. It ain't as easy as that. She's a sorta fool. Not a real fool. She sews real good; does real nice crochet and has made her heaps o' quilts, all hard patterns, and the littlest stitches you ever see. But she's bin kep' away from young folks. They never let her finish school. I don't know why, 'less her ma wanted her to help at home. She ain't never worked in the cotton. I think she's right pretty, big and fat and heaps of yellow hair.

"Me knowin' all. I know, I feel awful sorry for Emma Lee, and her troubles is only jest begun, too; and her without a friend to talk to. You know what a body's family is, a time like that.

"Well, you see where the sun is at. Must be gettin' on to supper time, and here we sit a-chinnin'."

Yes, there we sat, our pails heaped with berries, our fingers stained, our backs aching.

Mrs. West fished deep into her apron rocket and brought out her snuff stick and can. "Pity you don't dip," she said, "it sure is a heap o' comfort."

Going back home we climbed fences rather than take the longer way by the wagon road that ran through the cotton patch over toward the pond. Mrs. West needed no help from me. She could climb up and swing over the top rail as well as I could. Once when my dress caught on a long splinter, she made as if to go on and leave me perched there, but turned back, chuckling, to loosen me and help me down.

"Don't ya never tell no one what we bin a-talkin' about," she cautioned me. "Especially Avis. She don't think I know all them things. I'm jest Ma to her. She's my baby, and I don't aim fer her to get to studyin' about me no different."

14. I See Those Big Eyes Peeping

IT WAS EARLY SUMMER NOW. The weather was muggy and hot, with no breeze; like being in a hot house all the time with the sweat running down your back. Even in a tub of cold water the sweat poured out of you. It was terribly weakening.

A woman on the place had slow fever, and the rest had all had their turns at malaria and chills and fever. We were taking our quinine twice a week and hoped to avoid the malaria.

Everything seemed against you here. Health, wealth, and happiness— there was none of it. It was actually pitiful to see the little children that had to be taken out of the poor schools to work in the fields, but, as they told me, if they didn't pick the cotton or hoe it, who would? Everyone was working. Most of the alfalfa and oats were in the barn. The oat ground was now being disked for peas.

The melon vines out in the melon patch by the house had millions of tiny cream-colored blossoms and small inch-long melons. There was no fruit growing in the bottom lands. Sauerkraut was a treat, home-canned in large quantities in fruit jars.

Wayne was working awfully hard, I thought. The climate did not agree with him even as well as with the rest of us. He did more work than any other two men anyway, all the time, and did all the planning and figuring, too.

"Geneva," began Mrs. West one day as she sat by the well, drinking water from the long-handled dipper. "Avis done sent ya a message by me. She wants to know if'n you-all will promise to come up and be by her when her baby comes."

"Oh, I couldn't," I interrupted her hastily. "I don't know anything about that. I've never been where a baby was born and wouldn't know what to do, really I wouldn't. Isn't there someone else who could go? Surely Mrs. Dawson or Mrs. Wood, any one of the older women, would be glad to go, and I understand that they are used to it and know just what is needed."

"It ain't that exactly," said Mrs. West, wiping the tobacco juice away from her lips with the back of her hand, "It ain't that there'll be a thing fer you ta do. Jest be beside her, and see that she kinda stays kivvered up, is what she means. Jest you promise to stay right along, no matter what happens, and keep the sheet pulled up over her. I cain't promise myself, fer I know in reason I'll give out. I ain't no good for them places like I used to be. Used to I could a-got along better than ary doctor. The whole thing sorta sickens me now. I'm liable ta hafta go and lay down in the middle of it. Had seventeen myself," she laughed, "and this is my baby girl now aimin' to have her one.

"You know, when we lived out in the hills, they was a family lived in a one-room house near us. Didn't have a mite of privacy, a body might say. Avis knows what I'm a-goin' to tell ya, that's why it's on her mind so strong, I reckon.

"Anyways, one of the girls was a-havin' her a baby. The mother pinned two sheets together to make a kinda curtain, and behind it she put the two smaller children to bed real early so as to have them out of the way.

"Towards mornin', when the baby done come, and the doctor was there and several neighbor women, who always flocked in to see everythin' was goin' on, and the baby was a-layin' there kickin' on its grandmother's lap, naked as day, everyone looked up when they heard one of the young uns, a brat about eight years old, callin' out to the new baby in an admirin' tone and wavin' her hands, 'I see them big eyes peepin'!' There she stood, up on a chair, where she'd bin a-standin', her head pokin' over the top o' the curtain. She'd bin there, for the longest, takin' it all in.

"I coulda laid her out. Avis heard about that, and ain't forgot it, seems like. So that's what's on her mind. She ain't aimin' to be no show."

I promised, but very reluctantly, and before the week was out I was sent for, early one evening.

Avis lay in bed. Her mother was busy finishing up the kitchen work. Ben had folded several quilts and made a pallet for himself out on the front porch, where he later went to sleep.

Finally the doctor came.

All night the mosquitoes sang in the corners of the room. All night the lamp flared in the hot wind. The doors and windows were opened wide, but the air was oppressive.

The doctor smiled when he saw me. "What are you doing here?" he asked. I didn't know just what to say, just how to explain what my purpose was, but anyway, I was pressed into service. A can of ether was given to me later on, and I sat by Avis's head and adjusted the gauze pad on her face, and poured on the ether under the doctor's directions. He seemed satisfied that everything was all right, and had Mrs. West make him up a pallet, too, and went out on the narrow porch beside Ben. But he didn't stay long. His ministrations to Avis were all carried on under the covering of the big top bed sheet, and I was glad to assure her that she was well covered and could stop worrying.

My own nerves were tense and I was quicker and alert to obey each order of the doctor, for I remembered my own appalling ordeal only too well, and held myself braced to bear up under whatever dreadful conditions might arise.

Toward morning Mrs. West, who had been sitting quietly in a corner, left the room, and I heard her making up a fire in the kitchen stove. Later the doctor left me alone with Avis, bidding me pour on all the ether I thought necessary while he had a cup of coffee. I tried to act wisely, but I was torn between sympathy and ignorance, and kept pouring the ether on the pad over Avis's nose and mouth whenever she moaned.

Relieved at last as the doctor took charge again, I stood up to stretch my aching back, for I had sat nearly all night, leaning over Avis's head. I found myself staggering with dizziness. All I noticed was a sudden spurt of blood that seemed to fill my eyes, for I passed out completely and dropped down onto the floor. Rousing myself, I went on crazy footsteps through the kitchen and out the back door. I knew that the doctor was holding the new baby, and I heard him call something to me, but I only wanted to get away from there. I kept on, my eyes half shut, reeling down the sandy road. I made about twenty steps before I fell. I woke in my own bed hours later, with the shades drawn and Wayne sitting by me sick with worry. He said I had etherized myself. He had been plowing in the field nearby and had seen me fall, so had run and picked me up and carried me home.

Still, none of this was very important. A sense of life being full and big and flowing all around us kept us from taking our daily accidents too seriously.

As time went on, though, I began to see what really was important, what was in everyone's thoughts, and controlled his every action. This was cotton. To plant it, cultivate it, pick it, worry about it, talk about it, so filled their minds that the final step—to cash in on it—was a blessing of such rarity that it was almost taken for granted that either total failure or at best a meager reward would be the common lot, and nothing more was expected.

I never knew any share-cropper who actually figured on definitely having money in his hands from his year's work with cotton, and I saw very little that money had bought. Some had the essentials of common clothes and necessary food, but little else.

I began to sense the hypnotic spell that hung over the farmers. Everything seemed reversed from the usual business attitude that I had known.

I seldom heard the words "I want," or "I am going to get," or "I will have." All the getting and having was submerged under the heavy concern for the seasonal welfare of the great crop that should have brought a reward, not only of necessities, but of comforts and luxuries, too.

Avis expressed a general feeling when she said, "Somehow it didn't seem right to be getting so much money. Didn't do any more work than I'd been a 'doin'." She blushed and added, "I felt right guilty, like it was stealin'." We had been discussing the difference between the high wages during the years of the war and the fact that now the common necessities of life were almost luxuries. She couldn't understand my explanation of the corresponding high prices she had been obliged to pay then for shoes and food. All she knew was that she had been doing the same work for much more money and it seemed unfair to her.

When it came to looking after Avis and her baby, I didn't feel like the Good Samaritan or Lady Bountiful, either, for the conditions were so wanting in every essential for comfort or sanitation. Avis's mother did all that was necessary for the first few days, but she was needed in the fields to chop cotton, so I went every day for a while and washed and dressed the baby and did all I could to make Avis comfortable.

After a week had passed, Avis announced that she "aimed" to get up next day. All my explanations and protestations were in vain. "Guess I'm just as good a woman as my mother was.

"She never laid a-bed no thirteen days. No, nor five. Besides," she said, her blue eyes shining, "I'm jest a-frettin' to keer for my bebby myself. Ruther have you than anyone I know of, but seems I jest can't abide to see even you a-lovin' my little bebby. Want to be up a-doin' for it my own self."

Her poor mother. I could never forget our conversation in the dewberry patch by the river. Yes, up in three days or less, often, she told me. Up and back to long days of hard work. The mother of seventeen children; Avis was the youngest. Not one of all her children was able now to give any help to the aged parents. Mrs. West still stood square and straight, with her keen blue eyes full of smiles and her face crinkled with pleasant lines. She was usually chewing on her snuff stick and the brown juice stained the corners of her mouth. She was always clean and neat and cheerful, and always busy hoeing cotton, picking cotton, mending overalls, washing and cooking, lending a hand in any neighbor's house, when help was needed, or walking miles to visit or borrow.

Her children were always welcome back to whatever shanty she lived in through all the years after she lost the fine old hill farm. I never could quite figure out which of the men and women I met at her house belonged to Eric West and which to her first or second husband, but it never seemed to make any difference among them all. Ma, she was. A wonderful ma.

One afternoon I heard a repeated "thump, thump, thump thump," over and over again. "Thump thump." The noise sounded like someone hammering. It seemed to come from Avis's house. "Thump, thump."

"Maybe she is hammering," I thought, "maybe she is putting up her shelves, as I suggested."

I decided to walk up and see Grandma, and at the same time call on Avis and help her with her shelves. So I got my bonnet and started out, though I dreaded the walk up through the sand, for short as was the distance, the plowed loose ground on one side, and the bushes growing thickly along the bank on the other, made it absolutely necessary to literally wade through the hot loose sand of the narrow road. The soles of my feet were burning as the heat came through my slippers, and the loose sand fell in around my ankles.

When I got to Avis's, the thumping became louder, and I heard her humming out on her front porch. The little two-room shanty, sitting high up on the precarious piles of rock that uphold the corners, faced the thicket of bodark, where the bank fell away to the line fence. The back door of the house faced the big feed barn, and Grandma's house.

I went around to the front porch, and climbed the steps.

"Shh," said Avis, "he's asleep at last." I stepped across the old loose boards, and seated myself on the low board that was nailed for a seat between the two-by-fours that held up the porch roof.

Avis continued to hold little Roy tightly in her arms, his little head damp with the heat, the yellow baby hair plastered flatly on his skull. When he aroused for a moment and slowly opened his sleepy eyes, she hugged him tighter and, bracing her toes on the floor of the porch, lifted the front of the old cane-bottomed chair beneath her, bringing it down each time with a heavy thump—the sound I had heard away down at my house. "Bang, bang," went the chair, and the baby closed his heavy-lidded eyes again.

"Been a-thumpin' him for most an hour," said Avis, petulantly. "Don't know what's got into him. Must be this heat. It's so hot, ain't it, and sticky. I didn't hardly sleep any last night, and Roy cried a heap.

"The mosquitoes has been botherin' him a lot, I think."

At last, when the baby showed no signs of re-awakening, Avis tip-toed into the house and laid him on her bed, covering his face with a small piece of mosquito netting.

"Aren't you afraid that the leg of your chair will go through one of those knot holes when you are holding Roy? I noticed that your chair travels a lot over the boards when you bang it that way. I'd be afraid. Ben could take out those boards that have the big holes and put others in. He can get the lumber from the barn lot anytime he wants it, and there's lots of nails over at Grandma's."

"Oh, I wouldn't bother," Avis answered, "I'm used to it. Kinda know where the holes is at, and can watch out. It's all right."

I knew better than to ask such foolish questions as why she didn't have a rocking chair, or a crib, or even why she couldn't improvise a baby hammock. It was the same old story. No money for a chair, and no energy or ideas for anything else.

"I aim to try and get Ben to make me a little bed for the baby," said Avis. "It's so hot at night with him sleepin' so close to me."

"I could help you rig up a hammock that you could swing him to sleep in," I offered eagerly. "Those two posts would do and I have a lot of canvas. I'm sure there's plenty of rope around the barn. Let's get it and do it this afternoon, while I have the time to help you."

"Oh, I don't think I'll bother," said Avis. "He's used to thumpin', and Ben wouldn't take to no new notions. I guess I'll jest leave be." I felt defeated, but made one more attempt.

"I'm good at driving nails. Do you want me to help you put up your shelves? Anytime you are ready, I can come up."

"Well, I put that little one up on the wall there near the stove where there was a piece stickin' out to sit it on. Kinda figure it's all too much trouble. I'll jest keep my things in my boxes, and go on like I be, I reckon."

Soon after this, Mr. and Mrs. West went out to stay with Vena and Parse in the hills and left the little house to Avis and Ben and the baby.

"I'm goin' to miss my ma and pa without no doubt," Avis told me, blinking her blue eyes rapidly. "But at that it will be kinda nice to have the house alone. A body does kinda like to be by theirselves sometimes," she added in confusion, her face turning a rosy red.

Indeed I thought so. I had marveled at the nonchalance with which her mother had insisted that I step in and sit a spell by the fire, that evening in the spring when I went up to their house with a message from Wayne, concerning a trip down into the lower bottom the next day to get a load of oak wood.

Besides the two regular beds in the front room, there was a pallet on the floor in a corner, and here Parse and Vena were already in bed. Ben and Avis were in one of the big beds.

The other was for Mr. and Mrs. West. Mr. West still sat musing over the smoking cottonwood sticks in the fireplace. Mrs. West sat opposite him, her snuff stick in her mouth, her large efficient hands, that were so wonderfully shaped, clasped in her lap.

No one seemed embarrassed but myself. At home, except in case of illness, one always avoided intrusion on people after they had retired for the night; but here conversation became general, Parse and Ben finally sitting up in bed in their flannelette night shirts to wave their arms and settle the question as to which one should drive the team of mules and just which team should be taken.

I was so ill at ease that I left as soon as I could, and reported what I had seen to Wayne when I got home. "That's all right," he assured me. "They're used to that.

"I'd like to see your face when you see how many move into the new chicken house we're building, when time for cotton picking comes this fall."

15. Fleshpots of Egypt

ONE SUNDAY MORNING, warm and sunny, we arose late and break-fasted late. Wayne, for a change, glad to lounge about the house and enjoy a Sunday of rest and reading without interruptions, wore his beaded Indian moccasins and was taking great pride in his new pongee silk shirt, beautifully made, and sent by my mother for his birthday.

I saw Mr. Wood coming up the path toward the house, so Wayne went out to greet him.

"Din't want nothin', Wayne. Just thought I'd walk around a bit, being it's Sunday. Pretty weather we're having, but a feller never knows. Might rain any day now."

Then, as his position shifted on the porch steps where he perched, "What be them things, Wayne?"

"Them things" were the moccasins. Wayne was embarrassed. "Oh, just some old slippers I've had for years. My feet got to hurting and I thought I'd dig them out and put them on." A silence. Then, "Well, I ain't never seen nothin' like them before." Silence and a long contemplation of Wayne's slippered feet. "No, I ain't never had nothin' like that on my feet." Another deep silence. "No, and I ain't aimin' to."

I had been digging down into one of the large packing cases looking over some things that were still stored away. I took out my long silk Japanese kimono, a beautiful sea-shell pink, with a large stork heavily embroidered on the back in black and white silk threads, and threw it over a chair. My other kimono, a black silk one, padded and warm, I put on, and settled down to write letters. Soon Mrs. Wood and a neighbor of hers from further down the road called in to borrow some baking soda. I saw them staring at my robe, and eyeing the kimono on the chair and felt the need to explain my unusual attire. I turned around so they could admire the pink chrysanthemums that were embroidered on the back and sleeves. These two robes had been presents to me, souvenirs of a gay purchasing in San Francisco's Chinatown. When I explained this, my remarks were met with a dead silence, but I caught the look of suppressed opinion that was flashed from Mrs. Wood's eyes to those of her friend. Mrs. Wood's eyes were so very expressive, dark and piercing bright, and constantly darting everywhere. Nothing escaped her. As she herself so often said, when telling me some scandal, or recounting some late gossip, "My big eyes sees everything."

The inner side of the black kimono was lined with quilted scarlet china-silk, beautifully rich and exciting. I had always taken much pleasure in this garment, and saw with regret that it was becoming rather worn in places.

The brilliant silk I now displayed to the women, hoping for at least one word of admiration or appreciation. But no! Just another look. A look this time that spoke volumes, as Mrs. Wood remarked icily, "Yes, I'd heard there was things like that was wore by some women, but I never thought I'd live to see any." That kimono and the pale pink one with the storks, which I was thankful had not been recognized as such, were ever afterwards kept safely in my trunk. Something was lacking in me, I'm sure, for I never could get up the courage while I lived on the farm, to wear either of those handy garments where any but my own family could see them.

I sensed a lot of curiosity growing up about me, centering on my large wardrobe trunk. What was in it? Avis finally got around to asking me. "Ain't aimin' to be nosing none in your affairs, but we was just a-wonderin' the other night what you-all kep' hidden in your trunk."

Hidden in my trunk!

I used to go in the bedroom and open my trunk when I was all alone. In one drawer lay several pieces of pink silk underwear, and a frilly, pale pink, silk bed jacket. There was my beautiful flowered challis dress with the violet satin-covered buttons. Ornamental hair pins, one jeweled. A fan. Silver evening slippers. A beautiful perfume bottle that had been my grandmother's, and a bunch of artificial violets.

These things I would stand and finger and dream over. They brought back memories of carpeted floors and soft lights and music and gaiety. Ah well. I wasn't going anywhere to wear them.

I remembered the happy evenings when I had worn a beautiful long white satin robe and white mortar-board cap, one of the costumes provided for the drill team of the Rebeccas. I had only been a member for a few months but was already taking an active part in the social affairs. Some of my old teachers were members, and I had looked forward with so much pleasure to all the activities of the Lodge.

But, there were times when I leaned on the trunk and put my head down on my arms and sobbed, hard and deep, for the memories that these things awoke were too strong. But I learned at last to keep these treasures and handle them as dear possessions only, to admire the beautiful soft pink silk, just for what it was—cloth. And, I learned to laugh at myself, mourning for the flesh pots of Egypt.

Little silk was worn in the South. Woolen clothes were needed in the wintertime, and cotton in the summer. Silk did not wear well in that hot climate and would stick to the body like wet paper. In town I had seen the women wearing beautiful voiles, and brilliant colors in dotted Swiss, but all the farm women around me wore gingham or percale, much of it in dark gray or dirty brown checks. At the crossroads store, I once saw a poor woman buying cheap sateen of Kelly green with some scarlet sateen for trimming, for her little girl of six. I wondered where such piles of cheap cloth came from; such cheap dyes and sleazy materials.

In town too, I had noticed that the ready-made clothes seemed to have been bought mostly for the Negro trade. These were cheap, over-trimmed dresses, with gaudy tin or brass buttons and cheap ornaments. Most of the white women went up to Little Rock to buy their better clothes, though many women sewed for themselves, and in nearly every home dressmaking was always going on.

How my eyes used to ache for the sight of a woman's foot in an attractive shoe. The farmer's wives, dressed in their dark cotton dresses, wore rough weathered shoes, shoes that were large, loose, and grayed from much wear. These women seemed a race apart, almost, from the women I'd known who wore only high-heeled shoes or shiny shoes of neat-arched appearance. At times it seemed that my very soul recoiled from the women themselves, whenever I caught myself staring too long at their shoes. Of course, there was a reason for them. Poverty and necessity. In the loose sandy soil or on the rough ground, fine shoes were not only uncomfortable, they were impossible, and all cheap shoes, even the comfortable low oxfords, were soon shapeless and ugly.

It made me feel as if it were an insult to all womanhood that these women should have to know so much of ugliness and drudgery, their birthright of possible dainty feminine beauty forever hidden and worn away in the toil of the cotton fields. Their life was only a long dull routine. Childhood spent at the heavy chore of chopping cotton, then the yearly hard grind as a cotton picker—dirty, back-breaking, heavy work. Maybe a few months of happy courtship, then child-bearing and sickness and more hard work—for years and years and years. Cotton slaves!

16. All Will Be Well

WE HAD A MELON a foot long at last, the only one of that size in the patch. They would be ripening about the middle of July. It was easy to see why melons assumed such an important part in the life of everyone when summer came. It was because other fruit was so scarce. Everyone waited eagerly for the first ripe melons.

Old Eric West had told Jimmy and Donald to take a young watermelon, one about three inches long, and place it into the mouth of a glass fruit jar, leaning the jar on its side on the round, and being careful not to break the melon loose from the vine—and then watch results. They did this and were quite surprised. In ten days under the impetus given by the extra moisture that collected in the jar and the concentrated rays of the sun, the melon quite filled the bottle, burst it, and grew rapidly. It was the first melon picked that season.

At last the watermelons were ripening and the cantaloupes and muskmelons below the bank were coming along. Those first melons brought forty cents apiece.

Oh, it was hot! At seven in the morning the day was already too hot for comfort. There was more alfalfa to cut, but the hay had all been baled. I canned peaches, seven bushels of them, brought in from the hills. They were a dollar and a half a bushel.

We were disappointed when we found that the fine orchards of Alberta peaches on the hill farms had been nearly all taken out, and the cotton acreage enlarged when the price was so high. Only here and there did we find a farm that still kept a small orchard.

Later I canned pears and apples. Beans, beets, and cucumber pickles, all from our garden, I canned, too. We were enjoying fresh sweet-corn, string beans, whippoorwill peas and tomatoes, and I was still baking "light bread." I got such a deep satisfaction from bread making. I looked with pride at the freshly baked loaves with their lovely golden crusts, the bottoms of the loaves showing that clean tan that proves the absolute rightness of the baking. And, oh, the wonderful fragrance that only good home-made bread has. I used the store-bought yeast now and learned to make excellent Parker House rolls, which were quickly eaten up. Generally my baking was done twice a week, eight loaves at a time. I was told by Wayne of the Southerner who returned from a visit to the North and reported with disgust that the folks had actually given him "light bread" for breakfast.

With the fire roaring in the little kitchen stove I could quickly make a pan of biscuits and they were really most delicious for breakfast, yet we all liked home-made bread. Fresh from the oven, spread with lots of good butter and jam, what could be better?

We were escaping malaria by taking quinine regularly, yet each day I felt so weary. It seemed the quinine taken every third day kept us from breaking down with a chill, yet the malaria was working in our systems and showed up in a sense of despondency and listlessness. Whether we worked or whether we didn't, we would be tired and our bones would be aching and we were always glad to lie down and rest. There was little incentive for anything, unless one happened to have it within himself, and I grew so lazy and totally indifferent that I could have lain down and died.

The children on down the road across the line-fence had the chicken pox, yet the other children in that family were still going to school. Many had hookworm too, but apparently nothing was ever done about that. Quarantine was unknown here. I guess I had a sure enough case of Arkansas blues. There is no known malady worse.

The times were hard. They affected the people right where they lived. Anything bought on credit cost so much more than a cash purchase. When fifty cents was added to a sack of flour, and that method was used in all the stores, it made about a hundred dollars more added to a two-hundred dollar grocery bill. How long could the people stand it?

Wayne sometimes talked of staying in the South for several years, but I couldn't see any pot at the foot of the rainbow, and when I looked around at the children that were growing up in the bottom, saw the lack of good schooling and proper training, with no chance for the future except to follow a plow as a share-cropper, the comparison between what they might have and what they did have made me feel desperate.

Some days I decided "to hell with 'em all," the poor white trash. They were in poor circumstances, surely, but deserved to be, too. I hardened my heart against them, and felt that I would hardly give them anything if they were starving. But, I realized, too, that this feeling was only my rebellion against a sense of futility and suppression and lack of emotional outlet. Grandma and Henry were not the ones to share my discontent and mental suffering, and I tried to hide it from Wayne, most of the time. I tried to keep busy and not do too much thinking.

Tom Wood said he knew his health would be ruined before he ever had a hundred dollars to get away on. He looked like a skeleton but plowed every day in the fields. All the young fellows he knew, he told me, hoped that some day they would be able to get out of that awful country. And it was awful. There are more important things in life than money, and though one must eat to live, some didn't even eat here. They were dying right there in the bottom every week for the lack of food and medicine. I heard that the druggist in town offered to supply quinine free to the "poor bottom folk."

There was one woman who tramped up and down the road barefooted, looking for work. She had two little children, but nobody had any work for her. Her husband had left her after pouring the little coal oil that was left in their can into their cornmeal crock, so I heard. That was done for spite because they had quarreled.

Since it was nearly time to "lay-by" the cotton anyway, there was really no place for her. The farmer they had worked for wanted the little one-room shack she had been living in to give to several cotton pickers who would be coming down soon from the Ozarks, for although it was such a tiny house it would shelter maybe six pickers.

When Grandma heard about this poor woman she immediately started in to help her by sending her bacon and soap and sugar and eggs and other needed supplies. Finally we heard that the woman had gone down into the lower bottom and had at last found a place to stay.

I first heard about all this one day when I answered a call at my back door and saw standing there the oddest person I had ever seen in all my life. He was old Mr. Batson. He was a little man—not exactly a dwarf, but little and short and heavy, with a large head of bushy, uncombed gray hair, a dirty straggling mustache and a long beard. His feet were bare. His blue chambray shirt was faded and torn, and his overalls, faded and dirty and ragged, were turned up halfway to his knees. His feet were indeed bare, but I gazed at them in fascination for they were unlike any human feet I had ever seen. His long toes were spread apart from going barefooted so much, and were very, very dirty. On closer scrutiny I saw that they seemed to have never known water, for they were caked and horny and coated with black dirt, and all the toe nails curled upwards.

He reminded me of a leprechaun, only those "little people" of Ireland were clean and gay and smart, so I quickly regretted the comparison that had come to my mind.

Mr. Batson chewed tobacco and his beard and mustache were stained with it, the juice running from his mouth as he talked. And he could talk. He stood out on the sand by the back steps, and holding out an old cotton sack begged me to donate some food, or anything, for the poor woman who was left without food or shelter or chance to work.

I remember giving him a lot of things to put in the cotton sack which he swung over his shoulder as he marched off down the road. I remember being dumbfounded when I was told later that he was a Holiness preacher. He was not regularly ordained, it seems, but he could and did preach when the spirit moved him, and took an active part in the services when the Holy Rollers started their meetings in the fall. I made up my mind that I would make an effort to hear these people, when they came, and I hoped that I could get to hear this queer little man preach when under the influence of the "spirit."

Ah, September at last! This was the first time for weeks that I had been able to "take my pen in hand." I had been having tonsillitis and a lingering fever with a high temperature. Wayne did the housework, and the children were my nurses, so we got along all right.

Everybody was in the fields picking cotton. It was now selling at eighteen to twenty cents a pound.

The weather was becoming more bearable, with cooler nights so we could sleep. But, oh, I thought daily—"Oh, to be in California, now that autumn's come!" I would chant:

Oh, suns and skies, and clouds of June,
And flowers of June together,
You cannot rival for one hour
October's bright blue weather.

There was a little of that tang already in the air; California weather, bright blue weather.

Mrs. Randall, whose boy, Buddy, had died of malaria in the spring, was having a singing down at her house. Wayne thought it was about time I saw a singing, for it was the common social gathering here among the neighbors in the fall and winter months, so he and I went down there one Sunday evening.

The porch of the house was about four feet wide, slanting down to the outer edge, for the foundation stones had slipped from place and the front of the porch was fully a foot lower than the back. Several young fellows were sitting flat on the porch floor with their backs up against the house wall, their feet braced to keep from sliding off. I saw Wayne join this group, and settle down without even a smile, though I noticed that he carefully avoided my eyes.

The little front room was crowded. Two beds, a tall phonograph, an organ, the heater and a rocking chair took up nearly all the space. I was given the rocking chair, for it seemed that I was considered the guest of honor.

The phonograph was a Silvertone from Sears Roebuck. I'll never forget that new, shiny cabinet in a room where the walls were covered with old, dirty, snuff-spitted paper, the floor bare, and rags stuffed in the holes in the broken window. Mrs. Randall was dressed in a long, dirty, dark-colored calico dress, with dirty bare feet and snuff stains around her mouth.

On the beds, among the piles of coats and wraps, lay several sleeping babies. Three men sat on the floor, squatting on their heels against the wall, wherever they found room. Everyone else stood up, for there was no place even for the wash bench which would have made an extra seat.

We waited awhile for someone to come who could play the organ, but time went by, and finally Mrs. Randall played the one record of church music she had, and then carefully covered the phonograph with a huge white Battenburg cover that hung down to the floor. "In ain't fitten," she said, "to play other than church music on Sunday."

As no one else came, the men finally slouched in from the front porch, trying to appear at ease, and everyone gathered around the organ, where there was a hymn book. The two lamps that stood there were without chimneys, and black smoke poured up under the low ceiling, but no one seemed to mind that. Chimneys cost money and were easily broken in these houses where the rains often blew in during the frequent storms.

I failed to recognize any of the songs that were sung, but I did enjoy the wonderful harmony and the earnestness with which everyone sang his part.

I noticed that the notes in the hymn book were the old-fashioned type, diamond shaped or square, and the four-part songs were faithfully and feelingly sung. But, oh, the haunting pathos, the deep sadness, the fervent longing in the cadence of those old Holiness hymns.

The last song they sang seemed to me to epitomize all the dreariness and barrenness and hopelessness of their lives, and most pitiful of all, their complete resignation to the fact that there was nothing in store for them except a repetition of the same conditions on and on through the years. The chorus rang out again and again:

> *All will be well, when life is o'er,*
> *All will be well, when life is o'er.*

17. You Forgot the Grease

WE WANTED TO SEE how the new tenants, the McGuires, were getting along, down on the lower farm, so we decided to drive down and visit them and arrange about getting out fence posts, too. John McGuire and his wife and four children had moved onto the lower farm by the Petit Jean after Mrs. Rollins left for Oklahoma about the time Ben and Avis came to live on our farm.

Jimmy and Donald and I sat in the back seat of the car, and Wayne and Henry in front. The children and I, as usual, were left almost in a world of our own, for we had little interest in their talk about the people they knew, or the endless discussions about acres and crops and prices.

This time we went a different way. We crossed a high swinging bridge that spanned the deep ravine where a little creek ran far below and took an old rutted wagon road through the woods. There were so many large boulders and half-buried roots and stumps in the road that at times it was almost impossible for the car to make headway, and we had to drive slowly and carefully to avoid the big mudholes left from the last rain.

At last we came to the tongue of land which was almost encircled by the river and climbed the rocky road that was cut from the hillside, stopping the car right by the big well at the back door of the little house where Avis had lived. We had brought no lunch with us, for Wayne told me that such a thing as anyone eating outdoors by themselves when there was a home nearby with a fire and food was just not done in the South.

Besides Mr. and Mrs. McGuire, there were three older boys and a little girl about six years old. She was like a little ghost, she was so quiet. Her yellow hair hung in beautiful curls down her back. After lunch as Mrs. McGuire and I talked, I found that the child would allow me to comb her hair and form the soft curls on my fingers. Mrs. McGuire promised to bring her to visit me some day when Mr. McGuire was going to Dardanelle.

Wayne had made me promise to eat whatever they had, and to act surprised at nothing. I promised, gladly, and with great curiosity. That dinner! It was the best they had and we were more than welcome.

A worn oil-cloth was stretched over a large table. On one side of the table was placed the wash bench, brought in from outdoors. There were only a few rickety split-bottomed chairs. The spoons were of tin and the knives and forks were old and worn, with yellowed bone handles. The cups were tin, too, and we ate off of tin pie plates.

The meal consisted of white corn bread, thin and heavy, made with salt and water and grease; pinto beans, boiled with a piece of fat salt pork; pieces of fat salt pork fried until they were transparent; coffee, strong, bitter and black, without sugar or milk; and then for a treat, a jar of peaches that had been canned without sugar, and a layer cake, thin layers stuck together with jelly, with that peculiar, good old-fashioned cake taste that is so hard to find nowadays. I was really hungry and made up my mind to eat some of everything, but the strong coffee, the fat pork and the tasteless beans were hard to swallow. At last, as we were all about to rise from the table, Mrs. McGuire happened to look across at the cookstove where sat a round, white bowl, full of the fat that had been fried out of the salt pork. This fat was now getting cool, for the fire had been out ever since the corn bread came from the oven. Around the edges the grease was hardening and was as white as the bowl itself. Mr. McGuire, reaching for the gourd that hung by the cedar water bucket, was stopped short by his wife's sudden cry of disappointment, "John McGuire, you forgot the grease!"

John turned at her cry and a comical look of dismay came over his face.

"Pshaw!" he exclaimed. "You folks sure been done out of your soppins. That there was fer the corn bread."

When the men went outside to talk about the post cutting, I wiped the dishes and set them up on the shelf for Mrs. McGuire, and she got out some worn overalls and began patching.

"You certainly have lots of wood here," I remarked. "I never saw such big hickory trees."

"Well, yes, there's a sight of trees here. It makes it real nice in summer. But there ain't never a stick of wood ahead, seems like. The trees are down at the foot of the hill and it's a job to have to carry it up after it's cut. Hard enough to get the boys to cut it," she added. "I had good luck for a spell, though. Once my two oldest boys and three neighbor boys went hunting. They got off early and left me no wood cut at all. It came on to rain and rained all day, hard. I made out on some cold coffee and corn bread. Then I laid the fire, with a bit of paper and two or three small chips, made up some biscuits and put them in the pan all ready for the oven, set on a fryin' pan of salt meat and put some beans on in a kettle of cold water. All was set. By dark it was a regular tempest. They all come in hungry and cold and soakin' wet, jest as I knew they would be. Didn't say much, when they saw what I'd done. Nothin' they could say. We et that night pretty late, I remember, for they had to go down to the bottom of the hill in the rain and split wood and carry it up. After that for quite a spell they allus left me some wood when they went huntin' and fishin'. Kinda give them a lesson. But I shore hated to do it."

Mrs. McGuire was middle-aged and colorless. Her hair was not yet gray, yet I could not determine what color it once had been. Her eyebrows and lashes were nearly white, her skin a sickly yellow, her cheeks pale and drawn. Her back was stooped from years of field work.

All her life she had worked hard, apparently.

"All I ever knew, I guess, was work," she continued. "At home Ma had eleven of us. I worked in the fields and worked in the house, too. They was ten brothers. Ain't nothin' gets ya like them great tubs of heavy overalls. All through these bottoms where I been a-livin' since I married, the water's been hard, and seems like, 'cept here, John allus rented a place where I had to carry water for the house."

Later on in the afternoon we left Mr. and Mrs. McGuire and went down the hill through the woods, checking on the trees that could be felled and cut into posts. This was a squirrel hunting heaven for those who lived near. Wayne and Henry had decided to do a little hunting as it had been years since they had had such an opportunity as this. They had borrowed rifles from Mr. McGuire. Gray squirrels and red fox squirrels were here in great numbers, scampering over the ground and racing up the tree trunks. I sat on a fallen hickory tree watching the men stalking off through the woods with the guns under their arms and their eyes rolled up to heaven, stomping carefully along, to avoid fallen branches. Jimmy and Donald were following them with great interest.

I suddenly realized that I was only one more among the mothers of men to be brought up sharply to the realization that all life is an eternal struggle between idealism and reality. Not long before, Jimmy and Donald had joined me in reading:

> *I've plucked the brown nut from the bush,*
> *The blossom from the tree;*
> *But heart of happy little bird*
> *Ne'er broken was by me.*

My thoughts raced ahead from those conned lines to the idea I was gradually having to face—that most fathers agree very grudgingly as to the value of such teaching as those lines suggested, where it concerns their own sons, although they might concede, even hope, that maybe at some future time in their sons' lives, their mother's straining after such a perfection of values would make for gentleness and pity in their sons' dealings with their fellow men; but most fathers felt, nevertheless, that all young boys should be allowed the manly exercise of such sports as practicing with a bow and arrows or an air rifle.

I considered all the arguments I had heard in favor of fostering the gradual growth in a boy's mind of ruthlessness and hardness of heart in wanton slaughter and bloodshed and suffering where only a bird or a squirrel was involved.

I knew, in the long history of mankind, men's dependence upon guns in the struggle for daily food and protection against lurking enemies, and even yet in our so-called civilization his apparently perpetual need for the ability to defend himself in battle. I wondered why mothers kept on shutting their eyes to all this. Why they blindly hoped that in some way, under the influence of their gentle admonition, and aided, perhaps, too, by the bits of appealing verse or story that the school books offered, their little boys might finally come, some day, to a time when they would not only say, with understanding gladness from their own hearts,

> *I am old, you may trust me, linnet, linnet,*
> *I am seven times one, today,*

but also, that they might as grown men have permanently instilled in their minds and souls qualities that would prevent them, through all their lives, from ever taking up a gun for the ruthless and unnecessary slaughter of birds and animals.

Yes, I wondered why.

18. The Ozark Mountain Boys

WE HAD FINE WEATHER that fall in which to gather our crops. All were in from the fields at last and we began preparing the ground for winter wheat. We had raised an excellent cotton crop in the strip from the lane to the river. Old-timers said we had the best cotton they ever saw on the farm. We now had thirty head of cattle, fifty hogs, and one hundred and fifty chickens. All these, and seven hundred bushels of corn, and two thousand five hundred bales of hay, besides good pasture for the winter. We share-cropped the cotton land, making forty bales on sixty acres, and got twenty-one bales for our part. Now if we should win the lawsuit we would be in clover. We were holding the cotton for twenty cents, as there was only a seven-million-bale crop in the South, only half of normal; so we were gambling on an advance in price.

After the big rain, it turned cold. We thought that we might have an early snow. Still, we had lots of wood and plenty to eat—both blessings.

I put in those cold windy days teaching Jimmy and Donald. They took great pleasure in writing letters to California and that served as spelling and writing combined. I sent away for a set of sewing lessons, so I could go ahead with cutting and fitting my clothes. Nearly all the women here, as in town, either made their clothes or ordered them by mail, for the quality of the ready-made dresses found in the stores left much to be desired, and the prices were far too high. One was certainly hemmed in with household cares, and I was busy from morning till night.

For three weeks I had kept boarders! They were four boys from the Ozark Mountains who helped gather the corn and sweet potatoes and worked in the hay.

These young men slept up at Grandma's house. An extra bed in Henry's room was set up, and also one on the porch, but they came down to my house for their meals, to save Grandma that work. Our daily menu was: biscuits, sorghum, coffee, sow-belly and gravy in the morning. Dinner: beans, biscuits, sweet potatoes, baked rice, coffee and sorghum. Supper: beans, rice, sorghum, coffee, sweet potatoes and biscuits. Variations: corn bread, butter and milk. I made ginger cake, but they "didn't reckon" they had ever eaten any and wouldn't taste it. Pumpkin pies were strange to them, although fried-pies made with dried peaches they thought were mighty fine. Once when I made hash of oatmeal with gravy, potatoes, onions, sage and seasoning, rolled out like biscuits and dipped in egg and fried, they ate nine apiece, and wanted me to cook some every day. Every morning I put on the big black pot full of pinto beans. After supper it was empty. At home in the Ozarks the boys lived on squirrel, they said. Asked if they ever had bacon or ham, they said, "No." They raised a few pigs, but they sold them for cash.

They needed the money. "Ain't never et no pig meat." My canned fruit and watermelon pickles they enjoyed, though they could not be urged to take more than a small dish of the fruit with their meals. They "didn't aim to eat up all my work," they told me, in explanation.

These boys were all about nineteen years old, but they were like children in many ways. They played leap-frog, and tag, and jump-rope, and were experts at horse-shoe pitching. These games they played at noon-time before they went back to the fields, and they played, too, all Sunday morning. They got "plumb tickled" about the least little thing, and their laughter could be heard across the fields.

Our little dining room was paneled with those wonderful pictures from the rotogravure section of the California Sunday papers. I got an interesting effect by using wooden strips to separate the walls into even divisions, and here I pasted the pictures. The illustrations of a motorcycle race interested the Ozark boys greatly, and they argued hotly as to whether the pictures could be real, for nobody could ride a cycle at that desperate slant, they knew—yet they told me they had never even seen a bicycle. They "aimed" to try and see Little Rock some day before they died, for their teacher had told them that a visit to the Capitol would be worth several months schooling to them.

I never ceased to be amazed there among the bottom people at the lack of what had been to us just general knowledge about ordinary things. I was often asked to explain what made the big breakers dash so high on the rocks in my ocean Panorama, and I exhausted myself trying to make clear about the waves and the tides, but I thought the climax was reached when, one day, a boy from down the road came in and, looking up at the deer's head, asked, "Who stuffed the cow?"

I was amply repaid for the extra work my boarders caused me, by their constant laughter and jokes and by the pleasure we got from the stories they told of life up in the Ozark Mountains.

The pecans and black walnuts were ripening now and it was such fun to gather them. The peanuts were all gathered now, too, and must be picked from the vines and dried. The black walnuts we spread out by the pump to dry, but since the brown stain from the green husks was so lasting, I devised the scheme of stamping off the husks, for I wore two pairs of Wayne's heavy socks in his old shoes. Yet the black walnuts and hickory nuts were so hard to crack. We worked at the job day after day, banging away with the hammer on a big rock. The most tedious job was picking out the meat from the crevices of the shells, yet even a small amount of these nuts added to candy or to cake frosting was delicious and well worth our trouble.

One Sunday we took our boarders and two of the farm wagons and drove down to the lower place again to get the fence posts the McGuires had cut. Ben and Avis and the baby were along, too, and this time we took a lunch and ate it down by the creek. Avis and I gathered several bushels of hickory nuts while Jimmy and Donald took care of little Roy.

When the Ozark boys left, I sent their mothers two large sacks of hickory nuts, peanuts and pecans. The boys seemed rather overwhelmed at this unexpected kindness, and made us all promise to come to see them in their mountain home if ever we got away for a trip.

When Hallowe'en came we had a little celebration for Jimmy and Donald.

Wayne made two lovely jack-o'-lanterns, with saw teeth, and I made pumpkin pies and peanut brittle and pop-corn balls. For supper we had fried chicken, sweet potatoes, biscuits and gravy. The peanuts, pop-corn, molasses, potatoes, chicken, milk, butter and eggs were all from our farm, and that made us very proud.

Wayne brought from town some California apples and grapes for a treat. The apples were five cents apiece, and the grapes thirty-five cents a bunch. High prices for California fruit here.

The boys enjoyed the jack-o'-lanterns and we turned out the lamp so as to get the full gruesome effect. After supper they carried the pumpkins up to Grandma's and had the fun of holding the grinning faces up to the window to scare her and repeating the same trick with Ben and Avis.

19. Wild Geese

I'LL NEVER FORGET the morning when I was awakened very early by a peculiar whirring sound that seemed to fill all the air, a sound as of a rushing wind or a loud continuous whispering. "What is it?" I exclaimed, rushing out to the porch in my night-gown. It was the first time in my life that I had ever seen the migratory flight of birds. They came from the northeast, passing high over the house and extending to the south in a black curving band that disappeared beyond the tree tops.

All day, without a break in that wide-spread band, the blackbirds flew overhead. Off and on during the day I watched, fascinated, unbelieving, thinking that the line would break at last, but darkness came and they still were passing, the beat of their wings fanning the air, and making me feel, as I stood gazing into the sky, as though I, too, might be caught up into that swift flow that seemed so tireless, so sure, so true.

The red birds hadn't arrived yet, but now every morning and evening the wild geese flew overhead with their strange cry, going south. I never ceased to marvel at them, as I picked out that formation of birds so high above me.

Shortly after that I was sitting by our well talking to Avis one evening, watching the sun set behind the line of hills and looking after a flock of wild geese. They made their outline far away and high up against the pink of the western sky. I was thinking of the verses of William Cullen Bryant and mulling them over to myself to see how much I remembered:

Whither, midst falling dew,
While glow the heavens with the last steps of day,
Far through their rosy depths, dost thou pursue
Thy solitary way?
He, who from zone to zone,
Guides through the boundless sky thy certain flight,
In the long way that I must tread alone,
Will lead my steps aright.

At last, ruminating out loud, full of emotion and a peculiar inner exultation as I watched the flight of the geese, I cried, "Oh, Avis, there is one of the wonders of the world in action. The budding of trees and flowers, and growing grass, all are wonderful and marvelous, but there goes something of creation itself, a living, roving wonder. We can actually see it. It ranks with electricity and the rainbow. Just look at those birds and think of the sureness and method with which they head south under their leader."

Avis interrupted my ecstasy with her slow drawl, "Did you never see no wild geese before?"

Our castor bean bushes were gone now. The seed burrs, changing from green to a beautiful scarlet, finally browned and dried and burst, scattering the large black shining seeds all over the sand. The clean russet and green stalks turned to brown, too, and grew dry and brittle, and the leaves began to drop off one by one. So we cut the bushes down and now once again the view was clear, down past the mailboxes, where the postman turned the corner with his horse and buggy, and in the other direction up past the whippoorwill peas toward Grandma's.

The pecan trees, too, were bare, and all along the bank the weeds were a gray-tan, a tangled mass to be chopped down and burned. Across in the meadow the Bermuda grass was turning yellow and the willows and sycamores were all stark and leafless. The cottonwood trees, too, were bare. Their leaves of gay yellow had brightened the meadowland, but were all gone now.

All day the blackbirds called with their liquid high sad sweetness. The trees were outlined darkly with their fluttering forms.

We found a wildly exciting pleasure in the fall days and we often left our lessons and rushed through the housework to be out-of-doors, poking at the bonfires we made and watching the smoke-spirals that lifted and eddied in the wind.

The school readers were full of beautiful verses. Some of these we sang to a chanting tune and repeated the words over with deep satisfaction and emphasis as we raked and swept around the yard:

> *Flowers in the summer,*
> *Fires in the fall.*

or

> *Goodbye, goodbye to Summer!*
> *For Summer's nearly done;*
> *The garden smiling faintly,*
> *Cool breezes in the sun.*
> *The winter pears and apples*
> *Hang russet on the bough,*
> *It's Autumn, Autumn, Autumn late,*
> *'Twill soon be winter now.*
> *Oh, Robin, Robin, Redbreast,*
> *Oh, Robin dear;*
> *Robin sings so sweetly*
> *In the falling of the year.*

These tunes had a real lift and a haunting melody that Jimmy and Donald delighted in. They sang lustily as they worked with me, accenting the beats heavily.

> *Robin, Robin, Redbreast,*
> *Oh, Robin dear;*
> *Robin sings so sweetly*
> *In the falling of the year.*

The moon rounded nightly into fullness, but the line of stars that we had been watching as we sat, out on the platform of the pump had changed gradually until we wondered at last just where they had gone.

There was always one daily task that we dreaded during the freezing weather. This was having to thaw out the pump in the morning. It didn't make it any easier to know that all around us our neighbors had the same task.

We generally pumped water at night and filled the big teakettle and the water pails and some odd buckets, but always when the pump was frozen it seems that extra water was needed. So we hurried to make a roaring fire in the cool stove in the morning to heat the water we had on hand. Then old rags or pieces of blankets and sacking were wrapped around the plunger and the boiling water was poured down the pump while one of us hopefully worked the pump handle. Standing out on the well platform the cold wind whistled past us shrilly. The ground was frozen about four inches deep. The heavy white frost lay on every roof and post and board.

Later in the day, even though the sun might come out for a while, either Henry or Wayne would have to take the axe and go down to the pond in the meadow and chop the ice so the mules and cattle could drink.

All about us was a barren world. The trees were without leaves. The little shanties that could now be soon so clearly back towards the bayou, had each a spiral of gray smoke, and on every side the flat stretches of the cotton fields, with the bent and broken cotton plants, and the broken and twisted corn stalks were all drab and lifeless and dreary.

At Christmas time we went out in the hills again and brought back a beautiful little cedar tree and stood it in the corner between the windows. Our decorations included the little celluloid cow, the two swans, the little canoe that had hung each year among the shining globes and tinsel; and the deer's head again had golden oranges hanging from his antlers.

Christmas Eve we opened the wonderful gift box which had come from California and which had been so tantalizing, and played with the falling blocks, smelled the perfumed soap, listened to the tick of the watches and went into raptures over the photographs.

I had made pans of peanut-brittle and molasses candy and filled little stockings made of green cheesecloth sewn with long stitches of red wool. We had spent several evenings threading popcorn and looping it on the tree and we put boughs of cedar over the doors and made wreaths for the windows. So our little house was filled with Christmas incense as we again renewed our yearly feeling of magic expectancy.

20. Seen Better Days

SLEETING! DROPS OF ICE the size of sago and the ground frozen. All day long it rattled on the roof like shot. We had lots of wood and kept the big kitchen stove going all day, but even so, it was cold.

Nevertheless, I saw that the men were out in the fields with the stalk cutters, cutting down the old cotton. Winter had really come at last.

Then the days began to get a little longer, but it was still very cold. It started to snow several times and stopped. Then again it would be warm enough so that we could sit out on the porch with just our sweaters for warmth. The wind howled and shrieked and the cold air lifted the carpet on the floor. All our activities were confined to the daily chores of living, for there was little comfort except right beside the stoves, and each day seemed only a time of waiting until the weather changed and our lives could expand.

Then came a day with ashen skies, a heavy gray pall that lowered hourly. As sunset came on, the smouldering color faded out over the tops of the western hills and turned to a pearly whiteness that cast a ghostly glow on the earth beneath, and as night fell, a dry powdery snow began to sift onto the cold ground. Soon the fence posts, the porches and roof, were holding the flurry coating of pure white. Our little house sat at the edge of a plowed field, with high piles of wood neatly stacked on the two porches. This time there were two inches of snow, only lasting a few days, but the ground was frozen deep and the sleet made the farm chores very disagreeable.

The milk was frozen in the pans in the kitchen, so in the mornings, with our hot biscuits, we had canned peaches piled high with thick sweet cream. We cleared a place outside the dining room window and spread cornmeal for red birds. The male birds were really scarlet, with top knots. The females were brown and sand-colored, with bright pink bills and feet.

But the snow melted so fast that it soon became past history, as everything soon got to be here. After that we had two weeks of dry cold, and then the weather became so warm that we took the extra quilts off the beds, discarded our winter underwear and slept with all the doors and windows open. This unusual warm spell was followed by another hard frost.

Then we had weather as hot as ever I felt it, followed by another frost. People could really do such a lot of talking about the weather here, for it seemed to change from hour to hour. This changing back and forth is what made everyone feel miserable and all seemed to catch cold easily.

During the Christmas holidays we had company from town. Three of my old neighbors came in and we enjoyed coffee and Christmas fruitcake as we sat and chatted. They seemed eager to see my house, although having come in through the back door they had already seen it all except the

"North Pole" bedroom. They didn't want to miss anything, apparently, so I restrained my feeling of resentment at their bad manners and serenely opened the bedroom door and invited them in. I was interested to see just what their reactions would be. They walked around the room admiring my heavy white spread, noting the old ivory toilet-set on the dresser, and finally examining with interest the large clothes closet I had recently contrived myself. I had boarded up one of the corners of the room, and had nailed up a long pole, and wrapped it tightly with strips of white cloth, and here hung our clothes all on coat-hangers, in a long row.

One of the women turned to her sister, ignoring me, and said with a sniff, in a patronizing tone, "Well, it's plain to me that she has seen better days."

She didn't know that Grandma had told me that she had been living in town only a few years. Before that she had lived out "in the sticks" in a little house that was lacking in everything but the simplest housekeeping necessities. Her kitchen there was like so many others I'd seen, where each year fresh newspapers were pasted over the old soiled ones to cover up the dirt and stains. Her father was considered "no good," and they were not from the kindly and comfortably-well-off hill families. When I knew her in town she had often spoken with contempt of the "poor whites." Besides that, my visitors dipped snuff, privately. I knew this because Wayne had told me to watch the town women who professed to be above snuff-dipping, for I would catch them placing two fingers perpendicularly on their lips as they interrupted their conversation to go to the fireplace or the edge of the porch, wherever they were sitting, to spit. Ugh! This was an unconscious habit, for in using snuff they took this method of keeping the tell-tale brown stains from the corners of their mouths.

But it was always a pleasure to me to have Wayne bring in any men to have a meal with us. The expression, "come and go home with me," so continually heard, was far surpassed by the other frequent invitation, so eagerly given and so gladly accepted, to come in and take dinner. If we were out driving in the country we would often pass a farmer who would be on his way home for dinner or supper. And he meant every word as he called out, "Come on home and have fried chicken and sweet potatoes." I learned that it was so easy here on the farm to have these casual visitors—just another plate on the table, maybe a few more biscuits rolled out, or more ham fried.

I knew that when men sat down with us for a meal there would be no surreptitious feeling of the tablecloth to appraise its quality, no significant side glances at each other and no inquisitive glances around the room. If they were interested in our pictures or ornaments they would ask frank questions about them. It was a pleasure to see their honest enjoyment of good coffee, and chicken and gravy and hot biscuits. While their conversation was neither witty nor brilliant, still it was always sincere and varied and very entertaining to me.

A traveling salesman stopped by once with a stock of medicine and spices and was delighted when I not only asked him to stay to dinner but practically insisted upon it. We had been staying close at home for several weeks owing to the bad roads and I was actually hungry for fresh talk that dealt with something beyond the farm work, the weather, and cotton.

21. I Never Mind the Weather When the Wind Don't Blow

IN THE SPRING we would have about a month of drifting sand, and the wind would rise regularly every day, to blow the summer in. And it was in the spring, too, that the wind definitely changed. All winter the winds were from the north, but there came a day in early spring when one suddenly realized that the wind had changed and was coming from the southwest.

It blew every day until summer came, sometimes from the north again, sometimes from the south, veering and changing back and forth suddenly, until at last it settled down and blew always from the southwest, until again in winter, when "The north winds do blow."

I often stood on the back porch near the pump and looked across the fields at the long reaches of green cotton rows, with the ground dry and dusty between. Every day the winds blew across the fields from the southwest, and the dust spirals rose and whirled, whirled and danced rapidly in the hot sun, racing across the fields and subsiding as quickly as they came. These spirals were about twenty feet high. They were smaller at the bottom and of almost conical shape, keeping their shape and moving in a mysterious manner across the fields, even when at times it seemed that there was no wind blowing.

Once I ran out into the field and stepped right into the center of one of these large circling whirligigs. I did this purposely, just to see what might happen. I found that I was nearly blinded by dry dust and could see nothing, so I threw my apron over my head and face, and the spiral moved away and left me with my mouth and eyes and hair full of dry dirty sand.

It was a sign of rain when these dust spirals danced across the fields toward the river, I learned, for in dry weather they always seemed to whirl away lightly toward the hills.

In the sandy land, after the cotton stalks were cut and the land disked and plowed, how the loose, gritty soil lifted and blew: Every day, about nine o'clock in the morning the sand began to blow, and only ceased when the wind died down in the late afternoon. This meant special housekeeping plans. The washing must be done early and brought in before the sand began blowing too heavily. The food must all be carefully covered and stowed away. The oven of the stove was the only place I found that was practically dust-proof, and just as I had been told, I had to keep my butter and biscuits in the oven all day.

It was irritating to go around the house most of the day with a cloth pinned over my hair to keep out the gritty dust. No sewing, no writing, nothing but the necessary routine work, hastily done, to be followed by an escape into a book or magazine. Drinking water had to be pumped fresh for each drink, for the water bucket would carry a film of dust in spite of its covering. The floors would be thickly coated, too, and soft piles of filtered dust would lie along the partitions of the rooms.

Underneath it all was the sense of waiting, waiting, day after day, for the winds to cease and summer to come.

Hardest of all to bear were the sweltering days of sudden spring heat, the stickiness of face and neck holding a coating of dust, and the never-ceasing wind that blew and blew and blew. "I never mind the weather when the wind don't blow"—that old saying often came to mind.

And in all those years, every spring, for about three months, while the wind gradually learned to settle down for the summer, the stove in the kitchen smoked.

The trouble was that the roof of the lean-to kitchen was much lower than the house roof, and to carry a tin stove-pipe up high enough for a good draft was impossible, as it would be blown down in the first high wind.

At last I thought I had the problem solved. I nailed a small ladder to the side of the house and Jimmy and Donald would climb up on the low slanting roof and turn the chimney, for I added an extra stove-pipe joint that bent over and carried the smoke away from the wind. Nearly always, turning this extra stove-pipe would allow the stove to draw, and the kitchen would gradually clear, for the cottonwood gave off a thick black smoke that smelled rank and stung the eyes, and the little room would be so full of smoke that it was hard to see even the stove.

Even though the coal-oil stove could be quickly pressed into service, many and many a meal was spoiled because of the sudden changes in the wind, for I found the wood stove far better for baking and cooking the daily meals, and the wind always came up suddenly and unexpectedly.

The boys delighted in the job of climbing the ladder and calling down for directions as to just how far to twist the pipe to control the smoke, and our weather eye was always out to catch the first change in the direction of the spiral of smoke that rose over our little room.

But I just can't express the thoughts that arose in me, when, with a cake in the oven, and the boys off somewhere from the house, I had to climb the ladder myself, again and again, trying to adjust the chimney so I could finish my baking.

In the spring days, too, the mules always interested me strangely. Down in the Bermuda meadow eight of them would gather by the gate under the big pecan tree. They seemed to sense a coming storm. The overcast sky, with the restless three-dimensioned clouds—the clouds back home seemed now by comparison to have been like flat painted effects on the sky—produced a peculiar blue-gray light that enabled one to see clearly for long distances, and seemed to bring the little houses, that sat far off by the road leading back to the bayou, so much nearer.

A strong wind seemed to drive the hot, feverish air. The trees would sway from side to side, their great arms raised and lowered beseechingly.

Then the mules would run and jump, and race madly, far to the opposite side of the enclosed pasture, and back again with pounding hoofs, their heads turned sideways as they threw themselves forward with terrific force and abandon. They would stand on their hind feet and paw the air and each other, then bray loudly and turn like a flash, and repeat their wild racing and pounding until the first heavy deluge sent them huddling back beneath the trees.

The storms of that March I will never forget. The daylight would change suddenly from piercing sunlight to dull gray glooms. The dead leaves along the bank would lift and eddy across the fields and garden. The bare trees would twist and sigh as the bleak winds assailed them. The rains now were cold, and drenched the roads and the meadows and lay in every depression. Not in this month did we look to the heavens to admire the white fleecy clouds with their rose-tinted peaks. Those monstrous clouds that billowed like snowy mountain peaks all glorified by the setting sun, came later. The crimson and opal lights belonged to later days. Now we watched the early morning skies of vermilion and ashy gray with foreboding, as we learned to prepare for another day of lightning and thunder and desolate screaming winds.

These are some of the signs of rain I gathered from my neighbors:

When the sun sets behind a low bank of clouds on a Sunday night it will rain before Wednesday.

When the crows gather near the house or barn and call incessantly it will rain in a few days.

When the star that appears near the new moon draws closer night after night it's a sure sign of rain.

Mockingbirds suddenly bursting into song in the daytime, or a strong wind in the southwest, are sure signs of rain.

Lightning and thunder in the west foretell rain, and when the melon vines wilt down in the heat during the day it will likely rain.

When the train whistle is heard from across the river clearly and the mules get frisky in the pasture, or if the quail whistle soon after sunset or before sunrise, it is liable to rain within three days.

When quail or chickens get out in the road to wallow and dust in the sand, it will surely rain.

These signs of rain amused me greatly, yet they all must have worked magic—for it usually rained.

22. Farming in Dixieland

THE BIG GUNNY-SACK of prunes my mother sent arrived at last and were a real treat. I gave lots of them away to our neighbors, but they all seemed to be a little skeptical about them, for strange as it seems, they had never had any prunes, and had no idea of how to cook them. I found it hard to keep the prunes for the damp heat caused them to become infested with worms. That's what happened to the corn meal, too, even though it was kept tightly sealed in jars, and our flour and all our cereals soon became full of weevils. Groceries could not be bought in quantity for fear of spoilage, so that was another reason why Saturday was such a necessary shopping day.

Up behind Grandma's house we fenced in an orchard and set out twenty-five fruit trees; peaches and plums mostly, for apples would not do well in the bottom land.

We now had six grown mules and a mule colt. All the mules went in pairs. Even in the pasture Mutt and Jeff stood around together, as did Joe and Jerry. They did much better work when they were kept together. Pat and Mat were mouse-colored mules; they looked alike. Mat was the best mule of that pair. Joe and Jerry were dark Missouri mules with a white stripe between their eyes. Belle and Rhoddy were a pair of small, fractious mules.

In this one year we had our own lard, soap, butter, meat, eggs, and vegetables. All last summer we enjoyed lettuce, radishes, onions, beets, carrots, peas, and turnips. The string beans and sweet-corn and melons lasted until October.

These days were sunny and warm. I turned out the two hundred chickens with eight Plymouth Rock hens to run in the cotton fields near-by, although the oats were already planted. We'd had them cooped up in the hayshed.

One big Plymouth Rock we called Old Hundred, for to our surprise she actually corralled over a hundred of the little chicks and succeeded in having them all leave their own mothers and follow her wherever she went.

Wayne and I were gardening, from five in the morning till dark. "From can't to can't." We only stopped to eat, and came in at night aching from head to foot.

I was always at the same work: chickens, mending, cleaning, cooking. The garden was up and doing fine. Radishes were a half-inch and peas an inch. Last year at this time we had ripe dewberries, but the weather was much warmer then than now. Wayne was busy fixing fences and soon would cut the alfalfa. He made a contract to sell most of the alfalfa right out in the field, so as to save baling and hauling. He planted casabas and five hundred more hills of watermelons.

Planted and fenced in, chicken-proof, were: cantaloupes, radishes, peas, onions, beans, mustard, beets, carrots, parsnips, lettuce, bell-peppers, sweet-corn, and onion-sets—eight gallons of them, and 'Tucky-wonder beans and cabbage. In the fields were one thousand hills of wa-termelons, seventy-five acres of cotton, fifty acres of corn, fifteen acres of old alfalfa and five acres of new alfalfa, besides the twenty acres of Bermuda meadow and the pasture by the river, on the "made" land.

Again the little castor bean shoots were up and growing fast, for every day came light showers. They would work wonders on the new garden. The little Indian peach trees were in blossom now, with their beautiful pink petals that fell so soon. There were no green leaves yet, but the mockingbirds were calling, the blue birds swept swiftly through the bare pecan and cottonwood trees and small flocks of red birds drifted by or settled in the brown bushes along the bank and flaunted their gay color. Around the hayshed swooned the swallows and the little Jenny-wrens hopped in and out of the pile of driftwood stacked by the black pot. One little wren built her nest inside of a short length of stovepipe up on a high shelf on the back porch. She peered down at me with her bright eyes, and seemed unafraid.

I was looking forward already to our lovely summer nights. I remembered last year reading out loud to Wayne while we sat on the porch in the moonlight. The hot weather would begin in June and our rocking chairs would be moved from one end of the porch to the other as we tried to keep in the shade. No one here, I think, ever saw a real sundial, but many people on the farms would glance casually at their porch boards and judge the time of day almost accurately by the certain board on which the shadows fell.

But well I remember, too, how we had to pump tubs of cold water each night so we could cool off before going to bed.

The mattresses and sheets would be so warm that in spite of the refreshing bath we had enjoyed, for we all took turns using the big galvanized tub on the back porch, we would lie awake for hours, sleepless, tossing and turning, trying to ignore the prickly heat that broke out on our backs, and trying to lie quietly enough so we could gradually sink into slumber and forget the mosquitoes and the oppressive heat until morning.

We were certainly overjoyed to know that my mother was really coming to visit us and the days were marked on the calendar. The boys began and ended each day with, "How many more days before Grandmother comes?" We went around singing, "Somebody's coming to my house in Dixie."

The sun was up early now and so were we. I attended to the chickens, and the milk from two cows, and had been sewing, and cleaning the house, getting all the odd jobs done so we could have lots of time to talk when my mother arrived.

I washed and finished all the ironing in one day so that we were very, very clean. My mother would get to see me wash in the black pot, all right.

Wayne brought home a big drum fish from the lower bottom. I baked it with a stuffing of bread and tomatoes.

The cotton was now being chopped out and the corn hoed. The garden was coming up in fine shape. We planted fifty hills of honeydews. Several of the watermelons had eighteen-inch runners. These were nearly three weeks earlier than last year and ours were the first ones ripe then.

We wanted them ripe by the fourth of July so they would bring a good price.

Even in these small houses, there was a great difference in the life lived where there was a fireplace and where there was none. Grandma's three rooms were undoubtedly crowded but withal there was a sense of warm comfort and use and hominess. Grandma spent much of her time out working among her flowers. She kept every pot and pail planted with cuttings that grew and blossomed for her as though by magic. Her small porch was really a bower of bright flowers and long green vines drooped over the rough boards and hid them from view. Touch-me-nots and ferns, petunias, smilax and geraniums lent color and charm, and she loved the zinnias which were planted in neat little rows by the steps.

In bad weather Grandma could generally be found by her fireplace, rocking away and knitting or sewing her quilts of intricate patterns or cutting rags for her rugs. Grandma was often in bed with a sick headache, for she suffered for years with migraine headaches, yet she kept up her interest in the chickens, the garden, the farm animals, and all the details of the daily work. She knew, too, the affairs of all of the neighbors. When she was feeling well she took a long walk each day that the weather permitted and was welcomed everywhere. Nothing escaped her observing eye, and no problem arose but what she had the sensible, practical solution for it. Gradually everyone learned to bring the daily plans of the farm work to her for her opinion and advice.

Her big Bible was always at hand and she read it daily, read it and believed it. Her religion caused her to look on life with tolerance and patience.

Jimmy and Donald spent a lot of their time with her for she loved to talk and reason with them.

She read aloud to them the items of world affairs that attracted her interest in the newspapers.

Henry, who lived with Grandma, and had the little side room for his own, subsided into a routine life of comfort and ease. He liked to eat and Grandma was a good cook. He was lazy and Grandma made few demands of him. He liked to read the papers and argue and Grandma was a good listener. He loved to brag and Grandma believed him.

Besides, Henry loved to make trips into town, and Grandma liked to go in, too, to shop or visit. The fire in their fireplace was seldom allowed to go completely out, and Grandma churned there on the wide hearth, dried some clothes there on rainy days, coddled a box or two of little chickens there in the cold weather, heated her irons there, and even cooked pots of hominy and baked sweet potatoes there in the hot ashes. She said it was easier to keep a fire in the fireplace than to be always bringing in wood for the kitchen stove. Besides, the fireplace was comforting in winter and cooler than the cookstove in summer.

Her huge feather bed, so high and billowy, stood waiting for her in the corner of the front room. I can see the picture yet—the firelight dancing on the bright velvet squares of her fancy top-quilt, and Grandma snuggled warmly in her flannel gown for the night, inquiring sleepily of Wayne and Henry, who often sat late by her fire, about the day's work, or wondering if the pigs or cows or chickens were all under shelter when the weather was bad.

About this time we read in the local paper that the dipping vats were being blown up here, there, and everywhere in the county. On account of the Texas fever tick, no one was allowed to ship cattle out of the county unless they were dipped, so the officials were on the alert to build vats and see that the farmers used them. But the vats began to disappear in the night. An explosion of dynamite, and there would be no vat close enough for many of the farmers to use.

We had hoped to raise a lot of cattle and so Wayne and Henry decided to build a vat over at one side of the meadow, and of course this would be used by all the neighborhood, too. We had a pump put in, for it was not very far down to water, and a small corral built and a deep trench dug out and cemented. This vat would be filled with a strong solution of creosote and water, and cattle in the corral driven into a narrow runway and urged along until they could be shoved off into the vat. They would jump in and frantically start to swim, then their heads would be pushed under at least once. They would scramble up the incline at the end of the vat. After that the fever ticks would die and fall off the cattle.

Our neighbors were as excited as we were about the purebred Holstein bull and the four registered milk cows that we had lately purchased from a friend out in the hills. They had been shipped to him from Dallas, but he found he couldn't afford to feed them. We traded hay for them and got them for seventy-five dollars apiece. He was glad to be rid of them although he lost considerable money on the deal. These cattle were among the first to be driven through our dipping vat, but they were already so infected by Texas fever that within a week, one after another, we found them stretched out in the barnyard dead.

It had been hard to get Grandma to understand just why our five cattle had died. I was asked to explain it to her as best I could—and a sorry hour we had, for Grandma cried and I cried, too. I told her all I knew about the Texas fever ticks and the vats and the dynamite. She had read some accounts of these things in the newspapers and wanted to know all about them.

I told her of the scene I had watched, down in the meadow one early morning. I had heard the cattle bellowing and men shouting and boys yelling, and had rushed out on the porch to see what was going on. A small herd of cattle was being driven into the meadow. Here they were gradually maneuvered into the small corral and beaten and kicked to force them along the narrow chute. Then what uproar as the cattle were prodded and pushed off into the stinging solution, where they struggled as their heads were rammed down beneath the surface.

No wonder they were scared and nervous. No wonder the farmers who owned just a few cattle, or maybe only one milk cow, objected to taking a whole morning from their work to come miles in the hot sun and go through all the excitement of the dipping. They claimed they just couldn't do it, for the cows fell off in their milk for a week afterwards, and as they weren't going to sell any stock anyway, the few ticks that got on the cattle they could pick off.

I didn't blame the farmers. I couldn't forget those gentle cows, made nervous and excited, being kicked and prodded and beaten by the men and boys who sat along the railing of the narrow runway between the corral and the vat, eager to participate in the show. It seemed to me that they relished this opportunity to show their own brute strength and authority, for they apparently enjoyed the fear and misery of the poor, bewildered creatures. One woman told me that she had walked four miles with her three children to drive their milk cow to the vat. Her husband had to use their team for plowing, her children could not be left alone and there was no one to take the cow but herself.

The life of a Texas fever tick was such that regular dipping should eliminate them. Native cattle were more or less immune and did not die as easily as cattle from places where there were no ticks. Our Holsteins had apparently made most attractive hosts for the ticks. So now, discouraged by our loss, we decided to raise more hogs!

23. Lidy's Baby

AT LAST WE HAD a telephone, a twelve-party line. I would turn the crank and try to get central, but there was always someone on the line. Conversation was often interrupted by someone listening who could give correct information. If I was trying to get my doctor, a voice spoke up, "He's gone out to Centerville, I heard Miss Webber say so. Her niece is awful sick." If I rang and rang to send a message to a neighbor, someone would say eagerly, "I'm going over to see Miss Sara after a while, I'll tell her you want her to come over."

Oh, it was very neighborly and kind, but one soon learned to use the telephone very, very discreetly. It was so funny, and so exasperating, to hear central say to the long-distance operator,

"Are you busy, dearie? I hate to bother you, but this seems to be important. Someone is sick and wants to get in touch with his folks." All as if it were a matter of personal favor; in other words, Southern fashion. They seldom said, "Number, please?" Just, "Who do you want?"

In a thunderstorm when the lightning was bad, we had to run to unfasten the telephone wire where it was hooked to the house wall, and throw it out on the bank, away from the house, until the storm was over. If we didn't, the telephone rang and rang and as it was only a simple arrangement of bare wires, it was a real danger.

We realized the value of the telephone most when someone was sick and needed to call a doctor.

Lidy Cowder, old Mr. Batson's daughter, had a new baby. Lidy and John, with their three small children, lived down in the same house where the Randalls had lived, on the Jay Cossey place, across the road from our line fence.

The first I knew of her illness was when I went to the door in answer to a knock and found John standing there.

"Please, ma'am, would you-all telephone to town for a doctor? The line is down again by our house and gotta be fixed. Don't matter which doctor. Lidy's awful bad. The baby's done born. It's in the bed, an' I jest know somethin' orter be done. I reckon Jay Cossey will stand good fer the bill—he said he would, anyways."

John stood by while I cranked the telephone and finally got a doctor to promise to come.

"Who is Cowder working for?" was the question I had to answer first, and the doctor being satisfied in his own mind that Jay Cossey would pay the bill, promised to hurry.

"Shore do wish you-all would come down, ma'am. Don't seem hardly right to leave Lidy alone. Ain't no one with her, and the baby orter have somethin' done to it. Shore would take it kind if you'd come." John was very worried and white looking. I was fearful of what I might be forced to face.

"Have you got a good fire going, and lots of hot water?" I asked.

"No. Ma'am, I guess we ain't. Been a long time since we et, and we was out in the field hoein' when Lidy took sick bad."

"Well, you've got wood and kindling in the house, I hope. Is there any water pumped?"

"No, ma'am, I didn't hev no wood up."

"No. I thought not. You ought to be ashamed of yourself. You come up here calling for help! Why didn't you have wood and water in? You've been through all this before. You have three other children and should know by now what is needed. You should have some sense by this time. Get down there, as quick as you can, and get things ready like I tell you. I'll come in a few minutes."

John ambled off, and I fairly flew out the back door and up the road past Grandma's, for I saw Wayne out in the barn lot.

He laughed at my fears and nervousness, and advised me to go ahead and do what I could. "You'd do as much for a dog," were Wayne's parting words, as I rushed away.

Well, I know I wouldn't, but I got the idea.

Back I went on the run, and arrived breathless at Lidy's house. Inside, Lidy lay on the bed in the corner.

I saw a pallet on the floor where the three children slept. John hadn't made up the fire in the tiny heater yet, but I could hear him outside chopping wood. I looked into the wood-box and saw only dirty rags and small chips, not enough even for kindling.

Lidy lay with her eyes shut. The pillow had no covering whatever and the ticking was soiled and stained. She wore the same dirty old dress she had been wearing out in the cotton field. I looked at her bare feet, for they were sticking out from under the quilt. They were just black with dirt, black and misshapen and calloused, for she never wore shoes except in winter. The few ragged home-made quilts were so dirty that I was amazed, but as there were no blankets and no sheets on the bed, I knew they used the quilts for all purposes.

John came in, carrying an armful of wood and a pail of water, and I asked him if Lidy had things ready for herself or the baby. "No, ma'am, she ain't. She was took kinda sudden-like—wasn't expectin' nothin' for a week yet. Don't reckon you'll find much."

I didn't. I could quickly see what there wasn't, for everything they possessed was in plain sight. There was no dresser, no trunk, no boxes. A few dirty ragged clothes hung on a nail, and a pile of dirty clothes lay in a heap on the floor. I peered quickly into the back room, and saw only the stove, the tall safe, and the kitchen table with two chairs and the wash bench beside it.

I knew it was high time something was being done for the baby as well as for Lidy, but I shrank from the whole situation. I knew, too, that certain things would surely be needed, so, taking a fearful look at Lidy's white face as I passed, I went out and ran again up to my house. It was heavy going on the sandy road, but I held my hand over my pumping heart and ran doggedly on.

The big doctor book had many chapters on childbearing. Hastily turning the pages, I saw the list of supplies a doctor would need, so taking my sewing bag, I began to collect what I could. Scissors, string, Vaseline, powder, and safety pins. I had them ready and was about to step down from my porch when I saw the doctor's car down on the road, speeding along to Lidy. I was really scared, almost panic-stricken. I had never had occasion to really know what to do, so I just went inside and waited awhile before I started down the road again.

As I entered Lidy's room this time I saw that the baby lay in the lap of one of the neighbors, a woman who lived down past the store. Evidently John had sent for her. She was dressing the crying baby, with baby clothes she had brought with her. There was no chance among the poor people here to save their baby clothes, for they were sure to be needed again by someone, if, indeed, they weren't already worn out completely.

With the experience and knowledge I have gained since I often wonder just what the doctor did. Attended to the baby's navel, and its eyes, surely, but there lay Lidy with closed eyes, her face so white that she looked as if she were dead.

The doctor was standing by the window, but turned when I entered. "She's going to need a bath," he said, "and clean bedding, and a gown. Do you think you can find any?"

"No, doctor," I replied in a mock-civil manner, for I was trying hard to repress my thoughts and not give voice to my anger and resentment. "I'm sure it's no use to look, but if you will drive me up home I can get what is needed."

So we drove up the little sandy road to my house. On the way we met Mrs. Batson, Lidy's mother. She came plodding along, with a gray flannelette cloth over one arm. The doctor slowed down.

"Lidy's all right," he called out to her, "but she'll need some care. Mrs. Batson nodded.

"Yes, I jest heard about it. I was kinda expectin' it to happen soon, even if Lidy wasn't, so," lifting up the gray gown for us to see, "I went up to Dawson's and borried a nightdress for her."

We drove on.

While the doctor waited I hurriedly collected sheets, slips, and towels. I knew she had no soap either, nor clean clothes of any kind, so I took those, too.

"Why didn't you bring the nurse?" I asked the doctor. "I happen to know there is one visiting at your home. You knew the conditions you were coming to."

The doctor was polite but apparently indifferent.

"Well," he answered, "it's almost impossible to get the town women interested in these bottom folks. They don't really know about the conditions among these share-croppers, and I'm sure they don't want to. I know they would hesitate to set foot in one of these shanties, even to save a life, and they wouldn't touch a woman like Lidy Cowder with a ten-foot pole."

"Well," I replied, "I'm not a nurse, but I've been trying to do my best among these people ever since I came here. It's like pouring water down a rat hole though, there's so much poverty and sickness. I've filled dozens and dozens of capsules of quinine, and advised them in lots of their common ailments. Our own doctor, knowing what I have been doing, suggested that I just give everyone of them a dose of castor oil as soon as they get sick. He said that was what they all needed, anyway. But it's money and common sense and a heart that's needed, also, I think.

"You know very well," I continued, "as all the people around here know, that I can always be called on. Of course I'll bring food. I'll bring clean sheets and clothes and fruit. I'll bake them fresh loaves of bread and kill chickens, too.

"But Lidy isn't one of our tenants, you know. Their own landlord won't help out in a case like this. They owe him too much already. You took it for granted that Jay Cossey would pay you because you knew he could. Maybe he will, I don't know. I understand why Lidy had no diapers ready, no baby clothes, no gown, and no sheets—just nothing. John Cowder is young and he works hard. He always has, for I've heard the men talk about him, but he hasn't got a chance to have anything."

I think the doctor was glad that the distance was short.

John was down on his knees when we returned, blowing into the front of the heater. A teakettle full of water sat on top. The baby was now asleep on the neighbor's lap.

At the doctor's suggestion I went out into the kitchen to see if there was any food there that could be fixed for Lidy. Nothing! The safe door was open. In it were a few broken dishes and a bowl of cold white gravy—"sop." John had brought in a bunch of collards, and the big fresh green stalks were piled on the kitchen table. Several pieces of broken corn bread and the coffee pot still half-full of black coffee were on the back of the stove.

In the front room, the doctor stood with his back to the bed, his hands clasped behind him, looking out over the cotton fields. Lidy, awake, now lay with a look of dumb anguish on her face, her two hands pressing hard on her abdomen. Old Mrs. Batson, backed up to the heater to keep warm, said cheerfully, looking over at her daughter on the bed, "That's right, Lidy, press down hard. That's what I allus did."

The doctor, it seemed, wanted to wait until his patient was cleaned up before taking care of her. I helped to fix the bed, and managed to get Mrs. Batson and the neighbor to undertake to get Lidy's clothes off and bathe her while the baby slept. I left soon afterwards, before the doctor went, promising to return soon with chicken soup.

Grandma helped me, and for several days we made trips back and forth taking eggs and milk and fruit, and seeing that John and the little children were well fed, too.

Yesterday, I was asked again to telephone, to try and get Lidy's father, who had gone across the river, for it was believed that Lidy was dying. Childbed fever had set in and the doctor, fearing to move her, had had to perform a necessary operation right where she lay.

I didn't know about it until it was all over, but I could imagine the risk of infection that had been taken.

Speaking about it to me later, the doctor said, "It's a funny thing. Any of the women in town, with good nursing and clean surroundings, would have died, no matter what aid they had, but I guess these poor people are immune. They just seem to live through anything."

24. Oh, Sailor, Come Ashore

OH, WHAT GREAT HAPPINESS we found in the phonograph and records my mother sent us! What a surprise that was! We enjoyed every record and Wayne and I thought we knew why each one had been chosen. We knew that Grandmother would never fully realize until she was with us how much music meant in our lives. The old familiar songs brought back bygone hours. Tunes that awakened sad, sweet memories charmed us hour after hour, although some of them seemed actually to make our hearts ache.

We really needed that music. Each day we turned afresh to the pleasure in store for us and it hardly seemed true that it could be there in our little house, so each day we felt like sending to Grandmother our heartfelt thanks over and over again. Jimmy and Donald never tired of the music, and our little house was filled daily with piano solos and singing and the throbbing of violins.

The boys were growing fast. They played out of doors a lot, chasing over the fields with Wayne, driving the mules to the plow or harrow sometimes, to their great delight, learning to milk, running errands, and studying their lessons. Often at meal times we played that game that Douglas Fairbanks told about in the *American Magazine*. We would write on slips of paper the names of subjects to be discussed. Then each of us would draw a slip and take turns talking on the subject that fell to his lot. Much to our surprise we found it taxed our fund of ideas to talk for one full minute even on a subject we thought we knew well, but we also found that gradually we expressed ourselves with more ease, so we kept at this from time to time.

Jimmy and Donald were eight now.

Jimmy always made me feel like the hen that had hatched out the duck. He was some duck. He came in one day from the fields and announced that he was going to be a rich man. "Nobody can make me poor," was how he expressed it. He had made up his mind early from seeing the conditions of the various cotton-pickers. Wayne made an arrangement with the boys to buy all the eggs they found out in the hayshed. They had three dollars in a short time.

I often heard Jimmy talking to Donald after they were in bed, asking him just how many eggs he had found, so he could figure out exactly the amount that was due them. Donald seemed like a willing henchman, following wherever Jimmy led, yet he loved to stay with Grandma and she took pleasure in regaling him with her great store of farm lore and tales of her own youth.

Fearful lest I might be making a mistake in keeping the boys from school, and wanting to see the new buildings that had recently been finished, I drove down to the schoolhouse one afternoon. A man and his wife were teaching. She had the smaller pupils, he the older ones, even in the one large room. A small house had also been built nearby for the teachers.

A lesson was in progress when I arrived. Sitting in a row on a bench up near the teacher's platform were eight grown boys and girls. They all seemed to be between twelve and eighteen years of age, and the teacher was discussing with them a reading project he had started.

I was warmly welcomed and given a seat and the discussion went on.

"Did you read the *American Magazine*?"

"Yes, sir."

"What did you notice particularly about that magazine?"

"Well, it seemed to be full of pictures!"

"Did you read any of the articles?"

"Well, I read two of them."

"Did you notice anything similar about the stories? Can you remember what they were about?"

"No, sir."

"Did you notice this article?" the teacher went on, holding up the magazine showing an illustrated story of a legless man who had nevertheless made a wonderful success of his life. "This one about the man who had no legs?"

"Yes, sir, I read that."

"Can you tell me anything about it?"

"No, sir, don't reckon I can."

Turning to one of the girls, the teacher said, "What magazine did you take home to read?"

She mentioned *Capper's Weekly*, a farm magazine.

"And just what did you like in that magazine?"

"Well, I liked the crochet patterns."

"Yes," the teacher encouraged.

"I liked the quilt patterns."

"Yes."

"Sometimes I read a story."

"They are love stories, aren't they?" the teacher asked.

"Yes, sir."

"Would you take some of my magazines home and read in them?"

"Yes, sir, I reckon I would."

Such were the labored questions and answers, as the teacher strove to awaken the spark of interest so necessary to kindle a desire for better things in these young people. After that class was dismissed I went to the other end of the room and listened to the smaller children read.

When school was over I talked long and earnestly with the teachers. They told me their greatest problem lay not with the children but with their parents at home. The children were very eager to learn, but there was a sullen opposition to the teacher in many homes.

"For instance," the teacher said, "Those six little children you heard reading know that reader by heart. They have been quick to learn and they go over and over the stories. They need new readers. The only response I get when I send word to their parents is that they are in the Third Reader and they should stay there until the term is up. That's the way it was when they went to school and that's good enough now. I often buy extra books and extra school supplies that I need, out of my own salary which is not large, you know. But it helps me, because I can hold the attention of those who would otherwise lose interest and get into mischief. If I had more to work with than just the bare textbooks, it would make all the work easier." How readily I agreed.

I had expected the teachers to be the ones to agitate for improvements as well as all the fundamental necessities.

"We have been teaching in different places in the country," they told me, when they saw how disgusted I was with the conditions they had to face, "and we have always found that there was little we could do. The lack of necessary funds to purchase new things has gradually created a state of mind akin to real opposition among the parents, even regarding real necessities. We think it was wonderful that this schoolhouse was even built, and the little new house nearby for us to live in.

"All these things take time, and most of the people think that teachers are never satisfied.

"We really don't dare to criticize anything; we'd lose our jobs if we did and we can't afford that. No, there's nothing that can be done. The parents aren't really to blame, and most of the other people are simply indifferent."

The sanitary conditions had improved, however. The Board of Trustees had compromised finally by building an out-house for the girls.

The boys could still go over the bank.

Wayne and I had sent to Washington and secured many valuable pamphlets relating to farm life. Our finer-meshed window screens and extra mosquito nettings were due to the suggestions of government experts.

They had sent, also, a simple plan for a fly-proof out-house, for the possibilities of contagion in diseases like typhoid and dysentery were not to be ignored, and more than ordinary precautions were needed in a district where such an idea as the flies being germ carriers was scoffed at. Our little new building, so well made and so screened, back behind the hayshed, caused much comment, and many of the neighbors, hearing about it, came and inspected it, evincing great amusement.

"You-all must have been living high-toned out there in California," one of our neighbors told Wayne when they were discussing our late improvements. "Years ago there was three or four of them things around the farm. Your father had 'em built. The winds blew some over, some just disappeared. Guess they made good dry kindlin' when fire-wood run short. Anyway, we've allus got along somehow."

I left the school that day more than ever convinced that I could do much better for my children by keeping them at their lessons at home than if I sent them off to school every day into that environment. The teachers had admitted that nearly every boy smoked or chewed snuff and tobacco, and crap-shooting and cards were the pastimes of all the boys old enough to begin copying the habits of the men. Since they all carried knives, for protection in the quarrels that arose at recess and after school, pupils were often seen with a bandaged arm or head. Truant laws were not enforced. When the children were needed to chop cotton, they were kept out of school, and when cotton picking began, it was only the few who came from the larger farms who were able to stay out of the fields and attend classes.

I'll never forget one time after I had spent the whole noon at school lessons with Jimmy and Donald and they had then gone into town with Wayne and Henry. I was sitting on the porch when down the sandy road came the two older Wood boys to sit on the edge of the porch in the shade and rest, as was their usual habit. It was a hot day and their shirts were soaking wet.

Our huge Webster's Unabridged Dictionary lay on a little table near the well. Tom eyed it curiously.

"Ma'am, can you read in that there book? Dadburn me, that's a heap of book. I ain't never seen such a book, never. Heap of words in it. You wouldn't mind no looking in it, would you? Gosh! Hit sure is full of words. Take a heap of learning to be able to git into that. Never knew they made books that big."

"Tom," I said, "what reading are you doing now?"

"Well, we're a-reading poetry, but nobody don't like it."

"Will you read your today's lesson?"

Tom took the reader, his cheeks flaming, and carefully locating each word with his finger as he went along, and accenting the "al" in the word "coral," he read:

> Oh, sailor, come ashore,
> What have you brought for me?
> Red coral, white coral,
> Coral from the sea.

"Do you know what coral is?" I asked.

"No, mum," he replied.

"Didn't the teacher tell you?"

"No, mum, he ain't."

"I have a piece of coral in my trunk, white coral. I'll get it and show it to you. You may take it up to the school tomorrow so the teacher can show it to the other pupils. You could take the album of postals, too, if you like. The class might like to see then."

"Yes'm. Sure be proud ta. And those picture cards, too. Sure be proud to take them. I aim to get outa this place, someday before I die. Don't know how. None of us ain't ever had nothin'. Read and write and figger, that's all the use I got for any schoolin' fur as I kin see. I can figger what I got comin' to me, and write me an order for groceries. I always make out the Sears-Roebuck orders myself now, for the whole family."

He stood there gulping. He was a tall lean boy, hollow-cheeked and sunken-eyed, his face misshapen from protruding teeth and uncared-for adenoids.

Tears came to his eyes. "Gosh, I sure would love to get away from cotton. Love to be where I could get me some fruit."

I don't want to remember those tears. Perhaps, after all, it was only the sun in his eyes.

Everything seemed to be against these poor people. They were doubly bereft, for their lives were empty of so many material things that made for comfort, even at times of enough warmth and food, and they did not have "fire enough in their brains" with which to build castles, or dream dreams or make plans.

Those of us who have long known that "hope deferred maketh the heart sick" would yet choose ever to cling to some vain hope, some ambition and striving, rather than to have only the blankness of "Ain't got nothin'" and "Ain't aimin' to have nothin'."

The schools were soon out, here in the bottoms, but Jimmy and Donald would study off and on all summer. Donald announced that he knew there were twenty-five thousand miles around the globe. The big new geography was fascinating to him, and he was always finding amazing and disjointed facts in it that took his particular interest.

Jimmy brought down to the house a little Plymouth Rock baby chick that he found wandering out in the barn alone. It was healthy and saucy.

We adopted it for a pet, and it slept in a basket of fluffy cotton, kept warm by the stove, ate food from our plates and grew at an amazing rate. It really surprised us to see how early it got its pin-feathers, and then grew a full coat of perfectly marked gray and white feathers. We named her Maggie and she was beautifully curved and developed for such a young chicken. We made a special box out on the back porch, where she learned to roost. She and Peggy, the big collie, were good friends, and when Peggy lay down on the porch, Maggie often perched on top of the dog's back.

Donald's dog, Looie, was one of Peggy's pups. Peggy was a thoroughbred, but her seven pups weren't. Looie was white and brown, and just learning to bark. Yes, we seemed to be surrounded by animals.

It had been simple to teach Jimmy and Donald some of the first principles of life and reproduction. They were deeply sympathetic and concerned over our poor mother cat who had been getting heavier and heavier until the day when we found a nest of beautiful kittens. The fact that the boys were not attending school and associating with other boys made it possible for us to extend the poetry and stories from their books out into our actual life. The animals around us seemed to represent those in their readers, the mother hens so concerned with the happiness of each chick, and the problems and pleasures of all our animals duplicates of those that we read about.

All the delightfully silly stories in the readers, full of such expressions as "Then the little dog laughed, and said—" we incorporated into our daily conversation, and we were always beginning, "Then the little hen came by with her nine little chicks and clucked and said—" or "Then the old cow mooed, and said, 'When are you going to put me in the pasture?!'" Both Jimmy and Donald were learning to milk and took that possessive attitude in their work that makes for success.

It was easy to make the transition, too, from poetry to real life, in all the changing weathers. Both Jimmy and Donald took it as a matter of course, when we stepped out on the back porch after a stormy night and saw the dazzling sunlight and the clean sweetness of the outdoor world, that I would quote casually to them

> *There was a roaring in the trees all night,*
> *The rain came heavily and fell in floods,*

or we would recite gayly together, as we tramped over the meadow, picking wild yellow violets:

> *As I wandered through the woods,*
> *On a summer's day,*
> *At my feet a blossom sweet*
> *Bowed across my way.*
>
> *Down I stooped to pluck it,*
> *But it seemed to say,*

'Would you pluck me from my stem
Just to fade away?'

So I raised it gently,
Roots and stalk and all,
Took it home and planted it
By the garden wall:

Every day I watered it,
Every day it grew;
Still 'tis there, and bright and fair
Blooms the summer through.

Some of the older neighbor boys often came over to our place on Sundays, to eat melons or to take turns with the boxing gloves we ordered through the "wish book." Wayne enjoyed teaching the boys to box, and encouraged them to play fair and to be good sports when one hit harder than was necessary.

But, unlike the families around us, I was strictly against letting Jimmy and Donald "spend the night" with their friends. We kept them busy and happy with us, and so had no jeers or criticism or questionable talk to counteract our efforts toward education and moral training. For good or ill I knew that their young years were being kept from the contamination of the evils that were so commonly taken for granted in rearing boys. Their minds followed mine eagerly and earnestly. My regret was that I saw no hope of any changes being made in the school.

As time went on I found myself more and more on guard against the evils that I saw in the life around us, and the greater evils that kept me on the defensive—those that I imagined were lying in wait to take my two fine, innocent boys and turn them into the kind of boy I saw so often around us; fighting, swearing, smoking, chewing tobacco, full of insinuating jokes, and entirely lacking in the qualities I most wanted to cultivate in my children. Above all, I sensed in every man, woman and child I came in contact with, a complacent, half-complaining acceptance of every lack and ill.

Yes, some of them disgusted me, I'll admit. Some were lazy, and shiftless, but worst of all I found that most of them were deeply conscious of the fact that they were only making a try at a living. They well knew that at the end of the year they would have nothing, owing more to the landlord than their share of the cotton came to. I heard a farm owner brag that it wasn't good business to run a farm and take care of tenants all year and then have to pay them cash, too. Imagine that!

I couldn't understand why I had to be the only one to complain. I never heard a word of criticism from any of those poor people. I couldn't understand the total indifference of the landlords to all the hardships of the poor share-croppers who were practically at their mercy and who, through long years, had carried on with such lassitude, having little, expecting little. Most of them could neither be stirred into taking any action toward improving the conditions where they were nor in moving to a better place.

I was finding, too, that the effect of malaria, besides lowering resistance to colds and minor illnesses, slowed down the very tempo of life. Just as I saw the men plodding steadily behind their plows and the women plodding up and down through the loose sand of the road, so I saw that they all plodded steadily, without ambition or hope, through all the days and the years of their lives, and it was from this mental stagnation that I would save my boys.

25. Split-Board Sunbonnets

AGAIN IT WAS a bad day for field work. The sun was out early, but the heavy clouds hung low and during the morning it rained hard several times, although the sky almost cleared between showers.

In the fields the hoe-hands were busy chopping the young cotton, for the process of stalk cutting, plowing and planting was finished, and the new crop was up, as were the weeds.

Up and down the long, long rows they went, for the sandy soil absorbed the moisture, here toward the river, and the work could begin as soon as the heavy rain let up.

After dinner, as the crowd of hoe-hands were nearing my house, busily swinging their hoes and leaving the tiny plants neatly spaced on the high middles, a sudden clap of thunder seemed to shake the earth, and a heavy downpour began. All the choppers rushed toward the house. I hurried to let them in, for the rain was dashing on the back porch and seeping under the door.

There were eight women and girls. Their cotton dresses hung dripping, their cheap shoes soggy. All that really concerned then, however, was their bonnets. I put several sticks of oak wood in the big iron heater, and their clothes soon began to steam, as they gathered by the stove to remove the strips of cardboard that were used to stiffen their sunbonnets.

Rows of stitching formed narrow spaces in which cardboard or thin strips cut from berry boxes could be slipped, so that, as in this emergency, they could be quickly removed and fresh pieces put back, thus saving the time it would take to dry out a starched bonnet. I supplied both cardboard and pieces of wood, and helped to make the bonnets wearable again. Sunbonnets were generally worn in all the cotton fields by the white women to shield their faces from the sun's terrific heat. I made up my mind to make a split-board sunbonnet for myself, as it was really a clever idea and I wanted to keep one for a souvenir.

While this repair work and drying was going on, the rain continued to pour down. The women helped me to move the lounge and the machine and the phonograph. Even the big flour barrel was rolled to another place. They seemed to enjoy the opportunity thus afforded to look at my pictures and books. I caught them exchanging glances at each other, as they surveyed my improvised writing desk and the many photographs.

I was going to play some of the records for them, as one of the older women suggested, but the rain had lessened, the sun was shining again, and the cotton must be chopped.

Soon they could all be seen, with bent backs, moving diligently up and down the cotton rows. They started together by the fence, and each one took a separate row, all trying to keep together so they could laugh and talk.

The whole afternoon was broken by frequent showers, and the air was close and muggy.

One day during this rainy period, when the rain slackened, I looked out the window and saw Ella Hedley, a cotton chopper from Jay Cossey's farm, coming up the road to the house.

She was coming to borrow something, of course. She walked slowly on the wet sand, her stooped shoulders and flat-footed steps making her appear like an old woman. Yet she was tall and gaunt, a girl in her twenties, with indefinite features, weak eyes, and an undeveloped figure. It was hard to get a good look at her face, for her head was usually bent over, but I had seen her up at Grandma's and wondered at her putty-colored skin and blank, listless expression.

I heard Ella's step on the porch, and waited for her knock. I hoped she would knock this time, but I knew she would not. She had been to see me before, to borrow, and had just stood silently on the porch, waiting. I couldn't understand it, but it irritated me so that I had resolved to wait the next time she came and see what she would do.

I finally gave in. I knew she was there. She must have known that I knew. Anyway, a cold wind had come up, and she was poorly clad. My heart relented, and I got up and opened the door, inviting her inside.

A big fire roared in the heater. "Won't you take this rocker?" I asked.

"No, ma'am—no, sir," she stammered, much embarrassed. "No, ma'am, I'll just stand." But later when I went out for more wood, I came back to find her comfortably seated by the heater, the steam rising from her old ragged coat, and her thin shoes drying out on the fender.

"What have you been doing at home during this rainy spell when you can't get into the fields to work the cotton?" I asked her.

"We just don't do nuthin'."

"Well, I guess you just spend your time cooking. That takes a lot of time with wet wood."

"No'm, we hain't been able to cook much. The roof leaks."

"Why, yes, I guess it does, but even so, can't you cook your meals?"

"No, ma'am, we et corn bread all yesterday, left-over bread. The water runs down the chimney and fills the stove. Cain't light a fire, seems like."

"Couldn't you fix the roof a little around the chimney?"

"No, ma'am, don't soon like we can, somehow, and anyway there ain't no wood. Was a little, but it got soaked through."

Ella was a tenant from the next farm so I really couldn't follow through about the wood.

A dead silence followed. I had been reading but I knew better than to speak of books. Ella sat looking at the floor. I wondered just what I could find to talk to her about.

"What do you do most of the time when you're not in the fields? What do you do after the cotton is laid by? Do you sew any? Do you make quilts?"

"No ma'am, we don't never make quilts. We had a lot give to us once, but it don't seem there's anythin' left of our clothes to make nothin' out of. They got wore out and raggy. Even the shirts jest sweat out and rot away. Overalls is generally all gone, time we patch all year. No'm, we jest sit."

Finally her confidence began to come to her and she stammered shyly,

"My sister is sick in bed. Took a notion for an egg. Was your hens a-layin'? Thought maybe I might borrow a couple eggs. She ain't et nothin' either. Been a-chillin' I reckon. Could hear the old bed 'most shake the house down at first. She's got some fever now, I know."

She got some eggs, a bag full, and canned fruit and fresh bread. I didn't mind helping out, but I resented the fact that Jay Cossey could be so mean, so cold, so indifferent, so stingy and heartless to his tenants. Just what kind of a man was he? I meant to find out. Lily Cowder, sending for me, and getting me to beg the doctor to come to her when her baby was born, was one of his tenants, too.

Ella had a brother, Arney, a boy about twenty, who was tall and thin, with bushy rough hair and a blank face. Just another product of years of malnutrition, and mentally the product of parents worse than he was. Arney walked always with his head hung down, too, and his long arms dangling loosely at his sides. He didn't use tobacco as so many of the men did. Yet he made a steady hoe-hand and he was honest and reliable. His folks had always lived down in the lower bottom in one of the tiny, unscreened, mosquito-infested shacks.

It seemed all he was able to do was just to put one foot ahead of the other. He was a joke among the field-hands. Whenever they saw Arney coming up the road—for he used our farm, which sat in the angle of the regular county road, as a short cut toward town—the men used to bet among themselves as to just how long it would take him to reach the peach tree or the big hayshed.

When I was coming in from the hayshed one day, with my apron full of eggs, I heard laughter. Four of our men were leaning against the shed watching Arney Hedley as he came plodding along in the loose sand. One of the men had a watch and he was checking the time Arney made, for often it took him all of twenty minutes to amble that short distance up from the line fence to the barn.

"Arney," one of the men called out, when the laughter had died down and the bet was settled, "Arney, what would you do if you had a million dollars?" The man looked around at the others with a knowing grin, as though he knew just what kind of an answer he'd get.

Arney was entirely unconscious of any sarcasm. His overalls and shirt were both torn and dirty. He looked soberly down at his worn-out shoes, shoes that were broken and ripped and gray from much use in all weathers, and considered deeply. Poor Arney! Like all the others, always bone-weary or chillin'. Hungry, ignorant, sick and dirty.

"Well," he drawled finally, "I guess I'd jest sleep and eat until it was all gone—and then I'd hafta go to work again."

I had seen with my own eyes how nearly impossible it was for some of these poor people to keep clean. Day after day at dirty work in sweat-rotted cotton overalls and shirts, little money with which to buy soap, no hogs to kill to get the grease for home-made lye soap. No towels, just ragged pieces of old cotton-sacks; and in their poor leaky cabins only ragged home-made quilts. No mattresses—just tickings filled with hay. No money for new cloth—no nothing.

Yet the women seemed to spend all their time when not at work in the fields, and when they were not chillin', in patching, washing, scrubbing their floors—hopelessly aimin' to have things like other folks.

Jay Cossey let the Hedleys go when crops were "laid by." All they did was eat and lie around sick, so he said.

I had become quite interested in the Hedley family and in my ignorance of the customs of hiring farm help I thought they had been treated unjustly, so I was quite surprised to learn that they had accepted their dismissal as just a regular occurrence.

Once I had been down at Cossey's store when Mr. Hedley came in. He had a little slip of paper in his hand and on it was scrawled a list of the things his wife wanted for the house.

Mr. Hedley handed the paper to the storekeeper. I watched carefully to see what would happen. The list was read, and then half of it was scratched out. "You don't need all them things," said Mr. Cossey. "You-all ain't a-goin' to be able to work 'em out, I know, and there ain't no use lettin' you owe me. This is all I kin let you have."

A small pail of lard, a bag of cheap coffee, which was rank with chicory, pinto beans, a bag of salt, and some cornmeal were reluctantly handed over to the gaunt hollow-eyed man, who made no reply, but shouldered the old flour sack into which he had poked the groceries, and dragged his heavy feet up the road.

This transaction between the storekeeper and the sharecropper surprised me and I kept finding out more and more about it, even learning that Wayne and Henry, too, were forced to do the same thing with some of our own workers.

When crops were "laid by," most of the landlords turned their extra hoe-hands off, for there would be no more work for them until cotton-picking time. Some of the men and boys might go out in the hills and find work in the saw mills. Some who had a little cash and had made a small garden might be allowed to stay in their houses. Some just begged or borrowed vegetables and fruit and milk and eggs during the summer from one nearby farmer after another. Some were forced to steal in order to live.

It was not unusual to find the ends of our corn rows and the whippoor-will peas all cleaned off when it came time to gather the crop, and so we know that there were some nearby who were forced by necessity to depend upon this source of supply for daily food. We often found that the ends of the rows were left alone, and the corn and the peas were taken from farther inside the field, for they had come back again and again, knowing that the loss would pass unnoticed.

Many families never expected to stay in the cotton land through the summer months but went back up to the hills where they owned their homes.

These were the real hill people. In the mountains they could can fruit and vegetables and pick wild blackberries and huckleberries. If they had cows or pigs or chickens, these would be left behind in the care of their elderly parents or neighbors, who sometimes kept the little children who were too young to go down in the bottoms for the cotton hoeing or pick-ing. But generally all the members of a family, big and little, old and young, left their places in the hills when cotton picking began in the fall, driving down through the bottoms from farm to farm, hunting for a good place to work where the cotton crop was heavy and not too much "in the grass," and where they could find a decent house with wood and water handy.

The fortunate ones were those who found a farm where there was extra work waiting to be done and the landlord could pay them in cash for work in the slack time after the crops were "laid by" and before the fall picking began. Ditches might be drained, fence posts cut, sorghum cut and made into syrup, or hay baled. But most of the farmers had no such work. They planted only cotton. One farmer had one thousand acres of cotton and nothing else. His tenants were not even allowed a spot for a garden.

During the hoeing and picking time a landlord who was taking care of his regular tenants allowed them to buy a certain amount of groceries at the little store at the crossroads, or even in town. Even in such cases the storekeepers charged more than the regular price for these necessi-ties. Sometimes a storekeeper gave credit to a tenant on his own account, through the slack time, figuring that the man would have enough from his share-crop to pay him, or sometimes just because the storekeeper was downright sorry for a poor family.

In giving credit, the storekeeper was taking a chance on the man's health, on the success or failure of the crop, the possibility of a flood or a fall in price, the boll weevil, the army worm, or rain that might come just in time to prevent gathering the crop or that might lower the price of the cotton by causing weather stains. Cotton unrained-on was the most valuable, for the weathering often left it gray or rust-stained or a dirty blue.

The cropper who was getting paid in summer for extra work generally bought at that time the shoes and clothes needed for his family, and anything extra that they wanted beyond the barest necessities. Vanilla for the cakes that were often on their tables when they could get the eggs and sugar was considered an item of real luxury.

Even though the share-cropper got the landlord to agree to carry him at the store, he had to take his weekly list of wanted supplies to the landlord and have it checked over and signed for. The landlord would mark off the list everything he thought the cropper could just as well do without. I saw many of these pitiful lists. They were scrawled on a bit of old paper, in almost illegible writing, even the commonest words misspelled. Grease, coffee, flour, cornmeal, baking powder, salt, sugar, chewing tobacco, pinto beans, sow-belly, overalls, blue shirts, work shoes, cloth for diapers, cheap cotton material, were the regular things listed. This had been the case with poor old Hedley. His landlord had scratched off most of his list, and then the storekeeper had cut it down, giving only the barest of necessities and these of course with small hope of ever receiving any payment. The storekeeper, as well as ourselves, borrowed money in town at the bank at high interest rates to carry on his business. A sorry system!

26. Doodle-Bug

DURING THE LONG sultry days of early summer, with housework reduced to a minimum and with no desire for any extra activity, we read our books and magazines, wrote letters, lounged around the house, or slept. We were daily expecting the telegram which would announce the date of Grandmother's arrival.

In the heat our steps grew slower as we went up the sandy road on errands to Grandma's, or back and forth to the hayshed and garden for eggs and vegetables, and so we began to take more notice of the insects that lived in the ground. Over toward the hayshed, where the young cabbages, lettuce and radishes were wilting on the hot sand—a sure sign that rain could be expected soon—we found big beetles, their hard backs shining as though lacquered, hanging head down in the holes they were digging, their ten legs working spasmodically as they kicked out the soft earth into tiny mounds far behind them.

Nearer the porch, where the sun was a hot glare in the eyes, long caravans of ants headed for distant holes. Jimmy and Donald would get interested in these ants and would trace them to their holes and there sprinkle sugar and squat nearby to watch the maneuvers that called in all the stragglers, while each grain of sugar was carefully carried down into the earth.

Ants, spiders, beetles, bees, flies, gnats, and June bugs, but as yet no mello-bug. I had been hunting for one a long time and at last Wayne and I found one over at the creek lying on the quiet water near the bank. It was a small beetle, brown and shining on top, soft underneath, with an odor like that of the Daphne flower, a sweet heavy smell, as though it contained the pure essence of all the odors of summer's ripeness; the heart of a ripe melon, the bursting black juice of the dewberry, or the oil from a cracked hickory nut. Not any of these, yet all of them. A mello-bug!

But later on, one morning as I went about my work, I looked out the kitchen window and saw Mrs. Batson smiling to herself as she bent her old back over the washtub, for she came every week now to do our washing.

I went out to take away the can of concentrated lye which we used in the black pot to "break" the hard water. I knew that Mrs. Batson would keep on adding more and more lye, believing that the lye ate the dirt and grease and not caring that it also ate up the clothes.

Mrs. Batson's face had now broken into a broad grin. "Did you ever see a doodle-bug?" she asked.

"A doodle-bug, what is a doodle-bug? No, I never heard of one. What are they like?"

"Well, I'll call one for you. Come over here with me. A morning like this is a good time for doodle-bugs."

So saying, the old woman walked away from the wash bench out to a clear space where there were no weeds. Here there were many small, round holes. I had always noticed these after a rain, but had thought that they were made by the ants.

Down on her knees went Mrs. Batson, her face close to one of these tiny holes. "Doodle-bug, doodle-bug," she called beseechingly, "Come out, doodle-bug, doodle-bug." I waited, not knowing what would happen, but sure that this performance was only the counterpart of the old childish trick of holding a lady bug while one chanted, earnestly, "Lady bug, lady bug, fly away home, your house is on fire, your children will burn," and believing, when the little bug finally spread its wings and soared away, that it had really gone to rescue its children.

"I see," I said at last to the old woman, for it all began to appear a little ridiculous to me, "I see what you mean, but do the doodle-bugs ever come out?" I tried not to smile, for she looked up at me so seriously. "Oh, yes," she said. "I've called 'em lots of times. You have to call just right, though. Can't never tell what they'll do. Well, I guess we'll have to try another time. I'll try at home and see if I can bring you one."

None of my jokes or jibes could move her. She insisted seriously that there really was a doodle-bug that would bring good luck if it heard someone calling and came up to the surface of the ground. But I'm still waiting to see a doodle-bug.

While Mrs. Batson was always glad to come to wash for me, often she could not come when I wanted her, because Abe Dawson, the farmer on whose place she lived, would forbid her going when he needed her to hoe in his cotton field.

Then I would send for Old Zula. Nobody knew when she left her hoeing and trudged the back way through the fields to do my washing. The fields over in the black lands were cut up into smaller pieces, amid the heavy growth of weeds along the fence rows; and the grass and young trees that grew where the bayou waters meandered through the low lands close to the foothills hid the narrow roads and the small houses from casual view.

Whenever we went that way to go to the hills to visit our friends we drove the Ford slowly past those little cabins, so that we might catch a glimpse of the life within. Great billowy feather mattresses tucked under snowy sheets could be seen, and always a kitchen safe and cookstove, table and benches. In summer the worn porches were used as extra living rooms.

Here if we drove past on a Sunday morning, we could see some Negro out on one of the porches acting as barber. A man would be lying back in a kitchen chair whose front legs were raised on chunks of stove wood, getting his kinky hair cut short. We never heard any loud talking or quarreling. Always the scene was one of happiness and peace. As many as a dozen men and boys would be waiting to be served, while maybe two or three men, their spotless white shirts gleaming over their clean blue overalls, would be perched on chairs playing their guitars and banjos.

A long line of shining lard buckets swinging by wires over their heads held drooping vines or gay petunias, and always there were clumps of cock's-feather and castor bean trees growing around the cabins, and gourd vines in abundance.

In all sorts of weather, Old Zula would come when sent for. I never heard her called anything else but Old Zula. She was a great-grandmother, though her woolly head was but slightly touched with gray.

Even when the snow was still on the ground, Old Zula would put the wash bench out where the fire had warmed the ground and melted the snow and go at her washing in the usual place against the end of the house beneath the kitchen window.

When dinner time came she would go right on working. She never brought any lunch and refused the food I offered her. Finally, she agreed to accept lunch if I would give it to her on a tin plate on the back porch. I just wouldn't treat her like an animal, and insisted firmly that she come into the house and sit at the table with the children and me when Wayne was in town.

We enjoyed seeing her black eyes shine and hearing her soft talk. She was very clean, and I could not see what harm there was in treating her the same as I would anyone else. But Wayne had cautioned me against this when I told him how I enjoyed talking to Old Zula. "You'll only make trouble. You don't understand how it is here. Southern people really understand the Negroes and respect them, but just have to keep them in their place. It's better to give her her dinner out on the porch. And for pity's sake, think what will happen if Hettie Wood knows that you eat with Old Zula. "

I didn't care what would happen. I hoped something would, for I certainly intended to do as I pleased, and that was to treat Old Zula like a human being, as I would like to be treated myself, all Southern traditions to the contrary.

So Old Zula ate her dinner with me many a time. She was full of fear and very humble and said I was "jest a fool gal" and wouldn't do myself any good being so kind to her.

Not only did she eat at my table but she was closer than that to me; she sat in communion with my very soul, many a day. She worked out on the west side of the house where the two big wash tubs sat on the long bench, and washed tub after tub of clothes—sheets, tablecloths, towels, and heaps of blue denim overalls, her frail old hands wrinkled and drawn by the lye and soap and hot water and her poor thin old body bent over the washboard in the hot sun. I would often take my sewing and perch in the shade and talk to her.

When summer came I finally persuaded her to move the bench and tub up on to the porch in the shade where she would be on a level with the pump platform and would not have to go up and down the two steps to carry the water for the rinsing and blueing. Even so, dozens of buckets of water still had to be carried down the steps and over to the black pot. Finally, with the last article washed and rinsed and blued, I prevailed upon her to let me take a side of the big tubs and so carry them away from the house to be emptied out on the sand. This offer was met at first with a firm refusal.

"You sure is kind. But been a-washin' all my life. Ain't nobody ever helped me before. Ain't nobody care for Old Zula less she kin tote her own load."

But at last I overcame even this scruple on her part, and together we carried the tubs down the steps and emptied them out on the sand away from the house, and it pleased me to see the look of plain human happiness on her black face when she looked at me.

"I done got me seven, eight boys, but jes' don't know where at they is. Ain't heard tell o' them this long while. One done sent me a five dollars; every month he sent it to me for the longest. Them days I didn't hev to keer about nothin'. Mr. Meek is awful good to us colored folks. I been a-livin' over by the bayou goin' on four year now. I'd a heap ruther do folk's work than field work, but they ain't nobody 'round has any to do. I usta do a heap o' washin' for folks. Lived over in Oklahomy them times. I'se traveled a good bit in my day.

"No, ma'am, you bin a-talkin' to me about me bein' colored. Well, I'm here a-tellin' ya, don't ya never believe what they allus tells ya. Us colored folks ud all ruther be white. Why not? Ain't nothin' in bein' us, is they?

Shore I'd ruther look like you, so would any colored gal. They'd give they eyes, yes, ma'am!

"When ya gets old, things don't matter none, but I was young, oncet. Yes, ma'am. I ken recollect a-honin' fer a nice fair skin. I ain't know jest what the dear Lord been about, some of us bein' colored up this way, but don't let none of these folks be tellin' ya we don't keer.

"No, ma'am, I ain't mindin' you a-askin'. You've got a good heart. I'd sure like to pass ya the word o' God, 'fore I leaves. You-all want me, ya jest send word over and I'll come, sure."

I felt pleased and happy and satisfied that the reactions of the Negroes were just as I had thought they would be. Yes, quite pleased with myself, until Mrs. Wood came up to see me one day full of wrath. Her thin body was firmly belted in its clean starched gingham. Her black roving eyes fairly popped, she was so excited. The skin of her face, always a dark olive, was now several shades darker. She could hardly be civil, but hesitated, not knowing just how to begin her complaint.

"I feel I just have to let you know this. There ain't any one of us bottom folks going to stand for you taking niggers into your house and making them so free-like. Soon we'll have to make another drive to clear 'em all out. We ain't got but a few families of the colored folks left here. They jest ain't welcome. Can't stop Meek from hirin' 'em, it's all the help he keeps. Parker is his straw-boss. But they all stay back by the bayou. It's a cryin' shame, I says, and him with all that rich land, a-pilin' the niggers in.

"Even if you have to do your own washing, you jest can't go to making Old Zula so free, I'm a-tellin' you. Maybe you didn't think I knew, but my big eyes sees everything."

I listened patiently, trying to think of the best answer to make to this, but Mrs. Wood waited for no reply. She flounced away from the porch steps and pulled her gray sunbonnet lower over her face as she plodded through the loose sand on the road up to Grandma's.

I was always eager to see what I could of the way the colored people lived in their own homes, my previous knowledge of them being only what I had read, and it did not disappoint me to find that their lives followed the pattern of the white people in all I had yet observed. Always in summer their cane-bottom chairs were placed out in the open hallways and here the women could be seen mending overalls, making quilts, shelling peas, or churning. The hallways made a much-needed extra room and the cool breeze drawing through was very pleasant. I saw the cedar buckets standing on the little shelves with a gourd dipper floating on them. The wash benches held the wash pan, and big wash tubs were turned upside down on the edge of the porches. Gourd vines ran riot over the fences. They called the gourds "simlans" and the large well-shaped ones were often cut down for salt holders and hung beside the stoves.

One night as we were coming home quite late from visiting friends out in the hills, we turned in on a road that ran near the bayou, through the black lands.

As we passed an old barn by the roadside we saw a small fire burning outdoors and heard music, so we stopped the Ford across the road and sat and listened. The air seemed pulsating with the surge of melody and rhythm.

Some Negroes were having a dance. The measured strumming of a banjo mingled with their steady chanting. We could see two long rows of the colored folks inside the barn facing each other, the men on one side, the women on the other. All were bent over, their hands slapping their knees as they advanced and retreated to the slow monotonous beat of the banjo chords, and all were stamping their feet on the rough boards, keeping time to the music.

Several lanterns hung from the rafters and in the semi-gloom we could see that every now and then a man or woman would leave the waiting lines and step out in the center of the barn floor and circle and prance and wave his hands and bend back and forth; then the chanting would begin again, louder than ever, and the lines would advance and retreat as before. It was the serious, sweat-glistening faces of the dancers as they came and went in the flickering light from the fire that made the weird picture I have never forgotten.

We were told that sometimes these dances were rather wild affairs. There was one special game or dance that they often played. A Negro would step gayly out from the line of men and advance toward a woman of his choice. Stamping his feet and patting his hands he would roll his eyes, prance around in a circle and yell, "I'se a-honin'." The woman would answer, "What you-all a-honin' for?" Then the reply, "I'se a-honin' for a good sweet kiss." Thereupon she would step out from the line of women, they would embrace each other, kiss, and prance up and down, stamping their feet and clapping their hands. When they returned to their places another of the women would dance over to the men, and standing before the man she chose, would go through the same routine. This went on and on, until all had taken part.

Meek Loupe owned a great acreage over by the bayou. He lived in town and hired Negro hands. His strawboss was Parker, a huge handsome Negro, smart and quick, and a great favorite with everyone. In the cotton fields, Parker led the hands down the rows, all swinging their hoes in unison, with Parker calling, "Bear down!" to the choppers, or, "Jar the earth!"

One time, so I was told, Meek was buying hogs to ship, buying them all over the county so as to make a carload lot. A man near town reported to a friend of his, named Bill, that Meek had stolen one of Bill's hogs, and had fenced it in with the other hogs in the big lot. So Bill went and looked and saw his hog in the lot with the others. He called Meek out. "Meek," he said, "I see you have got one of my hogs up." Meek said, "Guess not, Bill. That's my hog. That's Bet, I tell you. I raised that hog, that's old Bet." Bill said, "No, you're wrong, Meek, that hog belongs to me." The Negro Parker was over in the corner of the lot. Meek called, "Parker, come here." Parker came. "Parker, see that hog over there, that one with the spot on his ear. You know that hog, don't you?" Parker knew what was expected of him; he knew he had to back up old Meek in anything he said. "Yes, Mr. Meek, sure I knows that hog. That's your hog, Mr. Meek. I helped raise that hog." "There," said Meek, "See there, Bill, Parker knows the hog and the hog knows Parker." Then raising his voice he called loudly, "Pig, Bet, pig, pig." And that argument was settled.

Meek fired Parker once, laying every accusation against him that he could. Next fall Meek missed the good work that Parker had done and hired him back. The first thing that Meek did, though, was to make an agreement with Parker, whereby he was to charge Parker fifty dollars for stealing. That was to be all Meek could charge, provided he didn't catch Parker in the act of stealing. If he caught him, then Parker was to pay twice the price of whatever was stolen. Parker was telling all this to Wayne one day as they stood by the fence talking. Wayne said, "Parker, you and Mr. Meek are pretty well acquainted, aren't you?" "Acquainted," said Parker, "I betcha this nigger owes Mr. Meek five hundred dollars 'fore the first of the year, but Mr. Meek jest ain't gwine know it."

The story was told of two of Meek's Negroes who had hated each other for years. One was a huge burly fellow, the other short and thin, a runt of a man. The little fellow had vowed he would one day get his enemy, but as there was always a big sharp knife in evidence, that day was a long time coming. One hot day out in the cotton fields it seemed that the right time had at last arrived. The big man, who was then leading the Negroes down the rows, was barefooted and was wearing an old torn shirt and a pair of overalls so ragged that the outer pockets were missing and the others were worn to rags. Apparently there was no knife and no place for a knife to be hidden, so the little man walked over to his enemy and challenged him to a fight, swinging his sharp hoe.

Quick as a flash the big black man flourished a long-bladed knife, hidden on him somewhere. The little fellow, scared to death, jumped back and, rolling his eyes to heaven, beseeched, "Lord, hold me; I don't want to commit no murder today." The big Negro kept advancing, with his knife raised, stepping down the row and calling, "Turn him loose, Lord, turn him loose, I want to kill me that nigger today." The hoe hands that had gathered to see the fight separated the two men, and the little fellow took to his heels.

These stories, like all the others about the Negroes that I heard, only proved to me that the colored people when treated fairly were just like little children, easily managed, easily provoked, believing childishly but implicitly in a heavenly Father who watched over them. They needed only kindly supervision and sympathy, and this they repaid with a deeper loyalty and regard than many of the white people deserved.

I refused to become involved in the endless arguments I heard about the morals of the Negroes, or the many discussions that were brought up regarding the possible danger to white women from the colored men.

I always remembered the grim way in which Grandma used to sum up the whole matter. "Anyway," she had said, "in all my life I never saw a white woman nursing a black baby."

At last Ben and Avis announced that they were going to go to Oklahoma to try their luck, and the little house across from Grandma's stood empty for a while. Avis and little Roy and Ben all gone from the farm. I could hardly believe it. Ben had long been wanting to follow his mother, Mrs. Rollins, and his brothers, to Oklahoma, but stayed so Avis could be with her own mother when Roy was born. Now Ben was determined to make a fresh start, and hoped to find conditions in Oklahoma better than those he had known in Arkansas. Avis had made my days so bearable, for I found no irritation when with her. She was always as I first found her, kind, gentle and good.

27. The Square Dance

THE CHICKEN HOUSE was finished at last. Built by California speci-
fications, with a cement floor and good roosts and nests and windows,
it was inspected daily by all around. No one had ever seen such a hen
house. Many told me they would be glad to live in it, instead of in their
dark, crowded, leaky houses. The cement floor was a special fascination,
and the new lumber and good windows were things they longed for but
seldom got.

As I stood in Grandma's back door I looked over toward the chicken
house and saw that a crowd of cotton pickers had arrived and were busily preparing their supper. They had set up a small cookstove out under
the willow tree near the pump. Black smoke was pouring out of the short
length of stove-pipe. One of the women was frying slices of sow-belly. I
could smell the strong coffee and knew that in the oven would be a pan
of corn bread.

I remembered what Wayne had told me about the crowd that would
be camping there, during the fall cotton picking, so later that night I
walked with him to see Grandma. When we were ready to go home we
went quietly to the door of the chicken house and looked in. Just as I had
been told, twelve people were sleeping in there. Some of them lay up on
the slanting shelf where the roost would be placed later. The rest were
lying in a long row on the floor. They had come in three wagons, with
empty bed ticks which they filled with hay at the barn.

I looked, I thought, and turned away. Later I noticed them in the fields.
They were of all ages, but even the youngest, three boys about eight years
old, went every day to the cotton fields. There were three families among
them and they were the best pickers we ever had. They all worked "from
can't to can't." I often saw them stalking out toward the cotton rows long
before daylight, through the heavy dew-drenched plants, and again long
after dark, stragglers would be returning dragging a cotton sack half-full
of cotton that had been picked after the last weighing was over.

Two of the older girls came down to my house one noon and asked me
if I would invite them to a party at my house. They explained that they
meant could they have a dance, but that their folks were against dancing,
so they would only get to come if they could say it was a party.

I was glad of the opportunity to have a dance and obligingly sent word
to their parents that I was giving a party. That day the news spread through

our cotton field and also around to some of the other fields, when it got dark about twenty boys and girls arrived. They helped me to take down the bed and we rolled up the carpet and moved everything out on the porch except the chairs. We brought in the wash bench, too, but even so, most of the boys squatted on the floor around the wall on their heels. One boy played the banjo with a brisk dash, and sang "Birmingham Jail" many times by request. It seemed as though the melancholy hopelessness and despair of that song had a special appeal. Another young fellow sat on the floor in a corner with a blank, pale face, but when I heard him sing I was amazed. He reached for the guitar, tuned it slowly, and then eased softly into the drawling account of some heartbroken cowboy. In every song he sang it seemed that someone had left his little darling behind, or someone was far away from his home, out in the cold world, homesick and weary. The throbbing, staccato tempo of other cowboy songs that followed was accompanied by the rhythmic stamp of the heavy farm shoes on the floor and the regular accent of hand clapping. With the heavy pulsing of the deep chords of the guitar, the square dancing began.

Every face was eager and intent on every movement, alert for the exact second when they would turn in at "Swing your pardners" or take their places in the intricate weaving sets. The eight who formed the set took up all the available space, and all the others stood around and talked. Jimmy and Donald sat quietly by, missing none of the gaiety, and evidently quite thrilled with the jumping and prancing that took place. They told me afterwards that it was the funniest dance they had ever seen. Henry did not care for dancing, and Grandma was again abed with a sick headache, so this was our own party.

There was no time wasted in waits between dances. All was rapid and exciting. They had come here only to dance, and a fresh group took their places as soon as one set was finished. They called the square dances "playing games."

Before the evening was through both Wayne and I had entered fully into the almost childlike gaiety and abandon that prevailed. I suppose we made what is usually considered a fine-looking couple. We were not yet twenty-nine. Wayne was quite tall and very handsome. His brown eyes were especially good, with heavy brows above them. His eyes were pleasant to look at, sympathetic and glowing with a bright boyish eagerness. His forehead was high and smooth, his skin smooth as satin and slightly olive.

Some people thought he resembled Lincoln, and maybe he did. I was about five feet two inches tall, with dark auburn hair and blue eyes. I know now that at that time I was naive, girlish, sincere and happy. I like to think of that young woman I once was.

I can still remember the wild lost feeling I had of being breathlessly transposed from one side of the line of dancers to the other and of being swung about and placed in a new position before I had even a general idea of what was going on. I loved it, though, and laughed at myself and the others.

I found myself in a constant strain to learn the pattern of the set and be ready to take the proper hand of those that were reaching out as the dancers revolved about me.

I noticed that the bridge of the fiddle which one of the boys had brought was cut on a decided slant to make the chording easier, and the beats were all accented by the regular tapping of the musician's foot as he played. The calls for all the sets were sung in a peculiar chanting tone, with a nasal twang, and in a smooth rapid monotone, that made the words quite indistinguishable to me, try as I would to catch what was being sung. Once one of the larger boys took to sulking and pouting, for another boy kept on calling all the sets and would not yield the envied position as caller up to his friend, who wanted a turn at it, too. All the dancers gathered around and wasted much time coaxing and begging so as to get the dance started again.

The big four-layer chocolate cake and the hot chocolate I passed around for refreshments were refused by nearly everyone. This surprised me until Wayne suggested that maybe, like my boarders, they were not used to such food.

I got the words of some of the square dance songs that were sung from one of the girls later. She told me that they all had a fine time, and she seemed very pleased that we, too, had enjoyed the dance so much.

SQUARE DANCE SONGS

First old gent swing the opposite lady,
Swing her by the right hand,
Then your partner by the left,
And promenade the girl behind you.
"Oh, that girl, that pretty little girl,

The girl I left behind me,
I'll weep and cry, till the day I die
To see that girl behind me.

Oh that girl, that pretty little girl,
The girl I left behind me,
The girl I want, the girl I love,
The girl I am to marry.

Circle four, shutang,
Shutang, shu;
Do si lady, shutang,
Shutang, shu.
Round the next lady, shutang,
Oh, shutang, shu;
Back around gent, shutang,
Shutang, lady, shu.

Chase the rabbit, chase the coon,
Chase the pretty girl round the room.
Chase the rabbit, chase the squirrel,
Chase the pretty boy round the world.

Chicken in the bread tray
Scratching out the dough,
You should have had that other gal
A long time ago.

Two little sisters form a ring.
Back to your partners and everybody sing.
Gents to the center and back to the bar,
Ladies to the center and form a star.
All the men left, with our old left hand,
Greet your partner, go right and left grand.

Take a chew of tobacco and spit on the ground,
I thought I heard my old dog hound.
Swing your paw and then your maw
And then that girl from Arkansaw.

You swing mine and I'll swing yourn—
Give me back mine—I'll give you back yourn.
Meet your partner, meet her right,
We'll go home in broad daylight.

Devil on the hillside kicking up gravel,
Meet your partner and everybody travel.

The wild cat hollered,
The panther squalled,
The house cat jumped
Through a hole in the wall.

Right foot up, left foot down—
Grab your partner, go round and round.

Corn in the crib,
Bees in the gum,
Sal, let me chew your rosin some.

28. Grandmother's Visit

THE GREAT DAY came at last when we prepared a royal welcome for my mother, the grandmother from California.

Everything in and around the house, and everything on the farm was in its best possible order. All the cleaning and washing and ironing was finished. I had made a large beautiful cake and several lemon pies. The melon patch was carefully searched for the largest and best melons. Blocks of ice were brought from town and closely wrapped in sacks and canvas and buried beneath the porch in the damp sand.

We felt that we were on exhibition and wanted to earn Grandmother's approval and admiration. We set up the extra bed out in the living room, placing it across the corner between the two windows so as to get the benefit of every breeze, and here Grandmother slept, and here we gathered around her, early in the mornings and late at night.

Her perfume and powder and clothes charmed us all and we felt that our little house was honored by her presence. I, of course, always wore

gingham house dresses, and I was kept busy preparing the meals, gathering the vegetables, filling the lamps, pumping the water, carrying out
ashes, and doing the many chores necessary to keep everything in good
order. It was a wonderful pleasure to me to see my own dear mother, so
beautifully dressed, so sweet-smelling and so happy, rocking away in the
big chair, with Jimmy and Donald hanging over her, so attentive and so
delighted.

Grandmother had brought us two hammocks in her big wardrobe
trunk. Such a surprise, such a daily pleasure! How constantly we used
them, and wondered why we had never thought of having a hammock.
Nor did I ever see a hammock anywhere in the South at any time.

Our little rooms rang with laughter. Music was enjoyed early and
late. We not only played our phonograph and discussed the records, but
Grandmother said her voice had come back to her.

She suffered from asthma, but in our climate she could sing again as
she used to years ago. So in the evenings she sang to us many old loved
songs; some were ancient tunes from her own mother's day. She sang
several old "Come-All-Ye's" for Jimmy and Donald, much to their delight—ancient songs generally beginning with the words, "Come all ye
men and maidens and listen to my rhyme ..."

If Grandmother saw anything that caused her concern or unhappiness
she managed to hide it from us successfully, and through her eyes we
began to recognize our life as one of richness and warmth, teeming with
activity and comfort and plenty.

How the boys delighted in explaining to Grandmother our milk-pail
tree, and how she chuckled over it. It was a bleached young tree with the
branches cut short that had been washed up on the river shore. When
I saw it lying there, with its many short sturdy branches, I had an idea.
I dragged it home and set it firmly in the ground out by the pump, and
that's where we hung our milk pails in the air and sunshine.

I had been campaigning for clean milk pails. Old lard buckets were
used everywhere for milk and many people thought that if a pail looked
clean, it was clean. Some of the pails that had been used for years I found
with a lodging of dirt down around the bottom, inside, and most of them
had crusted dirt on the outside under the rolled tops.

Our tenants and the cotton pickers all went up to Grandma's house
to get fresh milk or clabber or to "borrow" meal or flour, and lard buckets were used for every purpose. I cleaned all I could get my hands on,

digging the dirt loose with a sharp-pointed stick and boiling them in lye-water in the black pot.

It was really amusing to see the traffic in old lard buckets. A kind of unspoken feud went on. Pretty pails that had a gold maple leaf or a blue band were much coveted, and in neighborly borrowing were cleverly changed when the owner's back was turned, the substitute being handed over with such a disarming smile that even though the rightful owner knew what had happened, something seemed to keep her from open objections—she just bided her time until she could do the same thing herself.

I had the children bring me all the empty pails they could find around Grandma's house, where most of the eggs and buttermilk and clabber were kept, and also had them pick up all the buckets from around the big barn and the chicken yards where they had been carelessly left. When these were beautifully clean and sanitary, they were hung on the milk-pail tree, glinting and rattling in the summer breeze. But gradually, most of the pails disappeared, I never knew how nor when. They just weren't there. Only three or four were left.

One day, as an experiment, I marked all of my own pails that I could definitely lay claim to, painting bands of scarlet around each one. I saw those painted lard pails in kitchens as far away as three miles during the years that followed.

Funny little milk-pail tree!

When Grandmother learned that the children and I had been talking of making a trip to visit my sister Elizabeth in Alabama, she proposed that we all go now, together, so after part of her visit with us was over, we started. What an adventure that was! We crossed the Mississippi River and reached Memphis, Tennessee. The water of the big river ran in a channel that seemed to us to be no wider than the Arkansas, but we were assured that the great stretch of low meadow land we had crossed, where the train ran over high trestles, and where we saw many little houses perched on high pilings, was the real bed of the river when the great Mississippi was in flood.

Our stay in Memphis was brief, for we were going down the Alabama River south of Selma, to visit an old cotton plantation.

This place was surely "deep South." The cotton here grew only about three feet high, with very scattered bolls. Our proud descriptions of the cotton that grew in the Arkansas River bottom, so high that a Ford car could pass along the roads through the black lands and be hidden from view, were simply not believed. We enjoyed our visit with my sister and her relatives, and Jimmy and Donald met their five little cousins, some of whom they had known in California. The seven boys had a wonderful time together.

We found that the field hands in Alabama were not the same type of Negroes we had in Arkansas. Ours were of African origin. Most of those were from the West Indies, and spoke with such a peculiar guttural tone that I could not understand their speech. My brother-in-law saw my puzzled look as I tried to understand a conversation he was having with one of the Negroes who was working in the garden. "Where is Mr. John?" he called to the black man. "Wha who?" was the black man's reply, for he had not understood the question. I was told that he was asking, "Where is who?" They continued to talk back and forth but I could make out very little of the Negro's words. We admired the way the big black women waddled back and forth carrying laundry in huge tight bundles on their heads, just as I had seen them in pictures.

Crops were "laid by" here, too, so the Negroes were not in the fields. All about us were pine forests, thin spindling trees that covered vast areas of sandy ground. Beautiful little creeks cut through the farm lands, wandering down to the Alabama River. Black muscadines hung from the cottonwoods in the creek bottoms, and we went one afternoon to gather these wild grapes and visit the big peanut patch down near the creek. I was surprised to see so much okra raised for table use. The yards were full of crape-myrtle bushes and pomegranates, and big magnolia trees grew on the Bermuda lawns, just as in town at home.

We drove about the country one day and I noticed many very large empty houses. Tall white pillars held up little verandas high up on the fronts of the houses over the front doors. Large trees and bushes and vines were growing so close that the houses, huge as they were, were almost hidden from view. These places represented deserted grandeur.

Nearly every afternoon while we were in Alabama the sky became overcast and a heavy rain fell, with much thunder and lightning. The rest of the day would be oppressively hot and damp under the bright glare of the sun.

One evening we all drove through the pine forest to a small Negro church. A big stick of pitch-pine was stuck in the ground near the door and was burning with a smoky flare. This served to light the way to the meeting house. The meeting had already begun when we drove up, so we stopped the car and watched through the open door. The men sat on one side, the women on the other. The preacher, wearing a long black coat over his work clothes, had a big Bible in his hand and was reading the verses he had chosen for his sermon. Then he picked up a big hymn book and read one line of a song at a time. After each line the men and women would sing, but they did not sing together. First the men sang a line, next time it was the women's turn. That went on, verse after verse, reading and singing line after line. It all sounded and looked so much like stories that were running in the *Saturday Evening Post* at that time that I could hardly believe it was real. What amazed me so much was that the men sang so low, and the women sang so high—and alternating as they did, in their singing, made it seem rather like a teeter-totter, up and then down.

These Negroes were all very black. In Arkansas the Negroes were of varying shades of tan, high-brown, and coffee-color, although some were quite dark.

Elizabeth told us many interesting stories about the Negroes. She said that most of the ones who lived on the farms had little cabins with a door but no windows. Summer and winter they burned pine knots in their fireplaces where their cooking was done, and this was the only light they had. Whenever one of the Negro women was going to have a baby she would hang quilts over the doors to keep out both the air and the light. Elizabeth told me that once, before her own last baby was born, and before the doctor could get down to the farm from town, she had had quite an argument with the old Negro mammy who had come in to stay with her, for Elizabeth insisted that the heavy quilts the Negro had hung over the doors and windows be taken down. The Negro, feeling that she knew better, refused to obey. It was a very hot day and Elizabeth suffered acutely until her husband's mother and the doctor arrived and took charge.

It surprised all our friends in Alabama to see my boys wearing blue denim coveralls for play-time wear. The five cousins wore little white suits at all times. I was amused to learn that this was done to impress the Negroes. My sister pointed to my boys and said to her husband, "You see, Jimmy and Donald are wearing coveralls for play." I gathered that there had been a previous discussion about the children's clothes. "How do you manage to get away with that in Arkansas?" my brother-in-law asked me. "Here in Alabama we have to keep the Negroes in their place. They've got to realize that they're working for quality folks. Overalls are considered the slaves' uniform, and no self-respecting white man will be seen wearing them."

This led to an interesting discussion of the many differences there were between the two states.

He told me that because the Negroes in Alabama greatly outnumbered the whites, they had to be kept from voting. The poll tax, which few of them could pay, was one hindrance. The other methods used were either to tell them how to vote, or to order them to stay away from the polls and to see to it that they obeyed such orders and did not get a chance to vote. Our discussion here touched on the rights of Negroes. I found I was on dangerous ground for this was the time when the Ku Klux Klan was in its great power in Alabama, and one felt that there was a mental quicksand underneath even ordinary conversation regarding the colored people.

The Negroes were all given strict orders never to come inside the yard gate after sundown. How dark it was there at night when there was no moon! Our nights in Arkansas were teeming with night sounds as the crickets and katy-dids chirped and sang, but here in the darkness I heard long drawn-out cries, and weird calls and the mournful, trembling, "to-whoo, to-whoo" of owls. Many strange large night birds flew past the house with a loud flapping of wings, the air seemed filled with fireflies and all around us every aspect of jungle-like growth was intensified.

Every evening, for long hours, we would sit out on the big porch in rocking chairs, looking at the sky, discussing the stars, watching the heat lightning, singing old songs, drinking cold lemonade, and trying to pass the time until the worst heat of the day was gone and the cooler breezes of the night would enable us to sleep.

Elizabeth had been sending her older boys to a nearby country school, but now she and her husband were making plans to leave soon for California, so when our visit was over Grandmother knew that she would see them all again before many months. When I thought of the uncertainty of my future, I was very sad when it came time to leave this sister who had always been so very close to me. She had lived here three years, her routine of life broken only by occasional trips to visit friends in the little country town. Her husband and her five little boys were her sole interest. They had tried to adjust themselves to Southern plantation conditions, but the old days of the white man's supremacy were gone forever. Her husband, a son of a distinguished old Southern family, had worked successfully in the West, but he had craved to go back to the South to the life he had known as a young man, a life that he now found existed only in memories and dreams. He had at last been faced by realities and discovered that no white man could live in so isolated a situation among the Negroes and attain either success or happiness.

Although Elizabeth had all the help she wanted, either for house work or for the care of the children, yet she was too much alone.

While the seven cousins played along the beautiful little creek one hot afternoon, and our mother was asleep, Elizabeth and I sat in the shade on the railing of the tiny bridge and talked. She was asking me about my life in Arkansas and about our friends in California. Her voice was low and sweet, but I, who knew her so well, sensed in her serenity and calmness a hopeless acceptance of uneventful days, a depth of joyless, dreary years, too deep for tears. She said once, "I feel like a pond covered with green scum—without life or motion! I must get away from this place."

— . —

Back again on our own farm, we spent some lovely afternoons with Grandmother strolling in the meadow, or wandering along the river bank on the gray, wet sand. We showed her the place where we had had our Easter-egg-rolling. Above our heads the mockingbirds sang loudly and sweetly. Over among the elderberries and the bodark thicket the mourning doves cooed in amorous ecstasy. Bright, large-winged butterflies flew ahead of us and the spongy meadow grass was a soft carpet beneath our feet. The tall gum trees and the pecans and cottonwoods cast their pleasant shade eastward, and the sunshine was golden and comforting, with the soft breeze in our faces, so pleasant and so caressing.

Wayne took many walks with Grandmother around the farm, proudly explaining his work, and anxious for her approval. We also drove away down the river road to see the levees and out into the hills to see the Bohemian settlement. Jimmy and Donald were so excited and happy they hardly knew what to do. They waited on their Grandmother in every possible way, and were with her from morning until night, listening to her stories, begging her to sing to them, and listening with great interest to her talk of the future, filled with plans for them when they would be older and could go to college. Looie, the pup, and Maggie, the special Plymouth Rock chicken, and Old Hundred, and the lame cat, Old Timer, with her thirteen kittens, were all introduced in turn, and Grandmother praised and listened and admired—until the boys were just bursting with pride. They brought her the biggest eggs they could find, proudly exhibited their milking ability, and were always trying to think of things that they could do for their Grandmother's happiness. I took pride in exhibiting all my canned fruit as well as the spicy watermelon and cucumber pickles and canned sausage and sauerkraut.

I remember one afternoon as I stepped out on the front porch, where Grandmother was lying asleep in a hammock, I saw her suddenly rouse up with a suppressed shriek and a look of incredulous amazement on her face. Old Mr. Batson stood near the edge of the porch, grinning, and looking at her with great interest. He apologized for disturbing her, but stood there, wanting to talk and waiting to be introduced. This I really enjoyed doing, for I wanted my mother to meet this most unusual-looking man. As usual, he was barefooted, with his ragged overalls turned up nearly to his knees. I saw my mother looking at his feet and I was amused to see her look of disbelief as she saw his long dirty toenails turning up like claws. Mr. Batson wanted to know if he could have a melon from the patch near our house. He said he had been out there in the field thumping them and knew that some of them were ripe, but he had not tasted a melon yet and could hardly wait to get one. While he was talking we could see the melon seeds that were caught in his dirty goat-beard. This amused us all, but we sent him away happy, with a big melon in his arms.

Mr. Batson's little girl Allie came slowly by, shortly afterwards. She was on her way down to visit her sister Lidy. My mother invited her to come up onto our porch, but the child was so bashful she would only hang her head and poke the sand with her bare feet and refuse to look up or speak. How Grandmother did it I don't know, but she got the child to talk at last and found out that she had no doll, and had never owned a doll in all her life. When I agreed that this was very likely true, Mother took the long string of beautiful garnet-colored beads from her own neck and looped them about Allie's throat. I fully appreciated the feeling which I knew possessed my mother, but I was becoming rather accustomed to the uncertain emotions that so often overwhelmed me as I tried to weigh and decide conditions and circumstances that seemed so hopeless of solution.

Because the time was broken by the trip to Alabama it seemed that the hot months of July and August passed very quickly. Mother's visit was over now and she had gone back to California, leaving behind her four people who missed her more than words could tell.

How she had enjoyed all our excited living! One night while she was still with us a heavy storm came up and we all rushed around in our nightgowns putting in the windows. Mother laughed herself into hilarious shrieks of amusement, for we put on our regular sailor act, bringing in the tubs and the water pails, rushing out to cover the chicken coops, fastening the doors securely, and moving the boxes and barrels and crocks away from the places where we knew the regular leaks would come. I was very glad for once to have a storm fulfill all our worst fears, for several times before we explained to her the awful things a storm could do; yet after a period of lazy thunder and a lot of lightning low on the horizon, the storms had moved off to the southwest and the stars shone out and the frog chorus again filled the air. Now, however, the rain fell in torrents. The wind fairly lifted the house from its rocky porches, the windows rattled, the paper on the walls rattled and heaved under the blasts of air, the lightning made the room as bright as day as the constant flashes came again and again, and the thunder roared and rolled and crashed until we could hardly hear each other speak. The roof not only leaked in all the familiar places but broke out afresh around the windows of the sitting room where Grandmother's bed was standing. The water collected on the ceiling paper of the dining room, too, where it had soaked through the loose tongue-and groove boards above, and when it gathered on the double thickness of builders' paper in a sagging weight, we took the butcher knife, as usual, and cut a gash in the paper and let the water spurt down into a pail, while Mother looked on in amazement. At last the rain even came pouring down from the high ceiling above, right onto her bed. We hurried to pull the bed out into the middle of the room, but another spot began to drip and soon the water spurted down in a stream, again on the bed. So I got out my umbrella and there Mother sat, huddled at one end of the bed in her pink silk nightgown with a blanket around her shoulders, holding the umbrella over her head and trying to keep her teeth from chattering as she rocked back and forth, enjoying herself to the limit, and already enjoying in anticipation the recounting of this night's performance, which she declared we could not fully appreciate. Her oft-quoted favorite, Bobby Burns, was hailed now:

> *Oh, wad some power the giftie gie us*
> *To see oursels as ithers see us.*

We ourselves were quite serious about it all, full of concern and worry. The night had been very hot before the storm began, and Wayne and I had jumped from our bed to begin all the activity in our bare feet. Now our gowns were damp, our feet were wet and cold from the wet carpet and the kitchen floor, most of the bedding was wet, and we could not light a fire nor a lamp, for the wind was too strong. Toward morning, when the rain was over, we made a big fire in both the heater and the kitchen stove and got dry and warm again, and later all the bedding went out on the line in the sun. We removed the windows, put the house in order again and soon forgot it all. But Grandmother never seemed to forget that night's experience and said it would have made an excellent motion picture, although she declared no one would think it could possibly be real.

We revisited Nebo one day, taking both the grandmothers for a picnic. We had a long happy day. We drove around the Bench to taste the sulphur and iron waters and look for ripe blackberries, visited Sunset Rock and the Pavilion and lunched beneath the trees at Lover's Leap.

Wayne undertook to tell Grandmother what he knew of the history of the place. Long ago, the man who owned this mountain had developed it into a famous resort. Wooden sidewalks were built beneath the trees, and a great wooden stairway went up from the bench to the high plateau where, at one end, many really fine buildings housed a rich vacationing class from the capital city. Those were the horse-and-buggy days, and when leaving the train at the depot, away across the river, the long trek was made in regular caravans of surreys and buggies, across the pontoon bridge, through the town, and on up the steep mountain road.

The rich merchants of the town built summer cottages near the springs around the Bench, and landowners who could afford to do so built their little houses at one end of the flat-topped mountain overlooking the river valley. Wheels within wheels, even in this celestial place.

Here, high above the fertile cotton fields, in this salubrious atmosphere, had reveled these people of culture. They had refinement and charm and gracious manners. Having so much leisure they indulged themselves toward their neighbors and friends with generosity and hospitality. Their lives held little knowledge of such a thing as a struggle for existence, for they had wealth and education and knew every comfort and luxury.

With all their advantages and privileges, they had a smug complacency, too, that they had acquired along with their sheltered lives, a superior feeling of being of a class apart, far above the poor mortals who had to work hard for a living. These rich people seemed to have ingrained in them that old false idea of the divine right of kings, only now, in modern times, it was they themselves who held the kingdom.

What more than astonished me was that I found remnants of that same old feeling of superiority still persisted in many of the present generation.

They were proud of being able to have enough land and money so that they not only had the black laborers beneath them but could also ignore and sneer at those they called the "poor white trash." They apparently ignored the fact that they were controlling a remnant of people who had sprung from family trees once flourishing and mighty, although now brought down so low in their struggle for existence that they were, in many cases, occupying a place far beneath that which the old Negro slaves had held. Yet it was these very ones, this class of poor share-croppers, who were continuing to make it possible for the others to live in comparative luxury and security.

All the while that Wayne was talking about the early days of Nebo I could not keep from thinking of the malaria, the dysentery, the mucky yards and poor hovels, that we had left down below us in the bottom lands. I could not keep from thinking of the time when I was discussing with one of the town merchants the need for good roads. The man said, indifferently, "Let the roads be. We don't care if they have to carry the cotton in on a mule's back, just so long as they bring it in."

I still remembered with deep resentment the time when I saw an article in the town paper in which some of the merchants were criticizing the farmers for using the mail-order catalogs instead of patronizing the local stores. Feeling that I could give several good reasons for the farm women using the "wish book," I sat down and wrote a letter to the editor of the paper.

I told the editor why many of the farm women found it much easier to order by mail than to make the long trip to town with their husbands, perched on the high seat of a rattling, swaying farm wagon, often with small babies in their arms and little children huddled on hay in the bottom of the wagon, enduring the heat and the dust in summer and the cold winds and freezing weather in winter.

There was no regular parking place in the town, I knew, so the wagons had to be left on the back street behind the stores, where there was no shelter either from sun or rain. I had walked along there and noted conditions myself. The flies were thick over everything. Little children, tired out, tried to sleep on the old quilts under the wagon seats. Mothers sat huddled in the dust and flies and heat, nursing their babies. Dogs ran everywhere, barking and fighting.

In summer the trip to town by wagon was long and hot. In winter it was long and cold. Even those who had cars had to wait often until the roads were passable, but the freezing weather and cold winds and the rain storms made the trip an ordeal even for the men.

I also explained to the editor, that the women found little satisfaction in having their man pick out dress patterns or certain colors of thread or gingham, even if the men were ever so willing to shop for their wives. Too often when a farmer brought home shoes for his wife or children, he disliked having to return the shoes to the stores if they proved ill-fitting. Or, the next trip to town might be so delayed that of necessity the shoes would be worn anyway, even though they were very uncomfortable. It was the exceptional case, I knew, where a country child ever wore well-fitted shoes.

My article carried the suggestion that everyone should vote for good roads through the bottoms, so that traveling would be quicker and easier for shopping as well as for carrying the loads of cotton in to the gin.

The editor printed my article under the heading, "Farmer's Wife Takes a Rap at Dardanelle." When next I went into town I called on the editor and pointed out to him that my article was no more a "rap" at the town than the words of any woman might be if she were pointing out to her family same much-needed improvements around their own home. A good husband, I told him, would consider practical suggestions and act on them if possible, but helpful ideas should hardly be termed "taking a rap." The editor finally saw the point I was making, but admitted there was little hope of anything being done about anything. He had been out to California on a visit, he told me, so he sympathized with me fully, agreed with me, and laughed with me about all I mentioned. I left at last in a very pleasant state of mind, although feeling somewhat baffled and ineffectual.

I couldn't forget, either, the time I saw the woman who lay on some old quilts in the bottom of a farm wagon, dying of pellagra. She and her family had come up from somewhere in the lower bottoms, and they had been hired to work on one of the farms near us.

Wayne told me that he had heard about the woman having pellagra from the neighbor who had hired them. He was now going to fire them all, because they were all "poorly." Even the men were too thin and weak to make good hands, and could not earn enough to pay for their keep. Wayne wanted me to be sure to see this woman when the wagons passed through our farm that afternoon, using our little road as a shortcut over to the bayou. Where they were going or how they were going to live, I never found out. One wagon had already passed the house, but I saw the second one put by the hayshed. They had stopped to ask directions of Wayne.

I hurried out and went to the back of the wagon and looked in. It was a rickety old farm wagon, pulled by two poor skinny mules. Four long poles had been nailed to the side boards in an attempt to make a regular covered-wagon, but the small piece of canvas and the old quilt nailed on for a cover were sagging down and gave but little protection.

The woman lay on some loose hay. I saw her face. I saw her hands and feet. A filthy old quilt covered her body. Nobody, I am sure, would think it possible that a living person could be in the condition of this woman. Her eyes were closed. Her head and neck were so thin that she actually looked like a skeleton. Her bare feet were just bones covered with dirty skin. Her hands were claws. It sickened me to look at her. I was filled with a feeling that I couldn't understand—a feeling of dread and fear. I know I stared unbelieving, not able to do any clear thinking, and I stood there staring after the wagon long after it had rolled along through the sandy ruts.

And so, while Wayne sat at ease in the sunshine in the big Pavilion and talked of the grand old days of pride and luxury, the old days that all Southerners still loved to brag about, I thought of the other side of the picture. To my credit let me say that for the sake of peace, I made no disturbing comments. I kept silent so that our two mothers would have a happy day to remember.

Our life here had affected me deeply. I knew only too well the humid days of summer in the tiny houses down in the bottom lands; the bleakness of the wintry winds and all the drab emptiness of the dreary years; the toil unending and unpaid, the battle never won.

Many long lazy afternoons my mother and I had spent together, each of us lying in a hammock. We talked, we laughed, we cried. We remembered this and that, sang songs, read aloud, but often just sat quietly, realizing, as the days went by, that Grandmother must soon go back to her own life, and we, all her children, would be left behind, here, with no certainty as to time or manner of leaving, no definite plans—nothing, in fact, but a clear realization that this was our daily life and we were certainly entangled in it completely.

They were sad, sad, those long hours when we fell silent, thinking so
many things that we dared not voice.

The music of the phonograph, Grandmother's music, we called it, was
doubly enjoyed during her visit, and then finally Grandmother was gone.
I still cherish the verses she wrote and left behind as a farewell.

EVENING ON AN ARKANSAS FARM

Fire flung high athwart the sky,
With golden clouds between.
A lake of blue appears to view,
With opalescent sheen.

The soft, cool breeze among the trees
Awakes the katy-did,
I hum a tune, while sails the moon
The fleecy clouds amid.

I'll soon be gone, but this stays on,
This scene so dear to me.
In distant lands, upon the sands,
Beside the Western sea,

I'll sit and dream, and it will seem
That I am here again;
I'll see the trees, I'll feel the breeze,
My heart will throb with pain

As it does tonight, in the moon's soft light,
When I think of leaving you
That I hold so dear, behind me here,
And bidding you adieu.

But the moon will rise, in Western skies,
Where better days will come;
And there we'll meet, when your dear feet
Turn back at last to home.

ABOVE: *Brick and her mother, Elizabeth Mary Seed Grant, Felton, California, 1911.*

BELOW: *Young Genevieve with sisters Gladys (left) and Elizabeth (right).*

Brick and Wayne, who eloped in 1912, were somewhere in the foothills of the Santa Cruz Mountains when this photo was taken. People in the bottom row include Brick's sister Vera Grant on the far left; her mother third from left; and on the far right her brother Douglas Grant.

Wayne, in a photo probably taken by Brick near the sea in Santa Cruz, circa 1912.

Wayne farming in 1917, probably in Soquel, California.

Brick, Wayne, Donald, and Jimmy (clockwise from left), probably on the farm in Soquel.

En route to Arkansas from Santa Cruz in Ford touring car, in 1920.
From left: Oscar Wadsworth (husband of Brick's sister Vera), Wayne's brother
Henry Sadler, Brick, Donald, Jimmy, and Wayne.

Dardanelle, Arkansas.
Sept. 8, 1922.

Dear Mama:—

We've been hoping for a line from you telling of your arrival home, safely if not in good health. I imagine Palo Alto looked doubly good to you this time. Seems something from a dream, your sickness & sudden departure and all. Wayne says to please give you to understand that those were Alabama chills & fever you contracted & not Arkansas— as you were in Alabama long enough to have caught anything. Are you taking quinine regularly to rid it out of your system?

I found everybody sick with colds when I got back. Wayne had to go to bed a day. Jim was in bed nearly a week. I stayed with Eliz. a week after you left, but that climate certainly has a bad effect on one. The children's legs are a mass of sores, & I was afraid I was going to have typhoid or

First page of one of the letters Brick wrote from Arkansas to her mother back in California.

Nancy Isabella Hixson Sadler in a 1907 studio photograph. Penned notes on the back of the frame say that she was born in 1857 and married in 1877. She died in Palo Alto in 1932.

Nancy Sadler in what appears to be another studio portrait.

Donald (left) and Jimmy "in a cotton crop in 1925 that was a total loss,"
according to the notation on the back of the photo.

Group photograph in cotton field, circa 1925.
From left : Jimmy, Wayne, and Donald, with cotton pickers and sharecroppers.

RIGHT: This photo of Brick was taken in 1922 at Dardanelle Rock, a landmark on the Trail of Tears' Arkansas River route. The forced migration of the Cherokees in 1838-39 passed through Arkansas along a water route and an overland route. The tank in the photo once supplied water for the town of Dardanelle.

BELOW: Donald (left) and Jim with their horses Dixie and Ginger.

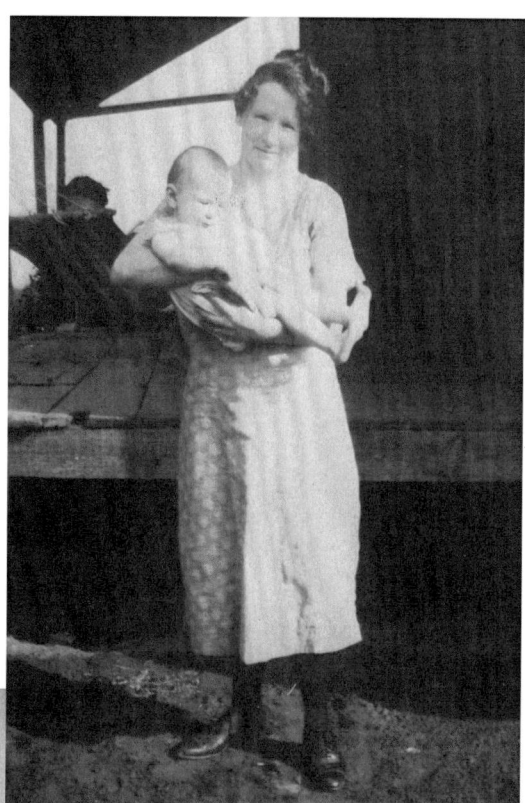

*Brick with baby Gareth
in 1924 or 1925.*

*Wayne with Gary in baby clothes,
bundled against the cold.*

From left: Jimmy, Wayne, Gareth, and Donald with snow on the ground in Arkansas, circa 1924.

Rufus Crispinius Sadler, Jr., born in 1850 in Logan County, Arkansas.

R.C. Sadler, Jr., on the far right, in 1904. He served as deputy sheriff in Dardanelle, which was and still is one of Yell County's two seats of government.

Elizabeth Candace Murphy Sadler, whose cotton farm in the hills near Fort Smith was sold to purchase land on the Arkansas River. The mother of Rufus Crispinius Sadler, Jr. (and eight other children), she was born in 1823, married in 1837, and died in 1910 in Santa Cruz.

This photo of Rufus Crispinius Sadler, Sr. was cut out and glued to the memorial cards handed out at the 1910 funeral of his widow, Elizabeth Sadler. Rufus Sr. was born in 1813 and died in 1866. In 1970, the fountain-penned 1852 ledger of transactions from his Shoal Creek general store in Logan County was rebound and preserved.

This pontoon (floating) bridge across the Arkansas River was the longest in the United States, stretching 2200 feet to connect Dardanelle and Russellville. It was constructed starting in 1889 and replaced by a steel bridge in 1929.

Brick with her hair loose. It was dark auburn, and she eventually grew it all the way to the ground. (She was 5 feet 2 inches.)

Wayne and Brick on October 25, 1929, in Palo Alto at the Stanford Hotel,
which they owned and managed after returning from Arkansas.
Wayne also went to work picking crops in California.

Together again in Palo Alto after returning from Arkansas are,
clockwise from top, Brick, Wayne, Donald, Gary, and Jim.

Brick in later years, circa 1950. She had the opportunity to travel widely with her son Jim, who bought and operated a number of residential hotels in northern California. As for her other children, Donald served as a career non-commissioned officer in the Air Force, while Gary served in the Navy in World War II, then practiced law in Washington D.C. and California.

This hope, you see, brings peace to me;
Ah, no I will not cry,
But hold this thought, or I could not,
I could not, say good-bye.

29. Indian Summer

IT WAS NOT quite cold enough to set the heater up again for the winter, yet the days were cool enough to be invigorating. The fall haze lingered in the air. The pecans were dropping, the melons nearly all gone and the sweet potatoes were being dug and taken into town to the drier. We would bale all the pea hay that was being cut. The ground was plowed again where the hay, corn and melons grew, and the fields were being harrowed for winter rye and oats. The big sorghum patch was being cut, too, and then there would be the interesting job of making sorghum. We were going to take loads of the cut stalks down to a neighbor's grinder and boil the juice there, too. It was an exacting process that ended when the raw juice came to the proper consistency and was finally poured into shining buckets.

The cotton fields looked odd, just the bolls with the white cotton hanging loosely on the bare stalks where the army worm had eaten all the leaves. The worms had gone at last but the weevils were still working. All the fall work was coming to completion. The corn was being gathered and the cotton picked. I would look up to see the high-sided cotton wagons rolling along to the gin. Cotton was down to twenty-three cents, but we weren't intending to hold any that year.

There had been no rain in all those weeks until Wayne's hay was cut and drying in the field. The pears and apples I had been canning were of poor quality owing to the dry weather, but at least I could cook and wash now without the prickly heat breaking out on my back and neck or the perspiration running down my face.

Up in the big barn were hundreds of bushels of peanuts, and three tons of peanut hay. We turned the hogs into the peanut patch and they ate about a hundred bushels that were still left on the ground. The peanuts had grown into a thick matted covering on the strip of sandy soil along the little road near the pecan trees. The men pulled the vines by hand, when they turned yellow, and the big bunches of peanuts that clung to the long roots were left upside down to dry in the sun. Later they were forked into the wagons and taken to the barn. Spanish peanuts, smaller than the Tennessee Reds, and many jumbos, with three large peanuts to the shell, we raised, too. Some were sold in town for seed, but most of them we fed on the vines to the cows and mules and hogs, for they made excellent feed.

When the wagons that were loaded with the peanut vines passed my house, I had the men put off a load on the front porch, and each afternoon I put on work gloves and stripped the peanuts into big sacks to save them for our own use. We roasted them in the oven in big bake pans and enjoyed them all winter long.

Below the bank some of the corn was twelve feet high. The blackbirds were beginning to gather, with much squawking and shifting, in the big pecan trees. All along the bank and against the fences the weeds were high. When we passed along, the gnats flew out in clouds, and the weeds gave out a sharp, nauseating odor that was almost overpowering. A growth of wild grape had succeeded in reaching far up into the highest bushes on the bank, and the strong twining vines were full of black possum grapes. I made some delicious muscadine jelly from the wild grapes we gathered out in the hills.

As the days shortened, the moon began to get full again and finally rose slowly like a glistening golden balloon to sail high over the pecan trees. The nights were really cold. The large box-heater had a small oven in one of the lower lengths of the stove-pipe. This drum-oven was excellent for baking. It was large enough for a loaf cake or a small pan of biscuits and became hot enough for baking in a very short time. Once I actually made a blackbird pie and baked it in the drum-oven, for Wayne shot several dozen of the birds. They had settled in the corn field and were eating up the corn. We stripped off the bird skin, feathers and all, and used only the plump breasts for the pie. Our mail order, this time, from the "wish book," included winter underwear and heavy shoes and wool blankets. I made several warm flannelette nightgowns, for the wind again roared in the chimney.

The carpenter was at home with chills and fever, we heard, so our repair work was held up for a while, but I took pleasure in looking out at the bundles of shingles and the pile of new lumber waiting for him. We planned to have a new porch floor, a large bedroom closet, and more shelves in the sitting room and dining room. Then the house would be jacked up, and the main roof shingled, and a new roof built over the back porch.

Some of the corrugated roofing that was put on the hayshed was second-hand and the sunlight gleamed through the nail-holes but it was rainproof. The shed was now full of baled hay. I often climbed up over the bales with Jimmy and Donald looking for hens' nests. One afternoon we were away up under the eaves, lying on the top bales, listening to the rain that fell like shot on the roof above us. I can still remember that wonderful odor of fresh alfalfa hay. The thundering downpour kept us prisoners until a slack came and we were able to dash with much laughter to the house.

I couldn't get used to having the time to enjoy such simple things as this; just as they came along. I remembered how, back in California, we always used to plan ahead to go places or do things, but here it seemed that we were living each day in a strange new world of "now." That was one reason I didn't mind not getting into town more often. Why should I go? I saw so little in the town life that appealed to me, and here on the farm each day was brim full, and we were all so busy and happy together.

Yet it was about this time that I wrote to my mother that, as Mrs. West used to say, I was quite "outa heart." I am sure that one result of my mother's visit had been to stir up afresh all my homesickness, longings, comparisons, resentments, hopes and fears, and make me realize as never before the situation we were in. Frustrated thinking, a baffled hopelessness, put me into a period of sick despair.

Here is the poem I sent to my Uncle Sid, in reply to the verse in his letter to me. Sometimes words just drop off one's pen point, and again, I found, they had to be dug up painfully.

I had just finished reading his long letter and sat down to answer it, but I burst into tears, in a flood of homesickness and the blues. So amid tears and sighs, I began and finished it in two hours, this, my first brain child. I sent it to my mother, knowing she would see it all and feel it, as I did.

> *Oh, I would love to sit with thee*
> *At rest beneath a pepper tree,*
> *I long, to hear the ocean roar,*
> *To see the waves rush toward the shore.*
> *I know a place where sunlight falls*
> *Dimly, amid the forest walls;*

Soothing, like twilight, yet its gleams
Seem full of life—and oh, the dreams
I've dreamed there, resting by a tree.

And I could guide you on a way—
'Tis heaven on a summer day,
Where roads wind in, and out, and round,
And there is scarcely heard a sound
Save meadowlark, or seagull's cry,
Sweet and yet sad; and there would I
Linger and listen, glad to know
A friend has also found it so—
Enchantment, known so long to me.

Our days were warm and pleasant now and not too hot. It was too cool for the hammocks though, and I had put the extra blankets back on the beds. The air seemed so quiet. In spring there were many cooing doves and mockingbirds singing. Now only the quail called occasionally and the cicadas whirred along the bank. Above us every night sailed the full moon.

One day I asked the boys, "Is that all the mail? No letters?" How funny that we should take letters for granted, and long for a letter every time the postman comes. I often got tired of writing, or I kept postponing writing, thinking I had no news, forgetting that the lives of everyone are really more or less uneventful. I became so aware of the surprise and pleasure of opening a letter that I resolved to write oftener to everyone, if only a short letter, so the postman could say, "Here's a letter for you."

Just doing the daily work, teaching the children, keeping everything clean and the mending done, filled the time. I seemed to have little extra strength. It seemed strange to remember the previous fall when we had the dance here in the front room for our cotton pickers, I had hardly expected to feel like dancing, but one gets keyed up, and pays for it afterwards. Still, we had a good time, and it was a break in the monotony.

Since the days were cooler the milk and butter were keeping well. I made the first beautiful pound of butter in the new churn. When Wayne showed me the new butter mold I was so glad to see the old-fashioned cloverleaf print on it. The canning was all done, at last, and my hands were getting white again.

I kept worrying about the partition between the kitchen and dining room. I was so tired of waiting for the carpenter to come that one day when everyone else was gone from the farm, Jimmy and Donald and I got the axe and hammers and started on it ourselves. The big wide kitchen shelf gave us the most trouble. It was nailed up to the wall boards, but when that was down, it was easy to knock the door frame and the partition down, board by board. So now we had one big room, and I would get linoleum for the floor. It would be more airy in summer, and we "aimed" to put a small heater at the end toward the garden, so that in winter the two stoves might keep the room warm.

The carpenter arrived at last from town and I watched him as he leisurely began the work of shingling the roof. After several days the main roof was finally complete, and he promised to come back later to do the roof over the kitchen and dining room. "No, you won't," I objected. "You could finish it today if you had a helper, couldn't you?" That he agreed to, so I became the helper.

I wore a pair of Wayne's overalls over my gingham dress, took my gloves and sunbonnet and soon was perching on the low slanting lean-to and holding the chalked string tightly at one end while he snapped it smartly to leave a blue line across the shingles. The old rotten shakes were easily removed, and I found I could lay the shingles and nail them on as well as the carpenter. Most of the time I finished my half of the row of shingles and waited for him to catch up with me. I was determined to have that roof shingled, and wanted no more leaks.

When he came back to work on the shelves, he thought I was very smart because I could tell him just how to space them. He had nailed them even with the floor, which was about four inches lower in one corner due to the foundation-rocks sinking into the sandy ground, and he looked worried when I showed him how slanting the upper shelves would be. He stood and scratched his head, puzzled, and was relieved when I told him to start at the top and put all the shelves the same distance apart even though the space beneath the lowest shelf would be quite irregular.

The bookcases were finished at last and all the roof and the well casing and a new roof over the back porch, and extra doors cut from the "North Pole" bedroom out onto both the front and back porches, with good screen doors.

The carpenter was amazed at the work we had done in getting the partition down. I had him put up a wide shelf beneath the two windows that looked toward "our" western hills, and build set of shelves in the dining room, too. Such style!

He seemed quite surprised when I told him I wanted the locks of the outer doors fixed, too. "Tain't the regular thing," he remarked, thoughtfully, "don't know what the folks around will think of you." I soon found out. One of the women who lived down past the store said one day, "Stopped by to see you-all the other day, but your door was locked, seemed like. Got to studyin' 'bout it. Ain't nobody here needs to lock up nuthin'. Couldn't figure what you might be up to, so I jest decided to go on back home."

I did feel rather guilty, and found it hard to explain my decided liking for privacy, but I found myself giving in and seldom using locks and keys except occasionally when we were all away from the farm.

30. Memories

Extracts from Letters to a Mother

I AM HASTENING to answer your letter. We are so glad you are better. You didn't need to have malaria yourself in order to verify all our accounts of it. I have been thinking of you lying in bed, but I have been sick with a cold and have been trying to "leg it," between the two houses, so have quite forgotten how to write.

The days are so short, that although I am always busy, I seem to get little done.

Grandma has been quite sick. She had the first stage of pneumonia. Wayne, of course, sits up with her every night; although we have a woman hired from town to take care of her, Wayne is more comforting.

We had one night of heavy rain, and yesterday it rained and blew terrifically, but today was beautiful. It is very warm, and the clouds are heavy above us, so we'll have rain again. This is early November, you know, and quite cold weather will come, even though our doors and windows stand wide open now. Weather here in Arkansas is always of interest, remember?

I got your nightgown finished at last. I sewed it all by hand, for my last machine needle broke and I had no chance to get another until someone went to town. The crocheting was my work, too. Your grandsons said they'd bet you would look lovely in it, and so you will.

The poor people ask after you often. As you say, "Kindness goes a long way." They have not forgotten you. Mrs. Batson's little girl cherishes her beads as though they were priceless, and is so happy about the doll you are sending.

Later:

Both of the boys were quite sick last week. They had colds, and such awful headaches, earaches, eye-aches, cramps, vomiting, delirium, fever and chills. I phoned frantically to the doctor who as usual prescribed quinine in great quantities with iron tonic, and said calmly that it was only a combination of flu and malaria. Each time we are sick I keep thinking of the other people here who are sick, too, and who have no money for medicine and generally no doctor, either.

How we enjoy the magazines you sent. They are splendid and are helpful in keeping one awake when sitting up with the sick. The boys just looked up from their checker game to report that there are forty-five baby pigs and to be sure to tell you that the fences have all been made hog-proof. This, l am sure, means to them that there will be fewer times when they will be called upon to help drive the pigs from some place where they shouldn't be.

The World Books you sent we use with our lessons, especially geography. Today we read of creeks and dams and looked up beavers. Sometimes we begin with oceans, and follow through with ships, whaling, blubber, ambergris, spermaceti and whalebone. I let the boys go ahead in this way as it encourages them to hunt out facts for themselves, and so they look forward with great interest to their lessons.

Last night we had a big rain and wind storm, presaging winter. The leaves are gone now from all the trees except the red gums. They are still so brilliantly beautiful.

At last I can tell you that the lawsuit is over, and we lost. We have been given the "privilege" of buying back the land. I cannot bear to think of it. You were right when you told us that the claim of paying taxes for twenty-seven years, and holding the warranty deed, should have held good, for I lately heard from someone connected with the courts that the lawyers knew all along that we were in the right.

So, on account of dishonest lawyers, we lost the money paid out for lawyers' fees, and the court costs, and now we can repurchase the land.

We must make all the money on the farms, and carry on the regular farm expenses at the same time. And cotton is not forty cents a pound now!

Later:

I am up, today, but so weak I can hardly stand. I have been sick for about two weeks with "flu" and acute bronchitis and malaria. The doctor said it would be two weeks before I'd be able to do much of anything. He doesn't know, though, how quickly we recuperate, even though we seem to fall sick so easily. He deals so much with the bottom folk, and they never get well; although I feel this morning that he could be quite right about me. Quinine and fever! Quinine and fever!

Later:

Your letter and package for Wayne's birthday came yesterday. We were glad to hear from you, and Wayne was delighted with the shirt and lovely cufflinks. I do hope you won't think we have simply neglected to write. But today I am only propped up. I've had two relapses since the last illness I wrote you about, with fever all the time. I can hardly walk across the room, but the doctor says I ought to be improving right along. Since you heard from me last, I felt pretty well for a few days and even managed to get all the dishes clean, and we were eating off a clean tablecloth again, but I went to town and got chilled. Wayne has been too busy and worried to write. He has been so good to me during sickness, for he is a natural nurse. He should have been a doctor.

I see the beautiful calendar that you sent hanging up, and the books for the boys, and the pile of magazines. And cards—Christmas cards—on the wall. But we think that you have made Christmas for us every day. Jimmy says you never go near anybody without giving him something, and well he knows that is the truth.

I see Wayne is cleaning a chicken in the kitchen. Jimmy killed it.

I hope I get stronger soon for I was in bed on Thanksgiving Day and you know that I would certainly not want to be sick at Christmas time. Wayne wants to go out in the hills again to get our trees. The time is so short, and I had planned on doing so much for Christmas this year.

During the worst of our illness a friend of Wayne's from town came to the farm to hunt quail. He evidently shot a good many for he came to the door and offered me several birds, introducing himself and thanking us for letting him hunt on the farm. I was so weak I had to hang onto the door frame while I talked to him, but I astonished him by refusing to accept the quail. Wayne had been in bed two days with terrible headaches,

Grandma was sick, and Henry was so busy. Henry even brought us bread from town, for none of us were able to bother even with biscuits. All I could think of when I saw the quail was the work they represented, and I didn't feel equal to it.

At the same time I was shocked by a fresh realization of the size of our little house and its utter simplicity, as I saw him looking around with interest. I saw the place anew, through his eyes. The same poor little house that had so disheartened me when we first moved in. Yet I felt a surge of loyalty fill me, for I knew, even if he didn't, how clean everything was. The porch was clean and was not used as an extra storeroom, there was no trash about the place, the curtains, floors, walls, and dishes were all as clean and neat as they could be. We were warm; we had our music and books; we had good food and we were comfortable. Yes, I was shocked, but somehow I felt on the defensive, and was glad, glad, to refuse his kind offer. It was an unpremeditated reaction. Are you surprised?

The weather is again very changeful. One night last week the thermometer rose fifteen degrees in thirty minutes. It was so hot we had to get up and take out all the windows, but of course you remember all that performance. The next night the ground was frozen. We are sleeping under all our quilts and between blankets and use the hot water bottles also.

How hard it is to get up in the cold and get the heater started in the morning.

I hope Christmas will be a lovely day for you in California and that you'll be feeling well and have a nice time. Somehow, it isn't a merry time we have, nor want—it's just the sense of happiness and contentment and peace that mean so much more.

Later:

The sun is shining and the door is wide open, although everyone says this weather is too fine to last. Christmas Day was lovely, too. It was warm enough for open windows and doors. The snow did not come, for the weather suddenly changed.

I was able to get the house cleaned and in order, and made a couple of cakes and some candy and cookies. We went out to the hills and got our little tree, and on it hung our decorations, the ones we've always had, and some fresh tinsel. We strung little round strings of popcorn and put the silver star and "Merry Christmas" at the top. Our deer's horns sprouted oranges, as usual, and the Christmas bells hung above the windows.

Christmas Eve, we put the boys to bed, put our presents on the tree and opened your box. Lo and behold! There were so many lovely packages in it. We were tempted to examine them, but we finally decided to wait until morning, as was our usual custom.

Next morning we all got up early and had the heater going fine and the candles lit. We spent hours examining and exclaiming and smelling and tasting. We didn't forget to think of how thoughtful you were about the "cup of the crature," that delicious green tea, and the figs and dates and raisins—and the pictures that represented other days and other times. The pink doll sat in the center of the dining table with animal cookies all around her. The baskets of sweet grass were lovely, and we all played with that big fine ball. So nice to hold, isn't it?

When we put on the records of those wonderful waltzes we could have cried, and I'll tell you why. They are so beautiful, but so sad.

Wayne said he felt like sending them all back to you, for they made him think of all the things he has planned to be and wasn't. We talked about the dances in summer down at the beach, where, when the music ceased, we could always hear the breakers on the shore. One waltz, especially, has such a mocking tone, jeering, as though to say, "Oh, yes, you think I don't know, but you are only pretending a Merry Christmas—whistling to keep up your courage." And we were, too. When I was setting the table for din-ner I laid a place for you, thinking that would add to our pleasure. But, alas, it worked out differently—it made us all quite homesick, although no one had such a pretty table and a nice dinner. Grandma was with us, but Henry was away visiting. Grandma is aging so fast. She has changed a lot since we came to Arkansas. Her memory is failing and she falls asleep so often in the daytime. Her last illness was very hard on her. She ate very little and went home early.

We played and played the new music to torture ourselves, but we re-ally enjoyed it after we cheered up a bit. Then we put the boys' new wagon together, and opened up the big Crokinole board and played games with Jimmy and Donald.

Later:

Again I have been spending the days with Grandma to care for her and help her with her work. We certainly do enjoy those records, now that the first shock has worn off. It's really a wonderful thing to be able to give another person a present of condensed memories and longings, isn't it?

Wayne is up superintending the building of chicken runs and the pouring of the cement foundation for the gasoline pump. The men are building a new house for cotton pickers. It is one long room with a slanting roof, large enough to house two families. Just a place to eat, and room for several beds. They will carry water from the pump out behind Grandma's house, for the new house is near the one where Avis lived.

Jimmy and Donald asked me to tell you about the cows. You took such an interest in them. Jimmy milks the Jersey and Donald the red cow. Lois is dry now. The Garner cow has a new calf and gives lots of milk.

We haven't had any snow yet, but are always looking for it. Yesterday we had one of those heavy rains that failed to materialize when you were here.

We have worlds of turnips and eggs and milk and butter and chicken and canned fruit and sweet potatoes and sorghum. Wayne had a sack of whole wheat ground by the same Bohemian who sold us the yellow cornmeal, so now we have whole wheat muffins for breakfast with our own honey on 'em! I have a loaf cake baking in the drum-oven and am boiling a pot of turnips and a kettle of black-eyed peas with onions and a ham bone added to them. The aroma fills the house.

Donald just carried in a hen and fourteen little new chickens to put in a big box by the kitchen stove, for the wind has come up now and it is getting very cold.

We celebrate all the holidays by having our supper table specially decorated; this makes a pleasant interruption in our daily life. Tomorrow they will connect the gas engine to the new pump so that the water can be pumped over to the huge galvanized tank for the stock at the barn.

There is going to be a "pie supper" down at the schoolhouse tonight but the roads are too bad to think of driving the Ford yet. Only those who live nearby will go.

Later:

I did not realize until last night that it was New Year's Eve. We were all too tired to sit up late so didn't see the New Year arrive after all. This morning, as we were dressing, by the heater, I said to Donald, "Happy new year." He simply smiled at me and answered, "I know it," so we just let it go at that. If the whole world knew as much it would be all right, wouldn't it.

Oh, yes, there was something else. Along about Christmas time we

*were talking about how much it really meant to us to have someone who
cared enough about us to listen to all our troubles, so, if sometimes we
seem to take your so many gifts and remembrances to us in the same way
that the flowers take the sunshine, don't think that once in a while we
don't stop and realize how few there'll be left to care about us when you
are gone.*

*The boys treasure the poem you sent them and delight in reading it
over and over. We read it aloud at the table on Christmas Day.*

From a Mother's Letter: A Poem for Her Grandsons

*Another year has been and gone,
A new one now is marching on,
The days are flying fast.
To get a little something done
We must not count on future fun
Nor ponder o'er the past,
But use the hours as they come
As though each one were the last.
Your loving letter just came in,
I've read it with delight,
It made me wish that I had wings,
I'd be with you tonight.
Instead of sitting all alone
As usually is my plight.
It made me laugh, it made me cry,
It made my sad heart ache,
It made me think of days gone by
And cakes I used to bake,
And pies, and tarts, and doughnuts, too,
And candy, and I'd take
And fill the stockings large and small
With goodies that I'd make;
And put the little ones to bed
And say, "Now, children dear,
Just shut your eyes and go to sleep,
Or Santa won't come near."
And in the morning, how they'd shout,*

"He came, Mama, he came!
Just look what I've got, oh, look here,
I didn't shut my eyes or sleep,
And he came just the same."
It seems but a few years ago
And yet these children grown
Are scattered far and wide today,
Not one of them at home.
But in the empty nest I'm left
With only memories, now,
While they repeat what I have done
And tell their children how
They hung their stockings, had their tree,
And how they'd plan and wish,
While Mama baked the cake and said
"Who wants to scrape the dish?"
Sure they were happy days we had
Together in our youth,
But these are just as happy,
If you only knew the truth.
For after all is said and done,
I'll tell you in a minute,
The pleasure we get out of life
Is just what we put in it.

31. Old Hames

ONE DAY I WENT out to the hayshed to look for hens' nests. At the far-ther end of the open driveway, squatting on the hay-littered ground, was Wayne; facing him squatted Hames Bowens, who had recently moved into the little house by the mail box. Ben's brother and his wife had been staying there awhile, but the first time they got paid, they took out for the hills.

Hames was called Old Neck. He had a long, gray, wrinkled neck like a sandhill crane, a reddy cast to his hair, little blue eyes, and a vacant face.

Both the men were earnestly engaged in conversation. As I neared them, they looked up, and then went on talking, so stood there, behind Wayne, and leaning on a bale of hay, prepared to listen in. I didn't keep my position long, for I was soon convulsed with laughter, and as I was standing where Hames could see me, I had to go around the corner of the seed-room to laugh every few minutes.

It seems Hames wanted to borrow a team of mules, so he could go back to his former home down in the lower bottom to get his milk cow and a calf that he had left behind.

Wayne agreed to lend him Joe and Jerry to make the trip. They were the toughest and yet the steadiest mules to handle. About seven miles further down the river the Petit Jean Creek cut through a small gorge. The bridge here, a wooden swinging bridge, was hung on steel cables secured to big rocks and braced with huge steel rods that were driven deep into the ground, high up on either bank. The bridge was safe and strong, but it did sway a little when wagons passed over it, and Hames had to cross it. A neighbor had brought Hames with his family and household goods in by another route.

Hames was asking the directions for his trip, and listening, with his head on one side like a bird, to Wayne's replies.

"Now, about this bridge," he said, when he had been told about it.

"What do I do when I get to it?"

"Why, you'd better draw up the mules." It seemed it was the management of the mules that was worrying Hames.

"How am I goin' to got them across the bridge?" he asked.

"Why, drive then across."

"No, but how do I drive them? You tell me just what I do."

"Well, old Jerry's the lead mule," began Wayne.

"Yes, lead old Jerry," Hames interrupted.

"No, Jerry's the lead mule. I mean by that, the side he works on—over here," motioning.

Hames said, "Then you don't lead him?"

"No, you drive him. Then hold your lines tight when you get to the bridge and hit old Joe a good wallop. Yes, hit Joe, then hit old Jerry, and drive right on across the bridge."

"Hit Joe, hit Jerry," Hanes repeated aloud, nodding his head at every word. He sat like a man stupefied, his eyes fastened on Wayne's face.

"The only danger will be if they start crowding together," continued Wayne. "Then one might crowd the other over the railing. It's about a fifty-foot drop into the creek, but you do as I say—hit Joe, then Jerry, and drive ahead."

"If they do crowd each other over the railing and fall in the creek, then what do I do?"

"Why, swim out. Never mind the mules. You just leave them and you swim out."

Hames squatted there, his jaw slack, his mouth wide open, his head nodding at every word. His hand was raised as if holding a whip, and he brought his arm down with a jerk at every "Hit Joe." His position was one of acute attention. He was actually trying to memorize the instructions and said vacantly, "Which did you say to hit first?"

It made me laugh, for the bridge was not dangerous, it was short and safe, and the mules were very steady. Hames's look and actions and my mental picture of him, probably drawing up at the end of the bridge and sitting there in his wagon, conning over the directions and trying to remember which mule to hit first, were, well—it was something one would have to see to appreciate.

They do say that where Hames came from there is a law against shooting before noon, as the natives are all up in the persimmon trees getting their breakfast. Anyway, some time later a man who lived down near the Petit Jean asked us what kind of man we had working for us, for he watched him crossing the bridge. First he unhooked the mules and led them across. Then he and his wife dragged the wagon across. Coming back, they did the same thing, leading the cow and calf over, and then hitching up and driving along home.

Wayne had given Hames one of the big cotton wagons with the high side boards, so he could put the cow and calf in the wagon, as the distance was too far for them to walk. Instead, we found out, the cow was tied behind the wagon, the calf left to follow as best it could.

Hames's wife walked behind, too, chasing the calf out of the woods and bushes along the road, and beating the cow's back with a stick to make her step briskly, for the poor cow pulled back on the rope all the way, scared and worn out with the long dusty trip. Hames sat up on the high seat, yelling at Joe and Jerry.

A few weeks after the scene by the hayshed, Hames rushed up the road to the big barn to get Wayne to come and help him with his cow. She was still kicking, he said, when he left her, but he needed help to get her on her feet.

When they got back the cow was dead. She had been tied out on the bank to eat the tall grass that grew so rankly in the shade, although she could as easily have been put across the road into the pasture. The rope had been tied in a slip-knot around the cow's neck, and the other end tied around a pecan tree, so the cow ambled down the steep bank, slipped and fell, and choked to death.

Wayne said he could have cursed when he saw what had happened. Hames could so easily have cut the rope on the cow's neck, for he carried a knife, and if he had given her a good kick, she'd have been up on her feet. As it was, the cow, worth about thirty dollars, had been more valuable than anything else Hames owned.

Afterwards, when Hames and his wife talked to Wayne about going back down to the lower bottom to live, the woman said, "I'll be sorry to be a-goin' back. Ain' never saw nothin' till I come here. Ain' never been into town yet. Ain' saw a train; neither a steamboat. I've heard 'em whistle and saw 'em blow, saw the smoke over the trees, but cain't rightly say as I've really saw ary one."

32. Arkansas Blues

Extracts from Letters to a Mother

WE ARE STILL *picking cotton, finishing last year's crop. Fourteen bales are already in and twenty more to pick. On the lower place they will have twenty bales.*

Everyone is predicting great floods this spring, due to the excessive rainfall in the upper Colorado and Arkansas River valleys. The weather is sunny but cloudy and it looks like rain, so I am going to wash today in the black pot before the wind comes up.

Jimmy and Donald are both in bed with fever and cold, again. The doctor says they have so much malaria in them that when they catch the least cold it brings out the fever. Arkansas is full of "flu" now, and the schools are closed at Fort Smith.

I've been out visiting a family who live in the hills. The mother is a woman from Kansas. She looks at Arkansas the same way I do. Her husband is a relative of one of the better town people. All her friends call this energetic woman "Mrs. Joe." She is elderly and gray-haired, with two half-grown sons, but she is so young at heart, and I am sure I'll have such pleasure in knowing her.

Mr. and Mrs. West returned from the hills and moved again into the house they had before. Mr. West is able to do only very light work now, but Wayne hadn't the heart to refuse to give him what work he could do, for he asked with tears in his eyes. So, with the Wests so close again, Grandma will have someone to talk to and sit with, and it will be good for her.

Ben's brother Claude and his wife, who have been down in the house where the Woods lived, have gone back to the hills, so now that house is the one that is empty. We intend to put cotton-pickers in it; one large family, or maybe two. "Pussy wants a corner!"

Today some friends of Wayne's called in to look at our California pictures and the book of views. These people are selling out to go to Texas where they have already bought a piece of land. It is near the Mexican border in the Rio Grande Valley, and, from their description, will be among fine opportunities, with good schools, good roads, and a healthy climate. Anyway, they got me well started on Arkansas versus California. We showed them all the pictures and cards we had. They stayed for dinner and then we talked some more. They asked permission to bring some friends of theirs to talk, as their friends are undecided between Texas and California, and it seems I'm to supply encouragement to the women, assuring them that far-off fields really are green. I'm only too glad to do so; only, when I'm through, I'm so discouraged myself, I could die. However, I feel that to put anyone out of this country is almost equal to saving his soul. It's saving his body, I'm sure.

Regarding your photograph: There's the bitterness of years around the corners of your mouth, and the sorrow of years around your eyes. It seems sometimes when I glance at it, as though you were going to tremble your lips and cry, and again as I pass through the room I fancy there's a dimple somewhere in your cheek that I missed somehow when I examine the picture closely, as though you might break out into a proud smile and say, "I know I'm surprising you to death." If I were to guess, I'd say that when you sat for the picture there passed through your mind regret at passing years, a few bitter thoughts, and then a calm pride. It really is a good picture, and I think represents you as I've known you in your best moments, which is what we humans all strive to be remembered for, isn't it?

Later:

Your letters have been received with great pleasure and a splendid package of papers and magazines which are greatly appreciated. Jimmy and Donald are enjoying The Youth's Companion. *From the "wish book" we have ordered* Scottish Chiefs, Pilgrim's Progress, Black Beauty, *and* Robinson Crusoe, *as the boys are in the* Fourth Reader *now.*

As seems usual, Jimmy is in bed with fever. I had a sore throat last week, even now I am hardly well, and he caught it from me. With malaria in his system his fever has been one hundred and four. I've been hovering over him all day, and was up with him all night, for he wanted so much water and had the nightmare. This fever! This weather!

I haven't written much, for every day I've hardly been able to get my necessary work done, and I often look over at my writing desk and wonder if it could be possible that I ever had time to sit down and write a letter. I haven't had answers to the letters I wrote some time ago, but I don't worry when people don't answer promptly. As Harry Lauder says, "Us never knows, do us?"

Those people I told you about are all back from Texas on a visit, more wildly enthusiastic than ever.

Jimmy says to tell you that he can't play William Tell Overture any more because that's the one record Looie doesn't like. Looie howls as though his heart would break, yet he is indifferent to all the other music.

Spring is really here, with warmer weather, so, from now on, barring windy days, we'll all feel better. The big wind storms seem about over and the days are generally sunny and calm and cool, with frost at night. We planted peas yesterday, and garlic, onions, beets, radishes, and lettuce. Irish potatoes were planted long ago.

The roads are still so bad that I have been into town only twice to visit my old neighbors, but all the acquaintances on earth can't equal one friend who talks one's own language. The three recent deaths in the family have affected me deeply. I can occupy my hands, but cannot find rest for my mind.

33. The Flower Garden

WE HAD PLANNED for a larger flower garden for a long time, and now that the chickens were up at the new hen house and the pigs running in the pasture across the road, it seemed the right time to go to work. I could not wait any longer for the day to come when Wayne would have time to help me, although he had brought a wagon-load of fertilizer from the barn lot to spread on the ground and be worked into the sandy soil.

I went ahead digging post holes for the fence, and with the help of Jimmy and Donald finally put in the posts myself and nailed on the chicken wire. The fence was about four feet high and we hoped it would keep out any pig or mule that got loose.

We had a plan for the garden. We placed a wagon rim on the ground in the center and made each corner an arc with flower beds connecting them. Wayne built steps down to the garden from the end of the front porch, and made a tiny gate at the back so we could carry water from the pump or the washtubs to water the flowers when it didn't rain.

I ordered packages of flower seeds from Fort Smith, and oh, with what hopes we planted them!

Around the edge of the center ring would be a border of white sweet alyssum, inside that would be golden California poppies and petunias. Phlox, four-o'-clocks, nasturtiums, daisies, morning glories, sunflowers and zinnias all had their special place, with more morning glories around the fence. We transplanted grape cuttings from the possum grape vines in the meadow, hoping that at least they might make a hedge of green.

So now we worked in our garden plot early and late, full of enthusiasm and hope.

A young elm that we dug up in the meadow we planted in one corner and it took root and grew well, and we carefully tended our little peach tree that was now four feet high. I wanted to show the boys how the columbine leaves would gleam like silver when immersed in water, and I knew they would enjoy the snapdragons that seemed like the heads of little puppies.

They had gone into raptures, when, on a trip out through the hill country the summer before, we had driven across a high narrow bridge near a sawmill and stopped to look down on a great expanse of yellow pond lilies and tuberous water lilies, all blue and pink and white. Wild iris, too, we had found, in a low swampy place where the wild onions and yellow violets grew, and great lush growths of velvet cat-tails.

We visualized a tall background of cannas and delphinium against our fence. I think our hunger for the flowers we had known was so deep that we were very greedy, and imagined a garden of verdant lush growth, full of every color and odor and beauty.

I often thought of the wildflowers at home in California that grew on the sunny slopes of the rounded green hills dripping down toward the sea. Blue, blue lupine, whole fields of them, splotches of gold where the California poppies burned, and acres of clear yellow where the wild mustard grew and the buttercups rippled in the April breeze.

I recalled the dark green spires of the redwoods and pines along the Coast hills and I seemed to hear the high clear exhilarating call of the meadowlarks. Such an abundance of glory there was in those spring days when we went on jaunts to the woods to gather hare-bells and gold-backed ferns and returned with branches of wild azalea, the sweet white blossoms tinged with pale rose and yellow.

Even among the sand dunes by the ocean we found the yellow sand-verbenas with their heliotrope fragrance, the rose-purple and pink flowers of the trailing abromias with their thick silvery leaves and sticky, sandy red stems, or we gathered the cerise geraniums, spicy and hardy in the salt air.

Ah, life had been so sweet, youth and Maytime and flowers, and if we did not fully appreciate all the joys we had, yet the fragrance of it all hung round us still.

And now at last we were planting our garden. We were sure it would be worth all the work involved, even all the extra pumping we would have to do. We had at last come to accept the pump as an inevitable part of our daily life, for we used lots of water.

Although we made endless trips from the pump around to the flower garden, after standing in the hot sun pumping tub after tub of water, yet the ground was so sandy and still so poor that the moisture seemed to vanish almost immediately. Jimmy and Donald were so proud of their work and so anxious for the garden to be a success that they pumped and pumped until the sweat would run down their faces and their shirts were wet and clinging to their backs. Then I would pump awhile. My hands were blistered from working that old iron handle up and down and helping to carry the big heavy tubs of water around the house.

Now, as might be expected, the point was choked up again with sediment and the volume of water was lessening every day. So we knew that once again the well would have to be "pulled." To "pull the well" meant all the process of rolling a big hardwood log up near the well where a long pole of elm or hickory was used for leverage. One end of a chain was tied around the pipe as low down as possible, and the other was fastened to the low end of the pole. Then the men all pulled down on the high end of the pole and the pipe was raised length by length, each length being disconnected as it came up and the chain put on the next length. When the last length hung free the worn-out point was hammered loose and a new one was screwed on. This point was a short length of pipe shaped like a pencil point, with holes along, the sides and copper screening at the end to keep out the sand. The soil, deep down, was a black sand that somehow ate the screening that sieved out the sediment, for this job of pulling up the pipe had to be done often. When the new point was on, some of the lengths of pipe were screwed together, an iron cap was set over the thread ends, and the pipe was pounded back into the well with a mallet. Then the other lengths of pipe were screwed on and the process repeated until all the pipe was back in place. Then the well platform was rebuilt, and we had water again.

I will never forget the anger I experienced on account of that pump.

The first time was early in our first year on the farm when the clothesline broke. The wind was blowing so hard that the sagging line of clothes was dragged back and forth through the dirty sandy soil. Some of the sheets were blown against the black pot and dragged through the hot coals and large holes were burnt in many of them. I ran out to rescue the clothes and was standing there in dismay looking at the wreckage of my morning's hard work and thinking resentfully of all the pumping that must be done again before the clothes would be fit to hang out, when around the corner of the hay shed came the farm wagon, loaded with a crew of men on their way to the fields.

They took in the situation at a glance and roared with laughter. I could have stood that, I think, if my husband had not been there too, for I saw that he laughed with the others. They did not even stop to help me put up the wire line, and as I could not stretch and tie it myself, it was left until the next day. I unpinned all the clothes and put them to soak in tubs of water which I had to pump. The sun was hot, and there was no protection from it on the well platform.

I was so angry, so deeply resentful, that I didn't care how much water I had to pump. I hardly knew I was working the pump handle. It took twenty strokes to get one gallon of water, just as usual.

Racing hotly through my mind was the keen realization of the utter indifference of most men to the things in daily life that make a woman either happy or miserable. I felt so hurt, so humiliated, so cheated. My very heart felt lacerated as I thought of the response I would get from any of those men if I should try to put into words the reason I was "mad." "Why," they would say, "it was funny to see the clothes on the ground and you standing there looking at them. Couldn't help but laugh. Didn't mean any harm by it."

Beneath it all was the thought that made me weak and ill—that my own husband was one of those men. Maybe no woman dares, in earnest criticism of her husband, to think of him as the man who really loves her. (Some day I'll follow that idea back and forth, and see if there, too, is not one more thing that every woman knows.)

Another time I was so angry was during my mother's visit to us. Wayne and Henry had an opportunity at this time to buy a lot of galvanized tin roofing with which to re-cover the big hayshed. The hayshed was old, and had been badly wrecked in the last big storm. The shingles were old and worn, many were gone, and when it rained, the roof leaked badly.

Now the shed was needed for the baled alfalfa, and that right soon, so Wayne put all hands to work building new walls and putting on the tin roof. Two brothers, who were old friends of the family and who were good carpenters, had been hired to boss the repairing, and the farm work was laid aside until this job was done.

I could see the men astride the hayshed roof, hammering and banging, laughing and singing, all busy and happy. Whenever Jimmy or Donald or I went out to the shed to find hens' nests, we tried to find where the half-gallon fruit jar of "chock" or Choctaw-beer was hidden. We had heard the men talk about it and knew that they had brought some with them. We were careful that they did not see us hunting, and we usually discovered the jar before long. Then we would empty it out and fill it with water and put it back where we found it.

This little game of finding the "chock" and substituting water for it was something like a game that we played whenever any of Wayne's and Henry's friends came to the farm for extra work, and nothing was ever said openly about it. I'm afraid that the men knew when they found the

jar full of water still at the tree roots or behind a certain bale of alfalfa
where they had hidden the "chock," that there was just no use ever men-
tioning it to me.

I watched them making the Choctaw-beer once. They borrowed my
largest crock, the one in which we "put down" sauerkraut, and they made
a mixture of water, sugar, hops, raisins, cracked corn, and yeast. During
the process of fermentation the crock stood out at a corner of the house,
covered with a white cloth and a board. The odor was sickening and even
after it had all fermented and the residue floated on top, the juice beneath still had a strong odor and a kick like a mule.

Now the August sun was very hot, and Grandmother suffered intense-
ly with the heat. We tried at first to hide from her the true condition of
our water supply and the hard work it meant to have plenty of water on
hand. Jimmy and Donald worshipped their grandmother from California.
They called her Grandmother, and Wayne's mother, Grandma. Toward
their Grandma they had a deep affection that was mixed with a sense of
protection and patience, for they knew how often she was ailing, and they
had become accustomed to helping her up and down steps and taking
turns staying with her at times when Henry was away. But just now their
glamorous Grandmother from California was the center of their life, and
they were anxious to make her visit with us a pleasant one.

But each day the force of the water lessened a little, and it became
harder and harder to work the pump handle. The stream of water grew
smaller, until at last it was only a trickle. The luxury of our daily bathing
was becoming something to worry us. Grandmother, though breaking out
with prickly heat, insisted on wearing the corset she was accustomed to,
and found it difficult to wade through the hot sand and over the rough
ground in her thin slippers and high heels, so if it had not been for the
relaxation she found in one of those hammocks which she had been inspired to bring us, her days would have been much harder to bear.

Finally the time came when the pump gave no water at all. The pump
point had filled up, and the well had to be "pulled."

Each day Wayne kept telling me that they would go that very day to
town and get a new point and stop work on the roof long enough to fix
the pump, but each day came and the pump was there in the sun, hot
and dry.

The men worked until dark every day. The two carpenters boarded
up at Grandma's, enjoying the full meals of ham and chicken and hot

biscuits and corn bread and luscious yams and jelly and pickles, and all the good buttermilk and home-made butter they wanted, so I knew that, though they were busy, and the new roof was very urgent, yet they were dallying along, too, enjoying every minute of their stay. It would have meant only an hour's time to go into town in the Ford and buy the new point, and if all the men stopped work and helped they could fix the pump in half a day. Yet, day after day, for nearly a week, Jimmy and Donald trudged off to the meadow to pump water there by the dipping vat. It was a choice of two evils—either to go up the hot sandy road to Grandma's or go down the road that cut through the bank and crossed between the alfalfa patches, over the big road, through the heavy gate, which had to be swung shut and fastened, and over one fence, to the meadow pump. At least the meadow pump was easier to work, and once in the meadow the pecan trees gave pleasant shade.

Nobody but Grandmother and I seemed to care about the two little boys who were required as a matter of course to go for the water. Time after time, day after day, during that hot, dry spell, I watched their young faces, thin and drawn from their recent bouts with malaria, as they returned wet with sweat and really tired out. Each day they, too, thought that by night the pump would be fixed, and each day brought fresh disappointment.

No, it wasn't such a terrible condition, it wasn't unendurable, it didn't last forever, but it did last a week—and such a week.

I soon found out that all over the South, at all the pumps, the women all faced this same hardship whenever the pump went dry and they had to wait until their "men folk" could get around to fixin' it again. The baffled feeling I had of enduring what I knew could be cured, my helpless rage, pitted against the smooth acceptance of the situation by Wayne and Henry and their friends, seethed within me daily. The soiled clothes piled up waiting for the next wash day. I tried in every way to use as little water as possible for cleaning the vegetables from the garden, but in washing the dishes I needed extra rinsing water because of the lye we had to use. I was trying to save water for the baths that were so necessary for even moderate comfort. All these things seemed to become major problems as day followed day, each one hot and muggy. When the housework was done we tried to keep cool and seek what comfort we could in the shade, and never before did we realize the large part that that old pump played in our lives.

To appear serene and calm before my mother and at the same time realize that our daily misery was due chiefly to indifference and procrastination, seemed more than I could manage, yet I kept my arguments and ill-nature until I could talk to Wayne alone—for I did not want my mother to realize how deeply I felt the injustice of the situation. Wayne was full of promises, telling me that it would not be long before they tackled the job of the pump, and anyway, we weren't really suffering. At last, however, shamed by Grandmother's nonchalant remark that she really could go down to the meadow, too, and somehow manage a bath and her necessary laundry, Wayne had the men leave the roofing, and they put in a day "pulling the pump" and putting it in good condition again.

I was a long time in learning that patience was the greatest virtue of all farm women, and that all things worked out if one but waited. Even yet, however, I remember the deep resentment and anger I felt at that time.

I know that throughout my years on the farm I never saw or heard of any Southern woman making a display of anger or indignation or temper.

We never entered any home where there was the least indication of excitement, irritation or bitterness. No one fumed or flew into a passion. There was no wrath or disputation or hasty touchy petulance. No grudges were held, and a wonderful friendliness existed throughout all their daily lives.

Even the men, although indulging in fist fights and the use of knives so freely, generally went about their quarrels calmly. They did not threaten and bluster or be loud and rowdy. If worse came to worst, they simply did what they were going to do and ended it. Whether the roof blew off or the crops failed, or the wood gave out or the rain spoiled a cherished plan, or sickness or accident or death, all was borne with quiet submissive acceptance.

I found out, too, at last, why the farm women were not the ones to stand admiring beautiful cloud effects or the artistic patterns of light and shade and color and harmony as the seasons came and went. Most of all this was kept from them by the dire necessity of viewing nature from another standpoint altogether. The changing cloud effects to them meant only worry about possible storms. The heavy rains or hail or snow were translated into terms of leaking roofs, scant clothing, and a daily concern to get dry wood for the fires. Dry weather or wet, wind or calm, meant the resulting loss or gain in growth to the everlasting cotton—and cotton meant their daily bread.

34. The Land of Can Happen

"TAIN'T A BIT LIKE JUNE." I heard that often. No, it was not a bit like June. Those five months since Christmas we had more rain than in all the three previous years. It rained heavily every day for nearly two weeks, and now the rain that was just over was a rain of eight hours, without stopping. Rain! Torrents of rain, with lightning and thunder and wind. We had hail and rain, and wind and rain, and cold and rain, and cloudy weather with wind. Hail in summer! I couldn't get used to it.

After a few days of quiet with overcast skies, it hailed again and rained a deluge and the wind blew a gale. It beat down the lettuce and spoiled much of the young melons and cotton plants. We were having the storms we used to prepare for the year before. The people all around were worried and disheartened.

I wrote a letter to my mother about this time beginning: "Your letter came saying you had received my stormy letter from the land of "Can Happen." Listen now to a tale of wind and rain and more wind and sand and rain. Rain that folks here call "gully-washers," "frog-stranglers," "clod-movers," and "chunk-lifters." How I hate the wind!

Over back of us, along the road by the Batsons', our cotton had to be replanted again. First the hail got it; then we replanted it, and a cold wave killed it; then we replanted it, and a "gully-washer" came and it just caked in that heavy soil. We thought it might be a total loss for it was "fixin'" to rain again, and that would prevent our turning the scratchers on it. Local rains would not matter so much, but if they come from further up the river …

Still, if we lost it all, we could still plant velvet beans and whippoorwill peas, Wayne told me.

The weather stayed cool. No one had cotton higher than six inches and on some of the best bottom land there was no stand at all. It was all to do over.

There was much that had been replanted three times. Of course, now the price should go up, but as there would be so little of it, it wouldn't do much good to raise the price.

The potatoes had come out so well after the last hail, and the lettuce was in fine heads, about one hundred of them. The 'Tucky Wonder beans were just stuck, and the green peas had taken a fresh start and were looking good. We had been having green onions and new beets and lettuce every day now. I saw that the peach and plum trees in the orchard were doing fine. Grandma kept her eye on them and expected to pick some fruit from her own trees this year.

Expecting a spell of dry weather and to insure moisture for the melon vines, we had planted soaked corncobs in the bottom of each hill. First it blew away the melon seed, which had been replanted after the first hail storm. Then it blew the very corncobs out of the hills, leveling the field that had been so carefully plowed and disked and checked for the planting. So now the melon would be planted for the third time. My mother was right when long ago she told Wayne he should be the leader of a lost cause. He was surely following his bent here.

Although Mr. West was not very strong, yet each day he shouldered his hoe and went out to the cotton fields. Mrs. West was an excellent worker and could hoe all day and keep up with the best "choppers" we had, but she, too, often had sick spells. One day Grandma asked me to go over and see how Mrs. West was, for she had not seen her going to work. I found her in bed, suffering from a severe chill. It was heartbreaking to see her big frame stretched out, convulsed with chills that swept over her, her face dripping with sweat and her gray hair stringy and wet. The bed would shake with the tremors of her body and then she would lie exhausted while the fever swept through her veins.

She asked me to give her some of her favorite medicine, 606, and a big tablespoonful of Black Draught, but quinine she refused to take. "Never had any use for it," she told me.

As it was nearing supper time I made up a fire in the cookstove and put some more chunks on the fire in the fireplace. Mrs. West asked me to dig down in the flour in the crock and get the biscuit dough she had buried there in the morning when she made the biscuits for breakfast. The dough was well covered with the flour and was still fresh and good, so I rolled it out and put the biscuits in the pan already for Mr. West to bake when he came in. This method of keeping the dough fresh even in hot weather was a handy trick that I found was practiced generally by many of the farm women.

Mrs. West told me that her daughter, Julie, and Mory, her son-in-law, were coming from the hills soon to work for Wayne in the cotton. They would live next door in the new house. We were glad to have them, for Mory was an exceptionally good worker, and the cotton was up now and all hands were needed in the fields for thinning and weeding.

One rainy evening, I remember, I had called in to chat with old Eric West. His eyesight was so poor now that there was little he could do except bring in the wood, and there he sat in his clean, faded blue overalls and chambray shirt, hunched over in the cane-bottom chair, with his hands slung down between his knees, watching the wet cottonwood log as it boiled and simmered in the fireplace.

I looked forward to these little visits with Mr. West, for more and more I sensed the warmth and kindness in his nature, and I always left with such a comfortable feeling that for days I could face all my problems with serenity.

I remember going out in the cotton field or the melon patch when Mr. West was hoeing near our house and walking along with him. He had a shy sweet humor that pleased me. He was proud of his ability as a speller and liked to have me challenge him with a word that thought might stump him. Then he would lean on the handle of his hoe and smile and blink his eyes as he carefully spelled out the word syllable by syllable. To my surprise he was usually correct.

"Jest, a-settin' here thinkin' of my old home in Missouri," he began, his little watery blue eyes blinking brightly as I seated myself near him. "Geneva, if you could have anything you wanted to eat, what would you have?"

"Well," I answered, "let's make a menu. Let's go right through a big dinner. What kind of soup would you like?"

"Like me some thick tomato soup; thick, pink, tomato soup like we always had back home."

"What kind of meat?"

"Well, guess you jest can't beat ham, now, can you? Great big juicy ham, baked brown. But say, I been a-studyin' about pie. Did you ever see real big gooseberries? I'd have a real gooseberry pie, like my mother used to make. I can taste it now. Real juicy gooseberry pie. Yep, that's all I'd want. Maybe a little more of that tomato soup."

Every day, year after year, since he'd lost his home out in the sticks, biscuits, cheap Peabury's coffee, "sow-belly," pinto beans, turnips, potatoes, buttermilk and corn bread. Butter, sometimes; home-made hominy, sometimes; sorghum molasses, sometimes; honey, one year, he said, when he kept a hive of bees.

Poor man, poor old Eric. He's dead now.

— · —

It was very hot, and the sun seemed to have an added brilliance, shining through the gathering clouds, and the humid air pressed down with a real weight on the lungs, but I was going to wash anyway.

I enjoyed my new sunbonnet. I made a split-board sunbonnet and a bungalow apron of pink-striped gingham. I was very proud of the bonnet, and Mr. West told me I looked very sweet in it. So there!

Jimmy and Donald wore straw hats and enjoyed going barefooted around home. There were too many poisonous stickers and sand-burrs and cockle-burrs for them to run bare-footed anywhere except near the house where we kept the ground swept clean and the weeds hoed down. The poison weeds caused terrible sores on their feet and legs; most of the children I saw all had bad scars and running sores on their legs, but they went everywhere bare-footed, for many had no shoes.

Two people in the bottom died with dysentery, so I heard. Dysentery was so weakening and they had so little strength anyway. Chicken pox, itch, hookworm and malaria were as common as dirt, and nobody did anything about it, for the doctors would not go to a share-cropper or cotton picker unless the landowner promised to pay the bill and many landlords would not do that. Of course, many of these people didn't believe in doctors, anyway.

There would be little fruit on account of the late frosts, but we had picked a few dewberries over by the river. It looked as though summer had come.

I read with interest about the May Day celebration at home. I used to follow those maps of auto trips in the newspapers, and constantly compared our California days with the life around me.

The children often said to me when I told them I was too tired to take part in some of their activities, "The trouble with you is that you need exercise. Now if you'd get out and run some every day, or come with us after the cows, or just come out and play ..."

I did go, often. I would go with them down through the Bermuda meadow to drive up the cows, but generally by the time my housework was all done I was simply exhausted. Although we were all feeling pretty well, still we tried to take our weekend doses of quinine, iron and strychnine regularly. Yet we just didn't have much pep. Some days we felt well and the next we didn't, with no apparent reason.

Jimmy and Donald kept at their lessons. They were doing good work in their reading, wrote fairly well, and were good spellers. They worked

hard over the letters they wrote to their Grandmother. They started selling melons to farmers who passed along the county road. They made
a big sign which attracted lots of attention, but the farmers paid only a
few cents for our finest melons. Often we sold fine cantaloupes for as
little as twenty cents a dozen. Jimmy still told me he was determined to
be rich.

Whenever I sat down to write to my mother I would thank God that
she had the disposition to appreciate sunsets, poetry, music and good
books. I decided that I would rather be as poor as a rat and be able to get
dry-drunk on a beautiful scene, or better yet the memory of one passed,
than to be owner of half the earth.

Oh, how I dreamed of summer in California, remembering the ocean
and the little town I knew so well, nestled in the rounded foothills with
the river winding through. I often woke at night and lay half-dozing, seeing that little town as we so often looked back at it from the wharf. Even
when I would be standing by my little low window looking out over the
cotton fields, I was really not there at all, for my mind was full of the
scenes I loved, as I thought of living again in California.

Julie and Mory arrived and moved their meager housekeeping outfit
into the new house. We found that Julie was what is called a "dope fiend."
That sounded exciting, but like most strange things, with better knowledge, the facts were often tiresome and dull. She did no field work, and
very little work around the house, but spent most of her time in bed. Why
she had ever started taking morphine, I did not ask. Grandma seems to
know all about it, but I never inquired into the facts. I know that one day
when she had kept to her bed all week, Mr. West begged Wayne to get
the doctor for her, and asked me if I would try to do something for her, as
she seemed in such misery.

When the doctor came he questioned her carefully, studied her bloat-
ed condition, and left with me a small vial of pills to be given to her so
sparingly that it was almost like torture to her. When Wayne left the house
he hid the little bottle so I would be able to withstand her pleas, and I
could honestly say I did not know where the bottle was. She could not get
anyone to stop work to go again for the doctor, so she had to endure what
she called our hard-heartedness, not realizing that we were only trying to
help her by cooperating with the doctor. But after two weeks of suffering Julie got Mory to borrow a team one Sunday and they went a-visiting
down in the lower bottom.

They were gone two days, and during that time we had one of the worst storms we had ever been through.

Never will we forget that afternoon when we noticed that the slate-colored clouds hung so flat and low, and then suddenly a wide splotch of clear pea-green sky appeared, as the heavens seemed to break apart. As we watched these changes in the sky we became alarmed, and Jimmy and I ran as fast as we could over the fields to attract Wayne's attention. We saw, however, that the men had already un-harnessed the mules in the fields and were hanging onto the reins while they raced up toward the big barn, so we ran back to the house.

We had put in the windows, brought in the wash tubs and the wash dish and had taken the clothes from the line, and were rushing out to fasten the big shed doors, when Wayne came running to the house. I could see that he was worried. Grandma, he said, was staying in her own house with Henry, but they, too, were expecting a big storm of some kind, yet they felt that they would be as safe there as down at our house.

Now all we could do was to go restlessly from window to window to watch the black rolling clouds. Several times we ventured out on the front porch, but the wind had a terrific force and the day had turned surprisingly cold. I noticed particularly that the clouds seemed so very low, much lower than I had ever seen them. Beneath, all the world look gray, yet there was a peculiar sharpness of outline to everything, and a sense of bigness to the trees and barns and houses, as though in some way they were magnified.

We put on our coats and some extra sweaters for warmth, and luckily we were inside when the hail began to fall. It came all at once, heavy chunks of ice that struck the roof like rocks with a steady bombardment; hail on the porch, the roof, and the ground, jumping and bouncing until it lay like a deep sago blanket everywhere, with wind-blown piles that had been carried with force in the driving gale and left in white heaps against the steps and the lower boards of the house.

We felt that the day had turned into night, for it was very, very cold. There was such a roaring as made speech impossible and the house shook and was blown so hard that it lifted one side from the pillars of rocks where it was perched, and seemed about to be blown over. There was nothing we could do, no other place to go, no move that could be made for further safety or comfort. We just stood together in the front room, against the inner wall, speechless; looking at each other and trying to see out through the darkened windows, feeling tense and expectant. No thunder or lightning this time. Just a big giant passing by, and we were waiting until he was gone.

Jimmy and Donald were as silent as we were. They showed no fear, but seemed to be impressed, too, with the realization that there was nothing we could do and nowhere we could go. We stayed close together and wherever one of use moved the others seemed to follow naturally.

At last we realized that the room was becoming lighter, the wind lessening. Finally we dared peer outside, and then we ventured out to look around. We were amazed at the big flat disks of ice, larger than dollars, and the irregular balls of ice like big marbles. Scars had been made on the north side of the house where the hundreds of hail stones had hit. We could see across the meadow that many trees were on the ground. The windows of the bedroom were broken and many of the shingles on the roof were split from top to bottom, and some holes made. Later we found young pigs with their eyes hurt and their sides skinned from the glassy ice, and the chickens that had been too late in scurrying to cover were dead in the barn yard.

We heard that up in town several men had their faces cut badly, for they had leaped from their wagons and stood at their mules' heads, holding them and calming them through the worst of the hail storm. Many who were caught out in the storm got under their wagons for refuge. Windows were broken everywhere, branches torn down, and small articles blown away. We had not been in the main part of the tornado. That had passed to the northeast and further down the river, where houses lay flattened and great trees were uprooted, and several people were killed outright. On some places sheds and barns were blown down and then the boards all blown away, so that there was little left to show that any buildings had ever been there. Yes, we were very fortunate.

When Julie and Mory returned she was apparently in good health again, laughing and pink-cheeked and happy. She opened her purse and showed me a big handful of pills that some doctor had been prevailed upon to give her.

Mory told us how awful the storm had been in the lower bottom. He said that on the day the tornado struck, the mules were in a pasture near the little one-room house where he and Julie were staying, but that he had hurried to put them inside a large barn nearby when he saw that the sky was growing blacker and blacker, and the wind was rising. The smell of sulphur was strong in the air, and the wind came in big puffs, dying quite away at times, and then suddenly sweeping through the trees and across the fields as though impelled by a terrific force. Then the whole world turned black around them. Only objects at hand could be seen. They saw that wherever the clouds broke, the sky beyond was a bright green, so they decided to get into the house and wait out the storm which seemed about to break above them.

They were no sooner inside the house than a blast of wind struck the side walls and they heard the wild beat of hail on the roof, and the house lifted and swayed. They lay flat down on the floor, up against the wall on the side from which the storm came, hoping that if the house blew over they would be sheltered to some extent. The walls of the house were made of wide boards running up and down from the roof to the ground, just as most of these poor little houses were built. These walls were held securely only at the floor and roof. Just when the tornado struck they did not know, for Mory said he and Julie cowered there with their hands over their heads, hoping and praying that they would not be killed.

They told me that it all seemed over in a short time. They staggered up and around and were struck by the big hail stones, some of which were lumps as large as walnuts and others large disks of ice with sharp edges. Yet their condition seemed of no importance when they saw the real danger they had been in. Just like a house of cards, the walls of the house were flattened—flattened and scattered far and near. The shingles of the roof could be seen far out in the fields around them. Trees were uprooted, brush torn away, the cotton ruined, the land flooded. The mules were unhurt, for the storm had just missed the big barn.

To my amazement Mory told us all this calmly. Then he added that had been through many such storms, but it looked as though this was going to be a bad year.

That hail put many holes in our new roof and as there were only three carpenters in town, I knew it would be a long time before I could get one to come. I was sure Wayne wouldn't have time to fix the roof himself. He was too busy replanting the corn, cotton and melons. Still, we had lots of cans and buckets to catch the leaks, and you couldn't buy any more shingles in town until they got some more in, anyway, they told me.

The little cotton that wasn't hailed to death was drowned. Then it turned freezing cold and killed most of it that remained.

Things were looking mighty "sorry." All the fine fields of cotton were either bare as a floor, or if any was growing it was only three or four inches high and full of weeds—"in the grass." Crops all over the state were late, though, so maybe ours would come out all right, if we got some dry weather, for our cotton looked better than the majority of crops around us.

Soon we had new potatoes and green peas from our garden, and I baked bread regularly again. Our horizon was limited to our work, the weather, and our thoughts.

I saved a copy of the local paper with the piece marked where Wayne and I went visiting one Sunday. We only walked across the fields the back way to a house by the gin, to try to get a man to come and fix the roofs, but it was all news here.

We rented a tractor to break up the cotton land over by the river, to plant peas, which should be put in soon after the rain. The tractor proved too heavy for the sandy land up above the bank, however, and we had difficulty getting it out of the fields so Wayne was still "turning 'em round." Out in the melon patch I saw six men working, scratching and hoeing, and two teams plodding up and down the long rows. Last year we were comparing melons with Donald's head by this time. Now the vines were only just in blossom.

Finally Mr. and Mrs. West left the farm to go back to the hills again to live with their son, Parse. Mr. West grew so weak that he had to quit working altogether, so they gave up the house, and we bade them a tearful farewell.

Julie and Mory were separating. He went somewhere down in the lower bottom land, and she went along with her parents, giving up the "new house" also. So there we were, with work for ten men crying to be done. Wayne and Henry had been out searching for hands without success. Then one day Mory returned, bringing with him the Bower family, who would board him. He was to sleep in the "new house," so now we had at least three good hands to work again. Cotton chopping would go to two dollars or two-and-a-half a day, we heard. Every cotton picker was hanging on to his job; the game of "Pussy Wants a Corner" was temporarily halted.

One night during the last storm period the thunder and lightning seemed much worse than any we had yet experienced. The boys were in bed asleep and Wayne and I had settled down in our rocking chairs for an evening of reading. Our evenings were so quiet that reading *was* reading. A big fire roared in the heater, the lamps were shining, and everything seemed snug and comfortable. I was trying to ignore the terrific crashes of thunder that grew steadily worse and rolled overhead with a force and vibration that shook the earth beneath. The lightning, too, was piercing bright and the wind was a constant fury, as though the air were being churned in every direction, not just blown with force against the house.

I began to lose interest in my book and was becoming very uneasy and tensed myself against the awful ripping sounds that preceded each terrible crash of thunder. Suddenly, I found myself across the room, down on the floor—not hurt, but rather stunned. To my astonishment, Wayne still sat reading. He had not noticed that anything had happened, and was quite surprised to see me get up off the floor, when he thought I was still sitting there reading the same as he. I never knew just what happened. I had no consciousness of being thrown down, yet the next day when we saw the long white gash down the side of the big pecan tree nearest the house, we knew that the tree had been struck by lightning and I had been knocked out of my chair and thrown to the floor at the same time. From then on I took little pleasure even in the heat lightning that had been so intriguing.

No matter how I tried to find comfort and peace in the seclusion of our little house away from the goings and comings of the various sharecroppers and cotton pickers, yet it seemed that it was impossible to keep myself entirely separated from these people. They represented problems and situations that I had never known existed, many of which baffled me, worried me, and sickened me. For instance, the new family that Mory had brought back with him. They were Bowers. Mr. Bower was the brother of Old Hames. There was Mrs. Bower and four children—a boy of seventeen, another about six years old, a little girl of four, and a baby of about nine months. Mrs. Bower took the baby with her when she went to the fields, carrying it back forth on her hip and leaving it on an old quilt between the rows while she worked. The little boy always followed along with his father. That baby was so tanned that it was about the same color as the ground. The skin on its body was tough and hard. I took care of it one afternoon when the mother was too ill to do any work. It was so dirty that I gave it a bath and took all of its clothes, bit by bit, and washed and dried them. Its little head was covered with sores, with the hair sticking to them. I had to actually conquer the feeling of revulsion I had when I looked at that filthy baby. At last, when it was clean and dry I rocked it to sleep in my arms, and sang a lullaby to it. Poor little baby.

The little boy followed his father and brother around the farm wherever they went. So—that left Zeddie.

Poor, wandering little Zeddie, with golden-red hair, pansy-blue eyes and fascinating freckles across her nose. An adorable plump little Peter Pan of the cotton fields. She was the most perfectly formed little child I have ever seen, but was sadly neglected by her ignorant father and a more ignorant mother, who was kept from even the ordinary maternal efforts toward cleanliness and care by a severe heart ailment.

Poor, neglected little Zeddie, born to poverty and dirt. Named in the hope that, like the last letter of the alphabet, there would be none to follow after. Someone had told the mother to call her Zeddie; maybe a doctor, for the doctors were always called on to supply names for the babies they attended.

Zeddie's rounded, ivory-tinted body was always grimed with dirt. Yet she was beautiful, from the top of her golden head to her perfectly shaped feet, as exquisitely shaded as some statuette lovingly created by a proud sculptor.

She wore only one dirty cotton slip, until Grandma, filled with pity, made little slips and panties and gowns for her and tried to get her mother to keep Zeddie decently covered. Grandma would often take Zeddie into her house and bathe her and dress her, comb out her silky hair and find a pretty ribbon for it, knowing all the time that her efforts would have no permanent effect. Zeddie would go back to the house across the road, carrying cake or a fried-pie or a large soda-biscuit filled with butter and jelly, only to wander again, aimlessly, when her father and mother left for the fields. Her older brother worked in the fields with his parents and only came back to the house to eat or sleep.

Zeddie could be found anywhere on the farm idling about among the field hands. She was given no training, no attention whatever, and performed her natural functions like any farm animal wherever necessity found her.

Wayne told me that once he met Zeddie wandering along with a beautiful smile on her face, munching a big chunk of popcorn-crisp that someone had given her. The popcorn, her dress, her face and hands, were all smeared with filth and the odor was nauseating, but of all this she was blissfully unaware. Big green flies, he said, buzzed around her, and crawled on the popcorn and on her face and hands. It was sickening to see her, Wayne said. He arranged to have the mother stay out of the fields so she could take care of her child. But that didn't help any. Although Mrs. Bower gained some relief from hoeing and had time to wash clothes oftener, her husband had her split all the wood now, since she wasn't making a hand in the fields. Mr. Bower could not believe that his poor wife was slowly dying of heart disease. He suggested that Mrs. Bower take over the job of milking the two cows, so they could repay their share of the milk. She was not working now, he said, and he saw no reason why she should spend her days in idleness. "Women folk were allus ailin'," he said. Grandma supplied them with milk and eggs, and Jimmy and Donald continued to milk the cows. I'm sure Mrs. Bower would have received more consideration if she had walked with crutches or had her arm in a sling. Something one could see!

Zeddie, of course, went about in her usual way in spite of all our entreaties and protests.

Now, in later years, as I think back on this poor child's unconscious accumulation of filth, I have coined the word "Zeddyism." Though her little body was smeared with filth, that was a condition due to ignorance and neglect. Now I am continually meeting men and women who relish mental filth far worse than any of the dirt found on little Zeddie. These are people of superior advantages and of seeming culture and refinement, who will, I find, for the entertainment of their unprotesting friends, be always on the alert to insert into any conversation, as a contribution of wit and humor, the sly use of the double entendre. They evidently do not understand the clean impersonal attitude of a doctor or a nurse toward the functions of the body, and consider it rather clever to descend even to the level of lewdness in their repartee of jokes alluding to love, courtship, marriage and childbearing. That pasture in which obscenity, vulgarity, immodesty and indelicacy grow rank, is one into which they seem glad to be the first to enter, providing they can be sure of blaming someone else for leaving the fences down.

My wonder has increased more and more at the number of people I find who, though basing all their own life's happiness on sentiments and emotions so closely related to the human body, yet can so readily thrust aside, with ribaldry and insinuation, all the true sacredness and beauty of their own physical beings, and relish in their jokes and stories a disgusting and distorted filthiness of mind. Apparently these people all have one common agreement: To them there is no such thing as romantic love.

All I can say of them, judging them by their own words, is that if they lack knowledge or experience of real romance, if real love has passed them by, it is because they have never deserved it.

— . —

Once again the clouds looked so very threatening that I feared it might be more than just an ordinary storm this time. I called Jimmy and Donald and together we set the pans and pails to catch the rain again, moved the beds into the middle of the bedroom, put away all the photographs, covered the phonograph, and brought in the cat and kittens. All of this was what the children call our "sailor act." Wayne and Henry were in town. We put on our coats and caps and locked the door and "took out" for Grandma's, to get her to go with us over to Mrs. Dawson's house, for we wanted to be near her big storm cellar.

It had been sprinkling off and on, and the sandy road was damp. When we passed Grandma's house we saw her fresh tracks leading around the barn and knew that she had gone to visit Mrs. Dawson, too. Grandma would never admit her fears, but she felt safer, we knew, on days like this, when she was near a storm cellar, too. Sure enough, there she sat, rocking and knitting and talking to our neighbor. They hoped the storm would not be severe enough to drive us into the cellar, although as we sat in the front room the storm seemed to be increasing. One of Mrs. Dawson's girls came in and said that they had seen a big snake out by the cellar and were going to kill it, so Mrs. Dawson went out and soon called me to come. There she was down in the cellar with a hoe clamped on the snake's tail, holding it securely. It was a big water moccasin. I was asked to get another hoe and see if I could cut the snake in pieces while she hung on and tried to drag it out of the hole down which it was trying to disappear. Between us we dragged it out and hacked it to pieces.

That snake had been in the storm cellar hunting hens' eggs. The hens laid there because the door was left open to air the place and it was cool and shady and made an ideal nesting place on a hot day.

Soon we heard the deep rumble of thunder, and the storm finally broke. The seven of us decided to take refuge. We had a lamp lit, and all sat, with our feet dangling, on the mud shelf that ran along the wall of the cellar on three sides. I noticed that there were a good many snake holes, so I felt very nervous, and could not be at ease as the others were, even though the elements raged outside. But I found that much can be endured especially when there is no choice, and my observations of what could happen in these so-casual storms made me glad to be down in that storm cellar, snake or no snake. I was now more anxious than ever to have a storm cellar of our own.

Wayne was bending all his energies to making all the crops a success, and it began to look as though they would turn out that way. He had all the farm in excellent condition; the best it had ever been, people told him. He was up at four in the morning, and came in just to eat and back again to the fields until after dark, and then to bed, dog-tired. He planted peanuts, both the Spanish and large jumbo, and a strip a quarter of a mile long was planted to honeydew and Rocky Ford melons. Whippoorwill peas, beets, turnips, parsnips and corn were all planted, too.

We were planning to get a day off to go up on Nebo, but each day brought some pressing work, or maybe it would rain, but we intended to go out in the low hills back of Dardanelle to pick wild blackberries anyway. There was a big pasture five miles away quite heavy with them.

Another large bundle of magazines arrived from my mother. I was glad it was raining for we could get a chance to enjoy them. There were so many splendid articles in *The American Magazine* that would almost convince one that with enough self-confidence and determination snowballs could be sold in Alaska.

At last we put up a mailbox just for our own mail, and our letters could go and come with safety. I was tired of having news on my cards reported to me before I got to read them myself. This new purchase surprised some of our neighbors who saw no necessity for that outlay of money.

There is a verse beginning: "Summer is coming! How do I know?" Well, I learned to finish that for myself. Because there was a bird building a nest in the locust tree on the bank near the house, and because the dewberries were ripening fast over by the river. We could not get over to the river banks though, for the mud and water were still over the meadow.

We enjoyed our music with every meal. We were accused of having a lot of high-brow music that nobody could understand, but I think the understanding of music is in one's own soul; if a certain appreciation is lacking, all the training in the world cannot supply it.

So far there had been only a few really hot days. It looked for a while as though we were going to have a drought, but finally it rained. The alfalfa was cut and the oats cut and hauled to the hayshed. One Saturday I went into town to visit my old neighbors. It seemed strange indeed to get away after four months of steady farm life. Jimmy and Donald wanted me to tell my mother when I wrote that the corn was higher than Wayne's head, the gourd vines which they had planted outside the dining room window were growing up over the roof, and we had eight new kittens. Wayne could be heard below the hill, shouting. "Get up, Joe." He was running the scratcher through the cantaloupes and casaba patch, for soon they would be ripening fast. Those generally raised were the Halbert Honeys, Kleckley Sweets and Rattlesnakes. There was a very delicious yellow-meated melon, too, but it was not raised in any quantity for it was not good for shipping.

We were still baling hay, but decided to go into town for the barbecue and public speaking at the school. They had put up booths and expected a big crowd. The Governor would speak and the District Attorney and the Mayor. About what? Why, against the good road program for which someone had been agitating. And they called this part of the United States!

The army worm was beginning to infest the fields, eating the leaves of the cotton, so there would be no top crop. That meant no second picking. When the cotton worm was working, a distinct odor like that of orange blossoms filled the air. But the farmers, instead of enjoying the heavenly fragrance, only groaned as they realized that their crops were being de-stroyed.

Doctors were already predicting a sickly winter on account of the ex-cessive amounts of water that spring. It was so very sultry that I had already begun to break out with prickly heat.

After all our efforts to have a beautiful flower garden, we brought in the one lone nasturtium bloom. We had planted so many of them. It's orange colored. Fancy treasuring one blossom. But the grape vines were beginning to twine on the fence and the cannas made a lush growth in one corner of the garden. A few of the sweet alyssum plants survived the blowing sand, but the blossoms were very scanty. Pale azure morn-ing glories and others of a delicate pink climbed on the window screen and ten golden poppies were in bloom. The four-o'clocks seemed hardy plants; they were growing thickly with their shiny leaves and filled all the space where the petunias should have been. Right in the middle of the garden grew three sunflowers, tall and straight, and scattered among the other plants were the zinnias, red and buff and orange, springing up in abundance.

Grandma had a beautiful plant in her garden—the cockscomb. That crimson plush-like flower grew on tall stalks, soft and velvety. The cockscomb grew especially luxuriant over at the Negro cabins by the bayou. The Negroes raised gourds there too, and they covered the fences and sheds and almost hid the tiny cabins. The gourd was a very common vine, yet here in the sandy soil even that failed to grow well. Still we were determined that we would yet have a garden spot full of beautiful flowers, and we spent many hours weeding and watering and hoeing.

35. All-Day Singin'

I HAD AN OPPORTUNITY at last to take part in an "all-day-singing-and-dinner-on-the-ground" affair. These singings took place in early summer all over the county after the schools were out. Usually everybody who could possibly do so brought a big picnic lunch and spent the whole day at a church or schoolhouse.

In the bottom, word was sent around that the singing would be held down the road in the schoolhouse for those who lived in our district, and on the following Sunday the singing would be held in the chapel, along the river road halfway to town, where dinner would be spread for everybody.

The first Sunday they didn't plan to have a picnic, so the singing began about one o'clock. The room was crowded, with boys even perched on the sills of the open windows.

There was a visiting singing-teacher in charge, and much of the singing was done by competing groups from different districts.

All sang by note and, though no prizes were given, a committee decided which groups won. They competed in harmony, sight reading, and the singing of four-part songs. Few solos were sung. All the songs were from the regular hymnals, and all types of religious songs were sung— "Mother's Gone," "The Wandering Boy," funeral dirges, and revival songs, one after the other. I knew some of them, but the others sounded unfamiliar and very sad.

The young people who attended Sunday-school joined in these regular singings, but, as always, there was a group of children who hadn't been long in the district, or who were thought too young to take part. Some of these children, I had been told, longed to have some part in the program, too.

One of the mothers asked me if I could coach these extra children so they could at least sing one song. At first I refused, as I had had no training for either singing or teaching, but at last the pleas of a certain mother decided me, and for several days I drove the Ford down the river road and collected the boys and girls and brought them up to the school. I never saw more eager, earnest children. I found it was impossible to teach them one of the regular songs from the hymn book for there were none in the group who could be depended on to remember the words. Finally I found that some of them knew the tune of the chorus to "Brighten the Corner Where You Are," so that song was decided upon. I wrote out copies of one verse and the chorus for each child, and how they enjoyed the practice. They were to repeat the chorus several times while I played the organ by ear. When at last the day for the singing came and their number was called, their parents were the proudest people I ever saw. The children scrambled forward, and I went to the organ with my heart beating wildly. But, oh, the applause we received, I can hear it yet.

I couldn't see the audience from where I sat, and I was very busy playing as loudly as I could to hide the places in the tune where the children forgot the words, while I struggled to make the old pedals supply enough air for the bellows. I met several mothers later and they thanked me profusely for being so good to their children. For them it was a grand event.

While I sat there, I smiled to myself to see the big circle I'd drawn with chalk on the rough floor, back of which the children, all scrubbed and starched, were now lined up in a stiff straight line to sing the verse. That over, with evident freedom and relief they turned abruptly, just as they had been taught, and circled around and around on the chalk circle, as they lustily sang the better-remembered words of the chorus, "Someone far from harbour you may guide across the bar; Brighten the corner where you are." They stamped their feet in unison to accent the rhythm, their eyes shining and a broad smile of pleasure on each face as they marched 'round and 'round and 'round.

No, I'll never forget that "all-day-singing" in the bottom.

Quite different was the concert that was given in the schoolhouse some weeks later. There were about twenty-five Negro men and women on the platform, all members of the "colored" church that lay back by the bayou. They were regular tenant farmers, most of them living on Meek Loupe's farm. A collection was taken up, the money to be used to buy regular uniforms, for they planned to travel around the country, singing, to raise money for their church, as they would not be needed in the cotton until time for picking.

A few of the men wore regular suits, but most them had on overalls with clean white shirts. The women, some tall and thin, some short and fat and shining, all wore percale or gingham house dresses. I caught myself watching closely to see if the heavy bass voices would come from the short fat singers, and if the sopranos would all be the tall thin women, but was lost in admiration of their white flashing teeth and gleaming smiles.

When they rose to sing, all this was forgotten. They used no books, and were not accompanied with music, but they gave us an hour of the most wonderful harmony I have ever heard.

Some of the songs were sung by the entire group, others were solos or duets. Such humming—four parts being taken; such heavenly tones, such deep, bell-like throbbing cadences that underlay the pulsing tunes. There seemed to be no apparent leader. Everyone was so well practiced that without hesitation song succeeded song, and the schoolhouse echoed to the spirituals and chants and the rhythmic harmony of lullabies and lament, all so well balanced with the sustained throbbing bass and the clear joyousness of the soprano as they picked up the air, that their voices seemed to meet and melt together. Of course they sang "Swing Low, Sweet Chariot," but the most interesting songs of all and the ones that called for a wonderful exhibition of coordination and exact harmony were "We Are Climbing Jacob's Ladder" and "Ezekiel Saw the Wheel." Never will I forget that wonderful evening of music.

36. Stringtown

AGAIN THE RIVER was high, just as it had been in June. We were told this was the highest fall rise in thirty years. Now it was raining again, but we put on our boots and coats and rounded up all the stock from the pasture.

Latest news! Still raining in Oklahoma. Some of the people referred to these days as the "time of trouble," the time predicted in the Bible for the last days. I knew that when I wrote to my mother she would say that these strenuous days of ours reminded her of the plagues of Egypt. "Let my people go."

The latest report from town was that the river was up twenty-four feet, and two more were expected. Down by the line fence near the mailbox the men were building up a dam; eight men with scrapers and slips were at work, with the water already six inches from the top of it. The river was already overflowing on the cotton across the meadows where we walked last summer and a tongue of water from farther down past our farm had already backed up to our line fence in several low places.

If the dam would only hold, it would save the alfalfa below the house. The pasture land further down the river, which we had rented for hay, was already under twenty feet of water. That would have yielded over a thousand bales of hay.

One afternoon we all walked down the road to the big new county levee and walked along its top to the old levee near the Mill Creek dam. The Arkansas River was running back into Mill Creek, but was still a long way from the new levee.

Then we heard that the river was up to the new county levee. If that didn't hold, they would be taking them out in boats down there, so every man was there, working, trying to hold back the water.

It was still raining, but soon everyone on our farm would be called out to drive the mules and cattle across to this side of the farm. They had been grazing on the meadow by the river. It seemed that we were now getting the benefit of those cloudbursts and tornadoes in Colorado and Kansas that the papers had been talking about.

I could see the meadow already glistening with water in every low place. Everyone on the farm went out after dark and ran the sows and pigs up from the meadow and fastened them up in the barn lot. I was afraid I would find myself sitting in the hammock watching the water come up over our garden.

Finally we decided to go down to the big vegetable garden and save all we could. We pulled up the beets and onions and dug most of the new potatoes, for it seemed now that the garden would be covered with water. The cotton planted in the strip by the meadow was finally submerged, and the water on the meadow itself stood several feet deep. This was back-water, for the pond overflowed and more water from lower down the river broke over the road bed and backed up toward the alfalfa below the bank.

All around us the same conditions prevailed. We were not the only ones, but that was poor consolation. They were looking for thirty feet of water in town. That would be the highest rise in years, although they always predicted higher water than they ever got. The men began to build up a dam at the corner down where the two county roads met by the mailbox and where the banks were low. The scraper and ditcher piled the loose soil higher and higher to hold the water a little longer off the alfalfa. It had been recently cut, although really too wet for cutting, and the mower had broken and time out had to be taken for a trip to town to get the mower repaired. It had been cut in the hope of saving the roots beneath it, for even though the alfalfa stood under three feet of water, if it went down soon enough the new growth would live.

All the excitement and uncertainty made me very nervous, but Wayne just ate it up. At last the dam was finished, and it held, so the garden was saved, for we had time to get the water drained away. Then the water began to recede and although it looked like there would still be more rain, Wayne began to plant pinto beans in the corn skips. We got most of the water drained from the alfalfa, yet a deep strip of water still stood below the bank. I caught a small fish there one afternoon.

Jimmy and Donald went along with Wayne and Henry and the other men over toward the bayou to catch fish in a small lake. Just as the Hawaiians do, a weighted net was thrown out in the water and slowly dragged shoreward, bringing in perch and carp and cat. I was told that the Negroes would be having "fish fries" over there, too. A group of them would go together to fish and cook the fish where they were caught, with corn bread made in skillets and lots of coffee, all eating together with talk and laughter. The Negroes always seemed to add happiness to any occasion, and they were extremely fond of fish. The overflow had brought in big fish in quantities. The men also caught fish in the meadow with a long-handled, two-pronged fork called a "gig." The fish there were left in the low places when the water went down.

It was no wonder we loved our garden. Turnips sweet and tender, carrots full-flavored and deep orange in color, large beef-steak tomatoes, and fine tender beets, besides the luscious melons. The ground was completely covered where the cucumbers grew and they were a daily treat. New to me were the whippoorwill peas and the black-eyed peas. The whippoorwills grew on low bushes in pods that were often six inches long. The peas were nearly round. We shelled them out green or let them dry for shelling later. The black-eyed peas were a larger, longer pea, with a black spot on the side. When fully ripe these had to be carefully handled for the pods would burst and the peas scatter. Cooked with a ham bone and onion and seasoned with salt and pepper, the liquid would thicken, and the flavor of either kind of pea was delicious. Butter beans and small lima beans were cooked the same way. The Negroes by the bayou seemed to raise lots of these beans, for they would plant them along their fences and in the rich black soil they grew wonderfully well.

Weeds sprang up everywhere, for the sun shone now every day. The men went back to their work cheerfully. I washed and ironed and we worked out in the garden. We felt cheered to see the bright sunsets again. "Red at night—the sailors' delight."

We still floated around nights in our gowns putting in the windows and taking them all out again. It hadn't rained now for about three weeks, only pretended it was going to.

During the overflow the water washed down a fine row boat with oars in it. Now we could go rowing over the Bermuda meadow, under the cottonwoods where there was still one large body of water by the rise. We went rowing one Sunday evening just when the sun was setting. The air was cool and the katydids were beginning to tune up, and the gnats and mosquitoes and bats were beginning to fly. I never saw so many bats.

About this time I made up some rhymes to tease Wayne:

AIN'T MADE NOTHIN' YET

Oh, he planted lots of cotton,
He worked it in so well;
The wind it blew, the rain came down,
You should see the hail that fell!

He built a good strong levee,
He built it long and wide,
But the Arkansas River backed right up,
The children they both cried.

It flowed out on the cotton,
It stood there eight feet deep,
It drowned out all the meadow,
And Wayne he could not sleep.

> *Chorus:*
> *And he ain't made nothin' yet,*
> *He ain't made nothin' yet,*
> *He came to Arkansaw for to get rich,*
> *And he ain't made nothin' yet.*

Wayne asked me to stick to "Whispering Hope." He said whenever he heard me singing that old song when he was out plowing, he felt better for he knew then that I was happy.

One evening in spite of the overflow, we drove down the river road a ways to cool off, for the moving car created a breeze. The oppressive heat had lasted all day, and although it was now after eight o'clock, it was still too hot to sleep.

Lights ahead of us at one side of the road attracted our attention and we drew up alongside a platform where an outdoor dance was in progress.

This platform was about three feet off the ground. Several lanterns swung out on long poles overhead. The usual cedar bucket and gourd-dipper hung under the big cottonwood close by. Three boys were sitting on split-bottomed chairs borrowed from a nearby house, one playing a banjo, the other guitars. Their fingers flew. The music was little more than varied chording, but the heavy "plunk-plunk" and "tink-tink" led the crowd of dancers through the intricate sets that the musicians were calling off.

> *Bow to your partners,*
> *All hands round—*
> *Back again and do-si-do.*

It all seemed so strange, out in the warm night air. A stretch of the overflow water lay close under the bank just beyond the road here, and in the pauses of the music the frogs could be heard shrilling in high excitement. Fireflies soared back and forth, away from the lights, and mosquitoes sang around our ears, where we sat in the semi-darkness watching.

It was just a crowd of cotton pickers from Stringtown. I had seen Stringtown. It was well named. It was a row of ten one-room shanties built beside the road on Cossey's place, to house the itinerant choppers and pickers in the spring and fall months. These tiny houses were of new rough lumber, with no fences, no yards, no porches, no shade, and just one water faucet on a pipe out behind the last house. I saw some broken chairs and empty galvanized tubs lying on the ground between the houses. Rubbish everywhere, tin cans and old rags. Hoes leaned against each house. A bit of barbed wire was strung along behind for a clothes line, and one large black pot lay on its side by the pile of ashes. No toilets, for the bank across the road was near enough.

The girls sat on wash benches placed around the edge of the platform, their heads shapeless with thick Mary Jane bobs that were held back in tight bunches with heavy barrettes. That was the new style. Vivid rouge and lipstick showed up plainly against skins that were tanned or freckled. All wore cotton dresses, long waisted and stiffly starched, with carefully pleated skirts. The men wore blue overalls and shirts and heavy shoes. Their hair was combed back from their foreheads, shiny with brilliantine. Somehow none of these people had the same appearance as the boys and girls I used to know, and I felt myself so alien to them.

Jimmy and Donald watched the performance with big-eyed wonder. It all seemed so strange and so apart from our lives. They were all big fellows up on the platform, and I was glad that my boys were as yet too young to even think of taking part in the dancing. They seemed quite content to watch, and expressed no thoughts of being left out.

I was struck by the rough appearance of some of the men and boys, their unshaven faces, the glitter in their eyes, and the look of tense determination on each jaw. Wayne kept explaining to me the social customs of these dances. No fooling here. There was an undercurrent of accepted procedure—"I'll dance with your girl while you dance with mine"—but Lord help the man who took a man's girl out of turn or kept her for more than the allotted dance. That's when the fights started. They weren't so concerned about their "main women" as to find a supposedly justifiable excuse for a quarrel. Every man had a sharp, long-bladed knife in his pocket, and a few of them carried guns. Those, I was told, were the ones with the wide leather belts, the bandannas around their necks, and the narrow-toed, high-heeled Texas cowboy boots. A bit of playing to the gallery, that was, but it was a costume that was very eye-taking and set a man apart in a most gratifying way, especially if he were good-looking. Those cowboy boots were the one thing that every young fellow coveted above all else, though few of them ever had a chance to ride a horse.

Wayne said that always a gun and better boots than ordinary were worn with a swagger by the men who sported the big Texas hats, and when these fellows attended a dance, the timid folk went home early, for the affair invariably ended in a fight.

In Dardanelle, on Mondays, many were the arms seen in splints, and many the heads wrapped in bandages. Black eyes were common. Nothing was said about it; everyone knew that the public Saturday night dances, in the bottoms and out in the hills, left the victims of the gun wounds and knife gashes to the care of the doctors.

No, we didn't dance. We sat and fought off the mosquitoes, dreading the extra doses of quinine we must take to pay for this indiscretion, and we were soon glad to get back home again.

It was hard for me to understand why so many families were found each year to fill up such places as Stringtown. It was hard to realize that there were hundreds of families who moved yearly from place to place always hoping for better conditions. They generally had one mule hitched to a light wagon, or several families would depend upon one man with a team of mules to haul them from place to place. But no matter how long they searched they always ended in some sag-roofed share-cropper's shanty, an unpainted unlovely hovel.

37. Malaria

AT LAST I "CAME DOWN" with malaria fever, not intermittent fever and chills, but real malaria with a fever at 103°. I was quite ill for a week. Wayne spent nearly all the week taking care of me and doing the house-work.

I became ill very suddenly. After spending hours in a semi-stupor I managed to get to the back door and ring the dinner bell. I was too weak to blow the horn which we kept for signals of distress. Wayne came hurrying from the field and when he saw my condition he helped me to dress and we drove into town. We found that our doctor was away so we went upstairs over a store where a strange doctor examined me, and informed me that I had typhoid fever. At the time I felt too ill and stupid to realize the odd procedure of his examination and I suppose that Wayne was too upset to consider carefully what was going on. The doctor scrutinized carefully two or three tiny red marks he found on my stomach and said they certainly indicated typhoid fever. Naturally we believed him. I was ordered to hurry home, cut my hair, keep an ice pack on my head, and take the medicine he prescribed. We got the ice and the ice cap and the medicine. I looked with pleasure on the lemons and oranges, for my fever was high, but I looked with distrust at the large bottle of black medicine. I began to think that I did not have typhoid fever, and I decided not to cut my hair.

That night I was out of my head and the next afternoon my fever rose high again and stayed up, and I was often out of my head. Henry, who was in town that day, met our doctor, who had just returned, and told him how ill I was. When the doctor came, he said I had real malaria fever. He gave directions for my care and left medicine for me. When Wayne told him of our visit to the new doctor we were told that this man was one of a school of quack doctors who were already on their way out. They had hood-winked people all over the country, but were at last being forced to discontinue their practice. Many of them had already been prosecuted. They had all been doing a flourishing business. That doctor we had visited, and his partner, had their offices up over the main street of town and had been posing as specialists in women's diseases. They had performed many operations right there in their offices. Their diplomas had come from what were called, at the time of their exposure, diploma mills.

Even in my sick condition, I couldn't help but notice the entire lack of animosity or even anger on the part of our doctor. The Southern touch!

Our doctor added greatly to our growing knowledge about malaria, a knowledge we had been gaining chiefly by painful experience. He said he found it hard to get some of the people to believe the facts about the tropical disease, or to persuade them to cooperate with him in ridding their own farms and homes of the mosquitoes. He admitted, however, the great difficulties entailed in the struggle for control in this scourge that was so widespread and so deadly.

So many still clung to the old beliefs that the low-lying mists, the miasma from the swamps, and the night air, were the cause of the malaria, rather than the mosquitoes that bred there. Ague, jungle fever, intermittent fever were all forms of the disease which was known to us generally in the chronic form, where the parasites destroyed the red blood corpuscles, causing the liver to be affected and producing a condition of anemia. Severe headaches, fever, aching of the bones, loss of appetite, nausea and dullness were all common symptoms.

If the parasites became so thick in the blood stream that they clogged the capillaries of the brain or the kidneys, death generally resulted. Even quinine was no prophylactic, although it was the best remedy known. Sometimes it failed to kill the parasites in some forms and stages of their development.

Our doctor told us the long struggle against malaria since the end of the nineteenth century, when it had been definitely accepted that the Anopheles mosquito, the female, carried the malaria parasites. This mosquito, easily recognized by the black dots on its wings and its habit of apparently standing on its head when attacking its victims, lived only about one month. Its bite did not cause severe itching nor large swelling, but one bite was enough to produce the disease.

We discussed ways of eliminating the mosquito. I learned that the mosquitoes could not live in sun or wind; they needed a quiet sheltered place in which to hide during the day. Beneath buildings, under bushes or in grass and weeds they could be found; even beneath the over-hanging banks of the flowing waters. The larvae or wrigglers could develop in any quiet standing water. Ditches and swamps as well as the usual empty vessels found scattered around barnyards and ill-kept home yards made excellent breeding places.

The doctor advised the use of oil to rid the farm of mosquitoes, as it would spread quickly on the surface of water; or Paris Green could be mixed with dry dust and scattered even on rain barrels since it was harmless but efficient if used in very small quantities.

Anyway, he said, he had been fighting malaria all his life, and since he saw little hope of completely eradicating it, he would probably be filling capsules with quinine until he died.

I got stronger at last when the fever left me, and soon felt it had been only one more thing to contend with. Wayne went back to the fields and the job of loading the melons to sell was now begun. Five wagonloads were picked at one time. The cotton was being picked down in the black lands but not near us. We would get only about a bale to the acre, owing to the weevils and worms.

Wayne made an excellent nurse. One night while the fever still raged he played "Meditation" over and over and over for me until I feel asleep for the night. It surely was heaven. I enjoyed all of our music so much at this time. John McCormack's Irish voice singing "Molly Brannigan" pleased me especially. One night Wayne carried me out in his arms and I lay for a long time in the hammock enjoying the balmy air. The same old moon was coming up, glistening over the pecan trees, and a warm breeze was blowing. There were a few little times like that that were so fine. Jimmy and Donald kept bringing me little bouquets from the garden and were so eager to do what they could for me.

Jimmy and Donald, although cousins, thought and acted like brothers, for they had been together since they were two years old. Jimmy was a little taller and a little stronger than Donald, and had a more aggressive attitude. I was glad to see that he usually took a protective attitude toward Donald, for when they were together I knew that Donald would cheerfully follow where Jimmy led, and I knew, too, that Jimmy seemed to have a wisdom far beyond his years.

As I listened to "Kreisler's Waltz" and "Russian Valse," it seemed as though I couldn't be the same woman who had danced alone in the moonlight.

Maybe I didn't take enough quinine. It's hard to take it and be made to feel so miserable from its effects when one feels well otherwise. It always caused my ears to ring and made me very nervous. I began hoping that this would be our last year on the farm.

At this time there were many stories in the *Saturday Evening Post* about Arkansas. They were so true to life, but we felt many readers might think them overdrawn.

Mrs. Barger had been insistent upon helping us during my illness, and as Wayne was really too busy to do much cooking, she often came in at meal times to help. We gave her several chickens, some to keep for herself and some to cook for us, so she appeared often with a big smile and one of the chickens boiled in a kettle of greasy soup with all the fat of the hen left in, all thickened with corn meal. A greasy mess!

While I was recuperating, the afternoons seemed so very long, and it was often a trial to live through them. I had too much time to think. I decided that a person had either to be a genius or a fool to stand Southern conditions and smile. But there's nothing like work to make one forget troubles and worries.

My little house again looked nice to me. Everything in order and all so clean and neat. I began hemming some sheets by hand, for the machine tired me so, and I got the mending done and again felt able to bake bread. We were enjoying green beans and cabbage and new potatoes from the garden and our own meat from the smoke-house.

On the Fourth, Grandma and Wayne and the boys and I went up to

Nebo and took a picnic dinner. We boiled our coffee near the big rock by
Lover's Leap. We took one hammock along, and enjoyed the cool breezes
beneath the trees and the view over the valley or toward the higher hills,
for I was still weak from my siege of malaria.

We gathered low-bush huckleberries, too, and ferns, and visited the
lower spring on the "bench."

Then it was August. It was too hot and sultry even to think. All the
fruit, like everything else, was late, but I made plum jelly and plum butter
and canned nine bushels of peaches with some marmalade and conserve.
We had lost our appetite for our regular food so we ate melons, drank
lemonade, and brought ice from town and used the milk for ice cream.
The days seemed to pass quickly. Wayne bought Jimmy and Donald two
saddle horses and a young colt. They named them Dixie and Ginger.
The colt was Spark Plug. Talk about two happy boys! They would saddle
the horses and race past the house like young Indians. They didn't really
sleep with the ponies, but I'm sure they would have liked to. Spark Plug
followed them around like a dog.

Looie, the pup that slept beside the sorghum barrel, grew up at last
and they trained him to go up and down the ladder when they had to
climb on the roof to turn the chimney for me.

Although I intended to have Jimmy and Donald attend the little Sunday-school regularly, it was they who persuaded me to give up the idea.
They did go three or four Sundays, after putting in a lot of time studying
lessons, but told me they would rather not go any more for the other
boys laughed at them when they answered all the questions and recited
the "golden text." They felt that it was silly to attend—none of the other
boys and girls took any part. We kept up our Bible readings regularly, but
after that they went down to the meadow on Sunday afternoons and rode
around on Dixie and Ginger in the sunshine, and were in good company,
happy and content.

One Saturday night we drove out to a platform dance "in the sticks."
There had been a ball game out there in the afternoon, and under the
full moon some men were still selling balloons and cones and hot dogs. It
was almost like a carnival.

We did not stay until the end of the dance so I do not know whether there were any knife fights that night or not. I noticed many of the
town people there, dancing, and the women all looked like pictures in
their long dresses, and there was such a variety of color. At home there

was never a time when outdoor dancing at night would be enjoyable, for the nights were always too cool. I didn't think I should ever become accustomed to outdoor dancing under a full moon—the brilliant stars so close overhead and the air so delightful. The effect of the light dresses of the women, the fiddle and guitars beating out the quick dance tunes, and the happiness so evident on every face making it all so beautiful and unreal.

Down in the bottoms during the heat of summer it was very difficult for anyone to arrive in town or at a singing all fresh and crisp and clean. I appreciated the efforts the young girls who lived down past the store made. I often saw them perched up on the wagon seats with the pleats of their well-ironed dresses all pinned together so that they would look nice when they finally arrived. Once in a while I saw an umbrella used for protection from the sun, but there were no parasols. Umbrellas were not used when it was raining, because there were seldom any soft drizzles. When it rained the deluge of water and the heavy wind would turn an umbrella inside out. When it rained, everyone, either in town or in the country, walking or riding on a wagon, was nearly always forced to take shelter.

Several times when we had gone out in the hills to dances held in a hall or someone's home, I would put on my dress and Wayne his white shirt just before entering the car, so that the breeze might check perspiration, and we could arrive looking presentable. After the first hour, however, everyone in the hall was flushed and hot, the collars wilted, sleeves plastered to arms, and a general air of disarray prevailed. No wonder the outdoor platform dances were popular.

I had been working hard at my dressmaking, getting patterns and studying them carefully, and cutting out my dresses with precision. But exact and painstaking as I had tried to be, I felt that I had much to learn before I would be satisfied with my work. I found some beautiful voile, in tones of gray and violet, when we were in Little Rock. I know I should have bought a good ready-made dress, but my vision of that beautiful material, in the colors that I never could resist, made into a softly gathered dress, so took my fancy that I could see nothing else.

I hoped to make a dress of which I could be proud. When it was done I knew it had been made correctly, and well-stitched, but just what was wrong I could not say. It had a certain look about it that displeases me, though Grandma and Wayne said it looked all right to them. I wore it into town one Saturday and went with Wayne to the ball game. We sat in the bleachers in the hot sun. Wayne apparently enjoyed the heat, but I was actually suffering both mentally and physically. The women I saw were all strangers to me, and the more they glanced my way the hotter I felt and the more I perspired. They all looked so crisp and cool in their thin dotted-Swiss and bright voile dresses. I became self-conscious and felt that my attempt at home dressmaking might be the cause of the laughter I heard. I envied the women their olive skins that took the sun's glare with neither sunburn nor freckles. Oh, the grimness with which I steeled myself to endure that long hour until I could escape to our car and get home again. The sense of my own inadequacy was a personal shock to me from which I suffered for a long, long time.

I had made many attempts to hire a dressmaker but the ones I knew in town were very busy. Our own farm work, our spells of sickness, and the bad roads, all kept me from getting into town to avail myself of any dressmaker's assistance at the proper times, and so I seemed faced with a problem that I never quite solved, and that filled me often with hot resentment.

38. The Cropper Speaks

GRANDMA ALWAYS SEEMED so happy just to be on her own land. Her sense of possession was very strong. Much of her time she spent cutting pieces for the intricate patterns of home-made quilts. For each of her children she had made, years ago, several beautiful quilts, Polk in the White House, Star in Texas, Double Wedding Ring, Southern Snow Ball, Rocky Road to California—she had made all these designs, yet even now a new pattern would start her anew to making quilt blocks for one more quilt. Between times she would make beautiful cushion tops and table covers, all embroidered in bright silk thread.

Her rag rugs, too, were works of art, the colors so soft and beautiful, and the work so carefully done. I deeply resented the countless hours that so many of the farm women put in "cutting up big pieces of cloth into little pieces and then sewing them together again," as some man explained the process of quilt making. I could see, however, the great difference between quilts made leisurely and with care, to give to some loved one, and the quilts that were heavy and dark and ugly and made from any available worn-out clothes; quilts so badly needed for warmth, so quickly worn out through constant use, the work undertaken as just one more piece of necessary drudgery.

Grandma had been quite well all that winter. Each day in good weather she would get out and tramp around the farm, or she would sometimes walk for miles up the river road to visit friends and neighbors.

One of these neighbors, Mrs. Varden, was a very dear friend whom Grandma had known years before. Mrs. Varden and her husband lived on a big farm close to the river. She was a large, kind, pleasant woman, with graying hair and pink cheeks. I always saw her dressed in clean gingham dresses, her hair combed neatly, and always wearing a white apron. Her house of five rooms was one of the few good solid houses in the bottom land. It was spotlessly clean and very comfortable. She had carpets on the floors of her two front rooms, and her rocking chairs, enlarged pictures and small organ gave her house a really fine appearance, greatly in contrast to the other houses in the bottom. Oval rag rugs were on the other floors. Her big puffy beds, so immaculate looking, that I saw in the bedrooms, had feather mattresses filled with down from geese she had raised when she lived out in the hills. Mr. Varden was an excellent manager, and their smoke house and cellar were full. From her garden came an abundance of fine vegetables, which were generously shared with all the neighbors. The big well was closed in with the porch at the back of the house. It was a wonderful well of fine water. Out beyond the garden there was a chicken house and a small barn for their mules and milk cow.

The big shining cookstove that almost filled the small kitchen seemed to be always in use. Every meal in her house was a treat to me. I have never eaten better looking or better tasting biscuits than the ones she made. Lemon and custard pies were her specialty, and the cakes she made were truly wonderful. She made a special cake that was rather firm, but not heavy, very aromatic, but not spicy, and with a fine flavor of good butter and eggs. Mrs. Varden always took the best food to all the picnics and singings, I was told.

One day I took Grandma in the Ford to visit Mrs. Varden. I had taken along a poem that I found in a magazine. I wanted to read it aloud to them. At first they seemed indifferent, but politely settled to listen to me, but when I finished reading and looked across the room, I saw Mrs.

Varden throw her apron over her face and burst into tears. Wiping her eyes, she apologized for crying, but told me that Grandma knew, even if I didn't, that the poem pictured just what her life had been like all through the years.

"All my boys are gone now to homes of their own, and I can only do for them as they come to visit us a spell." She went on, "Some of their wives won't let them come much. I wish we had been as comfortable then when they were all little and at home as we are now. Bud and me don't need much, but we keep on a-working and a-saving anyway."

This is a copy of the poem I read:

THE CROPPER SPEAKS
By E. E. Miller in The Nation

"Next year we'll buy a farm," we said,
My wife and I, when newly wed.
But next year came, and next, and next,
But always we were sore perplexed
To find enough to square the store,
And get a start for one year more.

I reckon somehow I'm to blame
That we have gone on just the same
For fifteen years; but looking back
I can't see where my work's been slack,
And we've not wasted what I made.
I know I'm not much at a trade,
And once or twice I've lost like sin
By letting someone take me in;

And twice a farm I've tried to buy,
But couldn't gather, low or high,
The cash I had to have in hand
To get possession of the land.
So, still we tend another's fields,
And pay him from our scanty yields;
From hut to hovel move about
Till all our plunder's plumb worn out.

At moving time, in years gone by,
My wife would fret and fuss and cry,
And say, "It's just no use to try
To keep things nice until we get
A home to stay at." "Right, you bet,"
I'd say. "Next year we'll have it, too.
I'm sick of this as well as you."
But now, we just pull up and go.
She says no word, because, I know,
She's too down hearted, tired, sad,
From giving up the hope she had.

It's hard for one to spend his life
Toilin' and moilin' in endless strife
With worms and weevils, grass and weeds,
For scarce enough to meet his needs.
It's hard to work for years, and then
Be just a slave to other men—
No home your own, no place to stay
If some man says to move away.
It's hard to feel men think of you
As one of a shiftless, thriftless crew—
"He's just a cropper"—it means, "no good;
He could do better if he would."

That's hard, but harder still is this:
To think of what your children miss,
And what your women-folks must bear
As you go drifting here and there.

What neighborhood has in its life
Place for a cropper's busy wife?
Who cares to have his kids about?
At school they're likely in and out;
They lose ambition as they grow;
They never set an orchard tree,
Or fix the yard up so 'twill be
A nicer place another year—

Next year they'll likely not be here.
So year by year they drift away
From folks with better show than they;
And year by year the wife grows old,
And less and less life comes to hold
For her, of things that women crave.
She, too, is nothing but a slave—
A slave to crops and a busy man
Who must keep going when he can;
A slave to toil that has no end
And does not help her lot to mend.

I tell you, it's no little thing
To take a woman's heart and wring
It dry of every hope she had
In days when she was young and glad.
It must be my fault that it's so!
I've tried and failed. But still I know
There's something wrong. I can't say what,
But what I've earned another's got.
A nigger cabin, a muddy yard—
That's my wife's portion. God! It's hard
To think of hopes that once she had
And keep myself from going mad.

Not long after our visit with Mrs. Varden, Grandma asked me to drive her up to the Chapel one Sunday evening. There was going to be a special meeting, a service of prayer and song. We went early so Grandma could have a little visit with Mrs. Varden, then later we walked together down the road to the little church.

Here were gathered about a dozen men and women. Although there would be no preacher they were expecting a friend from town who was coming to give a little talk and lead the singing. But as time passed and the lamps were lighted and the man did not appear, Mrs. Varden leaned across Grandma and asked if I would play the organ for the singing and give a little talk. I agreed to play so they could sing and reluctantly agreed to read something from the Bible, too. One of the men present who could have read had lost his glasses. For the others it would have been impos-

sible. Mrs. Varden said she herself just couldn't see in the dim light. I was
sorry that old Mr. Batson wasn't present, for this would have given him
an opportunity to exhibit his ability to preach, and I was still waiting to
hear him.

Anyway, I went to the organ and played as well as I could. It was really
hard work. I had all I could do to pedal and keep the air in the bellows
while I watched the notes, so I did not join in the singing. Some books had
been passed around and everybody seemed to be enjoying themselves.
I remember they sang "How Firm a Foundation," "Happy Land," and
"Shall We Meet Beyond the River." There was none of the quaint, plain-
tive singing I had expected, just simple old-fashioned tunes.

These older people seemed to be glad to be here together without the
usual crowd of young people, and a comfortable feeling of friendly communion descended on the little group.

While I sat at the organ I kept wondering what I could read this
friendly eager group, who were all so hungry for stimulation and inspira-
tion. I decided to read the Beatitudes, and I thought I could turn to them
without fumbling. Mrs. Varden nodded to me at last, and as no more
songs were called for, I stepped up on the little platform and opened the
big Bible that was lying on the old worn and chipped pulpit. I read verses
one to twelve, from the fifth chapter of St. Matthew.

"And seeing the multitude He went up into a mountain; and when
He was set, his disciples came unto him; and He opened his mouth and
taught them ..."

> *Blessed are the poor in spirit: for theirs is the kingdom of heaven.*
> *Blessed are they that mourn: for they shall be comforted.*
> *Blessed are the meek: for they shall inherit the earth.*
> *Blessed are they which do hunger and thirst after righteousness:*
>   *for they shall be filled.*
> *Blessed are the merciful: for they shall obtain mercy.*
> *Blessed are the poor in heart: for they shall see God.*
> *Blessed are the peacemakers: for they shall be called the children*
>   *of God.*

I was nervous, I was self-conscious, but I knew that if I did not read to
these people they would return home disappointed. What did they have
to go back to? So I read on—loudly, slowly, and with feeling. I did my

best, then closed the Bible and went to the organ and we all sang "God Be With You Till We Meet Again."

On our way home Grandma told me how proud she was of me, and that really pleased me. I told Grandma that it seemed to me that all the attributes listed in the verses I had read could well be applied to most of the people around us—the poor share-croppers. Poor in spirit, mourning, meek, merciful, poor in heart, peacemakers. And as I drove along I continued to think of that little meeting. I debated with myself the question of whether any of us really hungered and thirsted for righteousness, and in just what manner fell the persecutions and revilings. But, anyway, it was a good ending to a happy day.

39. A Jungle in the Temperate Zone

ONE WEEKEND Wayne and I drove to Little Rock to visit Wayne's brother Frank. We stayed overnight in a hotel and the following day drove down to Frank's cabin by the river near the Maumelle Mountains. While we were there, Frank took us for long drives through the surrounding country that was all so new to Wayne and me.

I had never seen a real tropical jungle, although years of varied reading had given me a vivid mental picture of the jungle as pictured by the novelist. But I had thought these to be only a most effective background for a story—dense, dim passages through moss-hung forests, and sluggish, infested waters, creating an effective impression of seclusion, loneliness and dread.

I am sure now though, since seeing so much more of the country here, that these descriptions of jungle growths were not exaggerated but were pictures of real places, for the semi-tropical conditions that I saw made me better visualize a real jungle of the tropics.

We drove past lake waters, shadowed and quiet, where the tupelo gums, uncanny-looking, utterly un-tree-like, rose with their vase-like bulges from the water's surface into true tree shapes, each hung with gray moss and tangled grape vines.

I caught glimpses of gloomy, quivery, shadowy lagoons, spreading under the trees. Farther along, the mellow haze of sunshine gave the waters a glint and a gleam. High up in the gums and cottonwoods that grew by the roadway, the Virginia creeper clung, the scarlet and gold leaves shining in the sunlight. Below them the muscadine vines fell verdant and heavy-leaved, in great masses and loops, their snake-like arms embracing the trees with tenacious coils.

We did not venture into a cane thicket nearby, for the canes grew so close together that it looked impossible for one to make his way through them. A short sally into the underbrush, among high weeds and vines and grasses tangled with briars and burrs, was enough for us, for the swarms of sour-gnats that we stirred up, and the sickening, penetrating acrid odor as of deadly scented pollens, sent us gladly back to clearer ground.

Beyond the dense bodark thickets, that stout tree growth, the lake merged into a bayou, with marsh-lands extending to the left and right. We drove under the trees, following a narrow road. Hard packed and damp was the ground here, sandy and slimy-coated in places, or covered with green shadowed water, where bullfrogs splashed with a distinct slap as they leaped from a log on the black mucky shoreline. Only an occasional bird cried in the loneliness.

We left the car here and made our way through the woods, climbing over bleached fallen branches and old logs, where huge mud-turtles basked in the sun and where long black-snakes slithered away across the musty black ground toward the water's edge. From where we stood on the oozy silt, we peered out under the gloom of the cypress branches across the sluggish bayou waters. There was a heavy odor of damp earth, though here the sun could penetrate in spots, and great lush patches of water lilies floated on the black ooze, while turtles' heads bobbed in and out of the water. Only the thousands of skater-bugs and a slithering water-snake kept me from reaching for one beautiful white lily. But I was content to admire the butterflies and white moths that circled in the sunshine. We stooped under the low-hung limbs, where dark green creepers clung, going through aisles that were dim and mottled from the interlaced branches above. Here were trees of various kinds—the oak, the hickory, the ash, the cottonwood, the holly, and the sycamore. It was very hot and muggy, a real hothouse condition, for the sky was becoming overcast and distant rumbles of thunder signaled a coming storm.

Frank and Wayne took as a matter of course all these conditions that struck me as so very jungle-like and weird.

Later when Wayne and I were on our way home, I marveled at the strangeness and beauty of all the various places we saw. Where the bottom lands merged themselves by table-lifts into the hills, and the hills rose higher into the Ozarks, we rested by clear sparkling creeks, icy cold, and found bubbling springs and waterfalls overhung with dewy mosses, which sparkled in splendor like dazzling showers of gems. The oaks and elms, the pecans and sumac, all green and glistening, protected every hillside.

But our road kept dropping and lowering into the cotton lands again, to the rich sandy loam and the black gumbo soil where grew the heavy cotton higher than a man's head, and the great fields of glistening heavy-burdened corn with huge black-green leaves, and the sweet potatoes and peanuts and alfalfa.

The land lying back from the Arkansas River was often cut into by young streams that finally settled down into lake beds or backed up on the low places to form swamps and bayous, all half-choked with reeds and rushes.

We saw that the jungle aspect in the lowlands was very close to the farms, too. The matted growth of Johnson grass, peanut grass, and crab grass, and the tough twining morning-glory vine and bindweed were all to be constantly fought from every cotton and corn field with unremitting labor.

The better farmers always looked with disdain on the shiftless farmer who was surrounded on all sides by a dense growth along his fences. In too many cases, sickness, lack of adequate help, or incessant rains prevented a farmer from winning in the competition with a growth so heavy that, in just one season, the weeds and bushes, the Osage orange, elder, goldenrod and sumac, and matted burr-growths, could grow so high and thick as to completely cover and hide a fence or small house. Many a promising crop of cotton had been left "in the grass" because the heavy rains made it impossible to get in on the black soil soon enough for the hoe hands to prevent the weeds from completely taking the crop. The most intense satisfaction I ever witnessed was the genuine triumph with which a bottom-land farmer announced that the long battle with the weeds was ended and his crop was "laid by."

On one farm near us on the Arkansas River, we knew of a farmer who planted three acres in sorghum so as to have his own syrup for winter and some to sell. The rains were heavy during the growing season and when the sorghum was ready to cut it was higher than a mule's head. The growth was so dense that the man who was to cut the cane on shares refused his part of the crop, for neither he nor his sons could stand the terrific heat and the awful humid condition which completely surrounded them, after they had cut their way into the sorghum growth to make a path for the mules and the wagons.

Buffalo gnats, dry flies, sour gnats, and mosquitoes bred amazingly in all the lush vegetation, and rabbits, quail, possum, squirrels, pigeons and doves made it their habitation, so that in the fall hunters were continually penetrating the jungle.

The quick, almost-tropic growth pervaded our farm too, for not only did the weeds grow rank but the crops matured faster than any I had ever seen during the warm summer days.

I had even heard corn growing! I sat once on a soft sandy ridge between two stalks of young corn and leaned my ear against a stalk. It was about ten o'clock in the morning, after a warm growing-night, the sky still overcast but with not the faintest breeze blowing. I heard a definite "squeak, squeak" as the corn growth moved itself where the leaves branched upwards and curled around the stalk. A gentle pushing sound which caused me to catch my breath in pleasure and astonishment. It was old Eric West who told me to sit down thus when all was quiet and the season was right, and listen to the corn grow.

That same spring, too, standing at the screened window that looked out on our flower garden, I saw the groping tendrils of my own morning-glory vine lift itself, and stop, and move closer, and stop, and move again, gaining about an eighth of an inch each time toward the wire screen where, at last, its delicate fingers clung tightly.

And in my garden every evening I had the pleasure of seeing my four-o'-clocks "be" four-o'-clocks. Just as the sun was setting in the cool delightful part of the day, the blossoms would open one by one, and I would watch the miracle of the tightly closed petals opening themselves into yellow or pinky-white blossoms, smelling so faintly sweet.

All around us the teeming life! I lay awake one summer night, listening to the various sounds of bird and insect life outdoors. High in the heavens, over the tops of the sycamores and cottonwoods and pecan trees, the full moon was sailing, round and burnished with a glistening copper sheen. No moon like a full moon in a summer sky in the South. In the grass the shrill cicadas, in the trees the crickets sawing, sawing—with a sound one was forced to become accustomed to in self-defense. Every few minutes their hum was interrupted by the katy-dids, with their "Katy did, she did, she did, Katy did," in dull monotonous tones.

I could see the flashing glow of the little fireflies, swarms of them that gathered and scattered, now here, now there. Against the screen, near my head, a mosquito "zeed"—mosquitoes, the menace from which we had to be doubly protected by the cloth netting over the bed, as well as by the wire screening at the windows and doors. Every now and then a beetle or a large June bug banged blindly against the screens, their wings making a rasping sound.

No such thing in the summer as going to sleep—in quiet darkness. No, one only succeeded in sinking into unconsciousness by making a definite mental effort to forget the ceaseless hum and stir of all the night insects, knowing that surely they would silence themselves, sometime, in the long night hours. And strange it was, to awaken, very, very early of a summer morning, and find the world so still. All the shrilling and zeeing and croaking and zooming gone, and in the soft half-light, to lie quietly and listen to the first faint sweet question of a sleepy bird across the bodark thicket by the meadow.

40. The Bargers and Cranes

AFTER AVIS LEFT we still called the little house across the road from Grandma's, Avis's house. The Bargers, Lem and his wife, Ann, lived there now. Her brother, Bunk, lived in the new house. Lem was dark of hair and eyes. He was a heavy-set man, with an underslung jaw and a drooling, open mouth for he was afflicted with overgrown adenoids. His wife was big and buxom. She was taffy-colored, her eyes were blue, her cheeks pink, her figure ample yet firm, and her step springy. Her tongue wagged incessantly. She had a small baby boy about a year old, whom she carried everywhere under her arm. She worked in the fields hoeing and chopping cotton, and the baby did not seem to make the least difference in her life. It was very strange to me how even the smallest children could be kept in the background, subdued and obedient, fitting easily into the rough daily life of their hardworking families.

Once when Mrs. Batson couldn't come to wash for us we hired Mrs. Barger. She insisted upon taking our clothes up to her own house, although she had to carry the water from the pump across the road at Grandma's. I'll never forget Jimmy's and Donald's faces as they came rushing in, breathlessly, to report that Mrs. Barger had hung our clothes on the barbed wire fence around the barn lots. Oh, yes, I was indignant and disgusted, but there was only one thing to do—send the boys to bring the clothes home, and wait again for Mrs. Batson, or Old Zula.

Hardly a day passed without Mrs. Barger coming to borrow something. She was always in a hurry to get back, so she never took a seat, but would stand against the walls while I tried to endure her dawdling talk. She skipped around from one person's affairs to another's so rapidly that it was useless to try to keep up with her. In almost every sentence were the words, "I says to Lem," and "Lem says to me," and with that introduction a fresh recital of someone's affairs was begun. She asked for anything her eyes saw in my kitchen. My spices, flavoring, cakes of soap, and baking powder, were carried away so often, that at last, thinking to check her, I made a list of the supplies she had already borrowed and hung the list up by the kitchen door. It took a lot of courage for me to go right before her eyes and list her latest borrowing, as I felt I could hardly face her embarrassment. My fears were needless. There was no embarrassment. There were no after-effects, either, and I still found it very inconvenient to run short of needed supplies on account of her "borrowing."

We had thought that with Mr. and Mrs. Barger and Bunk we had three good hands, as they were all very healthy and strong, but Mrs. Barger went gadding about every day in every kind of weather. She knew everybody for miles around and would not stay home. When their share of the cotton needed hoeing badly, others had to be paid to do her work. As was the usual custom, Barger and Bunk were to work out their part of the crop and then work for Wayne right across the field, doing whatever was needed to be done, hoeing, or scratching, or plowing the field. Wayne had agreed to pay them weekly for their work because they promised to be such good help.

They had a share-crop of about twenty-five acres of cotton. One-half would be theirs. What they actually would get depended upon the price when it was sold. About twenty bales would be the crop. The bales would be about five hundred pounds each, and the cotton should bring around twenty cents a pound. Their cash payment of one dollar a day was not the regular custom, but Wayne felt that it ought to work out in this case.

The Crane family was now down where Alva and Claude used to live, by the mailbox. Mrs. Crane waded the dry sandy roadway past my house every day, going up to Grandma's to milk the cows. She wore an old sweater and had a knitted cap on her head, pulled down low and almost covering her face. Her bare hands were wrapped in her apron. The lard bucket in which she would bring back milk hung on her arm.

Mrs. Crane was soon going to have a baby, but Wayne had hired the Cranes because Mrs. Crane and her daughter, Dove, a girl of sixteen, and the two young boys, were all able to work in the cotton.

Mrs. Crane stopped in at my house one morning to ask me a favor. She wanted me to lend her a towel so she could have one good enough for the doctor to use when her baby arrived. When I questioned her I found that, although she had some baby clothes saved from the last time, yet she had nothing else prepared. I sent down some new boards and suggested that she have her husband put up shelves in the bare front room where her bed stood. I gave her a lot of advice as to cleanliness and sanitation, to which she listened with a pleased smile. I suggested that she scrub the room thoroughly, while she was yet able, and keep the clean shelves for nothing but the things she would need for the baby and herself. I tried hard to help her by frankly talking to her about her condition. Above all, I tried to get her to shield and guard her body from the vulgar calculations and comments of the men who came and went around the farm. But she seemed to think I had the idea that her condition was something of which to be ashamed, and so was to be hidden as much as possible. It was impossible to make her understand that I thought her body was too precious, with its burden of new life, too sacred, too beautiful, to be so entirely ignored by herself that she might become the subject of jeers and ridicule, but I know she really did not understand me.

It seemed that she had nothing in the way of extra bedding. Although she had scrubbed and boiled some old cotton sacks and fringed the ends of them, she thought they were really not "fittin'" for the doctor to use. I gave her some towels and sheets and other things I knew would be of comfort to her. I found some old worn curtains, soft and white. These I boiled and dried in the sun one hot afternoon, and then I baked them in the oven and tied them up tightly in a bundle and put them away. Something seemed to prompt me to do this and it was fortunate that I did, for the day came, not long afterward, when Mr. Crane came frantically to the house to ask for some cloth or towels or anything that would do in the emergency. The baby was born, but hemorrhage had started, and all the doctor's supplies of gauze and cotton were used up. I had not known that the doctor had been sent for. The doctor thought it a miracle that I had this adequate supply of sanitary absorbent material ready. In Mrs. Crane's house there was not even a bit of clean cloth. My supply was the means of saving her life, the doctor told me.

Mr. Crane was a beastly man. Besides his general foulness he continually chewed tobacco and was always spitting. Once he recounted to me at great length the story of his life. We were sitting by the fireplace in the Cranes' house. I had gone in to get warm while I waited for the postman to come down the road. It was a wild tale he told of knifings and shootings and fights. Trouble with the law, as he called it, out in Oklahoma. To cap it all, he ended the stories with one about the time he had been taken out by some of his neighbors and hanged for something he'd done. He was left beneath a tree by the road, with his toes just touching the ground. Providentially someone came along just in time and cut him down.

"That was too bad," I said, but my sarcasm went unnoticed.

Between his two upper front teeth he had had a dentist bore a small round hole and rim it with gold, and he claimed he had won much money betting as to the distance he could spit, for through this hole he could squirt tobacco juice much farther than by the ordinary method, and had been Oklahoma's champion tobacco-spitter.

Once, Mr. Crane appealed to me to settle an argument he was having daily with some of the field hands. It seemed the men objected to his use of the expression, "Ay-God." They said it meant "By God." I told him that it sounded to me like a poorly pronounced "by," and I was certain that he was really intending it for cursing. He accepted my verdict with humble wonder at my brilliance, pretending that he did not know that he had been swearing; but I was not fooled by him.

When that point was settled, Mr. Crane began an argument about schooling. "You don't hold with all this here educatin' girls, do you?" he asked me. "Take Dove, now. She's able to do pretty good, getting along fine, the teacher says, in all the schools she's been. My stand is—you can waste a lot of time just goin' to school. Women should know cooking and taking care of kids. That's their place, getting married. Let 'em stay home, I says."

Even while I answered him, rather hotly, I realized that the broad smiles from Mrs. Crane and Dove were not only to hear me refute and override Mr. Crane, but were also an indication that they were enjoying the conversation as a sprightly break in their dull day. They knew that I could well defend myself, and Dove, as well as all women, for they had heard me talk before. I began to see that it was a waste of time to talk to such an ignorant man, husband and father, dulled to any possible spark of joy, so unthinking, so uncaring, so altogether un—everything. It would be a useless, hopeless task to try to teach him that scale of life's beauty which is based on the notes of cleanliness and order, or to show him the value of saving and preserving what he had, while he worked to make a home where peace and happiness and contentment should abide. I watched his ugly, jeering face, with its filthy tobacco-stained mouth, and saw all the calculated look of mean daring with which he still emphasized all his remarks, in spite of our discussion, with his half-hidden curse—"Ay-God."

The hens which ran loose around the big barn laid their eggs in the mangers and the hay mows in secluded places, and Mr. Crane would steal the eggs when he was feeding his team at night. He would fill the pockets of the long black oil-skin slicker he usually wore. When Crane, with a disarming grin, once asked Henry if he could borrow a few eggs to take to "the wife," as they hadn't had an egg for weeks, Henry couldn't resist the temptation of giving him a hard friendly push, crushing the eggs in his coat pocket, as he cheerfully told him to go ahead and help himself.

There was Crane's dirty self, and his stealing and general laziness— but to me the worst of all was the manner in which he was training his two small sons. One of the boys carried, in the top of his heavy boot, a razor, a long-bladed, old-fashioned one with a broken end. His brother had a large spool with a nail driven through it, the spool providing a good grip for his nasty weapon. Jimmy and Donald told me the boys met them at the mailbox and wanted to fight, but they saw the razor and the big nail in the spool and ran away. A few days afterward Donald came in with his cheeks red. He had been crying. He said one of the Crane boys slapped and beat him with his fists, because he told them he didn't want to fight. When Donald took off his knitted cap, the blood ran down his face where his head was gashed by the nail. Wayne told Jimmy and Donald they must not run away but must learn to fight to protect themselves. My blood seemed actually to run cold—as I visualized Jimmy or Donald with their faces and hands slashed by that awful razor in the hands of those young reprobates. It was a hard thing to do, to stand by and hide my feelings while my boys were taught how to fight to protect themselves. Wayne started boxing lessons again, and helped and advised them daily, so as to build up in them the combative spirit which they lacked, and which they sorely needed now.

The next thing we heard, Jimmy had come to Donald's defense, when the younger Crane boy ran into the house and came out with a loaded shotgun. Mrs. Crane stopped the fight—and came up and told me about it. She said that her husband had allowed the two boys to carry the razor and the big nail in the spool—and even to shoot with the gun. She admitted her boys were bad, just like their father, but she could do nothing with them, for he encouraged them in every kind of meanness; he had even helped them to steal things. That was one reason, she said, that they had never owned anything; they had to keep moving on account of Mr. Crane's orneriness.

Wayne told me that one afternoon he and Mr. Crane were over in the meadow clearing up the fallen trees and loading the wagon with wood, with the Crane boys helping, when Jimmy and Donald came along. The boys got into an argument and Jimmy appealed to Wayne to know what to do—for the older boy was calling him awful names and he wondered if he had to take it. Wayne told Jimmy to go ahead and fight—that he knew he could whip the Crane boy and he well deserved a beating. So Jimmy started in. He had only his fists and a righteous indignation. The Crane boy pulled out the razor he carried. Jimmy picked up a stick of wood and knocked the razor out of his hand. Then Mr. Crane, behind Wayne's back, pulled out his knife and advanced on Wayne, who turned just in time to see him. Wayne stepped back and ordered Crane to put the knife up and go on back to work. Crane, after trying to out-stare Wayne, put up his knife, ordered his boys to the house and went on with the job. There were no more fights after that.

Crane was a blustering, stealing rascal. We felt that he would even kill, but only if he had the advantage—never when he faced an opponent. He was always wrong—and he knew it.

About a week before Christmas, Mr. and Mrs. Crane came into our kitchen with Wayne, carrying in our supplies. Mrs. Crane took from her coat pocket a small set of aluminum salt and pepper shakers. She showed us these with a broad, happy grin, saying they were a Christmas present from her husband, the first and only present he had ever given her. We thought that he probably had stolen even these from one of the stores, but we said nothing to spoil her happiness.

As time went on I was more and more amazed not only at the dire lack of even enough privacy for necessary bathing and decent care for the women in the little farm houses, but especially at the general acceptance among the older men of a life that held an almost total disregard of even rudimentary personal cleanliness.

Among the farming class, generally, it had always seemed to us there was a general attitude of contempt for the more fastidious cleanliness of the town people. Perhaps, here in the bottoms, the deterioration that went on in the blood stream of those subjected always to the onslaught of the chills and fever caused by malaria also caused a general mental decline, a waning of ambition and a lapse into general bodily unwholesomeness. The younger people seemed to rally again and again, and seemed to re-establish themselves, if not to a hale, fresh, hearty and vigorous condition, at least to one where they still kept their self-respect; but among the older men and women I noticed more and more a general let-down of all effort toward more than superficial cleanliness. I used to look at the men, with their unshaven, tobacco-stained, dirt-begrimed faces; their broken, blackened teeth, their dirty hands, their long filthy fingernails and their evil-smelling bodies. So many were toothless or had decayed and broken teeth, so many had begrimed necks, and heads that were never washed. They were apparently unconscious of all their unwashed offensiveness. I marveled that they could think that their disgusting personal foulness should not be a constant offense to their wives. Their slovenly habits were altogether so repulsive to me that I never failed to be astonished and resentful when I met a smiling woman, wearing clean, starched clothes, her hands clean, her face shining with soap and water, and her hair combed neatly—and then looked at the man who claimed her as his wife.

I felt that a man to whom such things as bodily cleanliness and beauty in living were non-existent deprived his wife of the necessary appreciation and admiration she needed to make her daily tasks an unbroken chain of joyful effort. At last, it seemed, she either opposed him and struggled alone to "provide things decent in the sight of all men," or took the easier way of daily subsiding into an amiable lassitude that sank not only herself and her husband, but her children as well.

> *As the husband is, the wife is,*
> *Thou art mated to a clown,*
> *And the baseness of his nature*
> *Will have weight to bear thee down.*

It is no wonder that one of our milk cows was such a comfort to me. She was more gentle than the others, and made no effort to move away when I approached her. Her big dark eyes were so placid and full of understanding.

Many times, when down under the big pecan tree in the meadow, I stroked her head and leaned my face against her gently heaving side. She drew her breath so evenly as she turned her head to stare at me, even sighing with a deep heave of sympathy, as though she sensed that I was really seeking some solace from her.

And such solace! She was so clean and her breath was so sweet and her whole outlook so serene. I always felt rested and refreshed from this tranquil association with Jersey—far, far more than I was after striving to find some common ground with the poor share-croppers' wives.

41. Cypress and Yew

WE FOUND THAT the hay that had been cut and drying down by the next farm, where we had rented some extra land toward the river, was a complete loss, for the river had backed up and flooded the field, washing away most of the hay and spoiling the rest. Hay was eighty-five cents a bale, too. Yet we thought we would get around thirty to thirty-five bales of cotton, enough to clear all our debts that year, so we still tried to be optimistic.

The cotton fields were all brown now, and the gum trees were beginning to turn, and the pecans were dropping each day.

In town the annual circus had arrived. We were all going to go in and take the boys, for a circus is ever a circus, and belongs to children especially.

And now would begin the regular fall round of work. Six cotton pickers were working already, starting down the long rows. There was still some hay to bale, three acres of peanuts to rake up and haul to the big barn and store in the loft, and the winter wood to cut. That meant a trip to the lower place to get hard wood. Then some hogs would be killed for market. When the cotton was all picked, it would be time to begin plowing again for the spring planting.

During this time I was waiting one evening in the late fall for Wayne to come home from the gin. I wandered up the sandy roadway toward our "new house," and paused for a few minutes to visit the family of cotton pickers who had recently moved onto the farm. They had come down from the hills only a few days before, so I didn't know their names.

The sun had long sunk behind the hills and dusk was slowly changing into darkness. I could hear the crackling of the fire in the tiny stove and see the gleam of lamplight through the open door.

Only one of the women was in the house. She was cooking supper, turning the thick pieces of fat salt pork in the frying pan. A pot of coffee was boiling at the back of the stove. A kettle of pinto beans emitted a strong odor of just plain beans, boiled.

She seemed to be about sixty years old. She had graying hair, and her eyes as she glanced up at me were black and brooding, and seemed to have lost the ability to express emotion. They held only a blank, pitiful look.

"Sure proud to have you step in," she said, looking up from her pan of meat.

"Fine pickin' weather we're havin'," she continued, as I stopped at the door. "Gettin' over the ground in proper style. The men folk was a-sayin' that they ain't saw prettier cotton anywheres.

"I heard that you-all ain't been in this country long. Must seem strange, the cotton and all. I been a-wonderin' to myself if I could ask you to be kind enough to sing a special song fer me. I been a hearin' you sing when I passed by your house. Couldn't help but stop and listen. It seemed to me the best song I ever heard. I'd be right proud to listen. Was somethin' about growin' old."

So there I stood, and leaned against the door jamb, looking out into the darkness, and sang for her that beautiful old song, "Dear Heart, We're Growing Old." When it came to the last verse, and the words,

> And when we cross that shining bar,
> And pearly gates for us unfold,
> God grant that both may enter in,
> Dear heart, and never more grow old,

I turned and peered at the woman. She was still standing by the little stove, with one hand wiping her eyes on her apron while she busily turned the meat. "Oh, God," I heard her say, "If my life could only have gone on to that—"

A few days later, I was told by one of the other cotton pickers who came from the same place out in the sticks where they all lived, that this woman had lately found out that her husband, a man of her own age, had been unfaithful to her for years, and she had lately decided to leave him.

Before the cotton was picked, three times this woman came to my house and asked me to sing to her, and always she wanted to hear "Dear Heart." I don't know why, for it must have lacerated her deepest emotions anew each time. Yet I have always remembered the short lines from one of Mrs. Johnston's "Little Colonel" books:

> *This learned I from the shadow of a tree*
> *That to and fro did sway upon the wall—*
> *Our shadow self, our influence may fall,*
> *Where we can never be.*

Once in the fall, when the last loads of cotton were being hauled in to the gin near town, I rode in with Wayne late one afternoon on the high cotton wagon. Everybody had gone in early that day to attend the big tent show, where we would join them later. Over my dark suit I wore a long linen duster, and I took along a whiskbroom to brush the lint from my clothes, for I wanted to lie right down in the soft, loose cotton and look up into the sky as we went along.

The sun was just setting and the sky was most glorious. Puffy white clouds caught the varying colors and long golden rays streamed across the heavens. It was so comfortable to lie there as though drifting along under the sky, and I enjoyed every minute of the trip. As we passed the little house up the river where our former tenants, the Woods, were now living, Wayne called out a greeting and I sat up to see to whom he was speaking. Mrs. Wood saw me and was so startled that she did not know what to say. However, she waved her hand at me and called out, "Every dog has his day." Indeed, yes. I didn't know what she meant by that cryptic remark and perhaps she didn't either, but anyway, I still have the golden memory of that rare experience. I instinctively knew that if I did not get to ride on that cotton at that time, as I had often longed to do, the opportunity might never come again; and it has been proven to me many times over that, as Robert Burns so truly said, "The present moment is our aim, the next we never saw."

For years I had been hearing the story of the traveler in the South who sought shelter for the night with an old couple in a mountain cabin. He was very hungry and as it grew dark he became worried as he saw no preparation being made for a meal. At last the old man sat down at the table and the old woman began to dig in the hot ashes, and finally brought out some large yams which she placed on the table. The host then waved his hand toward the guest and said, "Stranger, draw yer chair and skin a tater."

I could well believe this story, although all the hill people I knew had an abundance of good food, yet golden yams, baked to a nicety, were always a delicious treat. They were so easy to grow, too, either in the bottom land or in the hills.

Out near the big hayshed, in the rich sandy loam, our sweet potatoes had been started. They were planted whole, close together, in beds built up against the shed so they would be protected from the wind. Each potato sent up many sprouts. When the sprouts were up about six inches they were pulled off and a load of them taken out to the field that had already been prepared for them. Some of the farm hands set the slips in the rows, others followed along covering them, and generally some young boy would have the job of carrying water from the barrels in the wagon and watering each plant. This watering generally sufficed to keep the sprouts from dying, for the planting was done with an eye out for a coming shower.

The next task was to keep the weeds down, and later the rows were plowed. By that time the vines would be covering the ground, and would be let alone until the potatoes were ready to dig.

To finish the process, they were gathered up in the fall, taken in to the big drier in town and left until cured, when they were sorted, packed, and shipped. Those wanted for home use were left on the farm, where they were usually cured out in the barns.

I remember that sweet potatoes were listed in the San Francisco papers as selling at two dollars and fifty-six cents a lug. Wayne received a check for twenty-three cents for half a carload of potatoes we shipped. Twenty-three cents! And they were fine big golden yams, the variety of sweet potato that when baked has a thin skin and flesh that is as sweet and soft as though mixed with honey.

Those potatoes cost ten cents a bushel for curing and the packing baskets were twelve cents, to say nothing of the routine work of raising the crop and paying for the work. Wayne calmly tore up that check.

Sunday had become a typical day on the farm. All day long men had been calling to see if we had mules or pigs to sell, or to ask about a location for the next year. And yet with all the interruptions we managed to put all in order and get dinner over. Wayne was reading in St. Matthew to the boys and was ready to begin reading aloud to us from *The Power of Will* by Frank C. Haddock.

I wrote to my mother that her grandsons had arrived at that enviable condition where their "cup runneth over." They had their ponies and a good wagon, a large game board, checkers and balls and books, and had remarked that there wasn't a thing more they wanted. I read the few lines that Wayne added to my Christmas letter: "Christmas is here and I feel better. A New Year is coming and I feel better for that, too. Experience is a great bird, but she lights mighty hard on the limbs of Satan."

I don't know what got into me. My mind was in a turmoil. I turned from pen and paper as though writing were the last thing I wanted to do. I guess I was making up my mind anew to another winter of seclusion and waiting.

While it was snowing one day Wayne came in with a stranger to sit by the fire and get warm. I went about my work, while they carried on one of the usual wearisome and desultory conversations that I had heard so often—about work, the crops, and the weather. First there would be a question from Wayne, then a long thoughtful pause, and the hesitating answers, spoken as though every word was of such import that it had to be carefully considered as to value and exact meaning. Then more silence. Then another mild question to urge the unimportant subject matter on a little further, and another slow careful answer.

Finally Wayne came out in the kitchen and asked me if I thought there were any shoes left from the clothing that had been sent us from home. I found a pair of very good, almost new shoes, far too fine for anyone to use for farm work. I had heard of people whose shoes were so worn that their feet were actually on the ground, but it had always seemed to me to be just an expression that was never intended to be taken literally. But here sat a man wearing an old pair of shoes whose soles were entirely gone. The tops were tied around his ankles with pieces of string. Some of the top leather was folded underneath the foot. He tried on the shoes and found that they were exactly the right size. He was wearing no socks. I was full of pity and so moved by this proof of poverty that I whispered to Wayne to try to get the man to accept a pair of socks, too. I needn't have hesitated to offer them. We gave him a pair of fine wool hand-knit socks that Grandmother had sent to Wayne from California. The man was so overcome that he was speechless. He took off the shoes and put the warm socks on his thin dirty feet and sat there staring at them. Had he been a woman I would have broken down and wept. He told us that in all his life he had never had a pair of socks. As a boy he had gone barefooted, and since then he had never had enough money to waste on himself. It took everything just to take care of his wife and children.

After he was gone I went around for a little while with a warm glow of satisfaction. I felt that we had done a kind and generous thing. As we discussed this man I learned that he and his wife and four children lived in a little house behind our farm on the road going toward the bayou. They had a small garden and kept a pig and a few chickens. When the crops were "laid by," and so many were idle all through the bottom land, he relied almost entirely on his garden and hens for a living.

But, Wayne added, we had been furnishing the corn for the pig and chickens all year. Wayne had ordered him out of the corn field several times, where he had been gathering the corn by the sack full. His chickens were always to be found over in our corn field along the fence. He had ruined four rows of corn on the side of the field nearest his house.

They had been eating our corn from the time the ears were in the milk. They grated it on an old tin grater and fried it. Later they boiled it, and grated some for rough corn bread. When the corn was finally gathered and in our corn crib, the man came there at night and stole it. He would loosen a board near the bottom of the crib and dig out the ears as far in as he could reach, cover up the hole, and return again and again, feeling sure that he would not be caught.

All this was too much for me, but after my opinion had been freely expressed, and I said that I regretted giving him the shoes, badly as he needed them, Wayne silenced me by blandly remarking that he couldn't blame the man, he had to live somehow. He couldn't make enough by working—so he "got by" by working a little and stealing a little. "And," Wayne added, "wasn't it in Proverbs that we read, 'Men do not despise a thief, if he steal to satisfy his soul when he is hungry.'"

I felt much happier when I found, in the big bundle of good used clothing that my mother sent to me later to distribute as I saw fit among the poor bottom people, a fine dark suit that had been my brother's. This I gave to Mr. Suggs, with a shirt and hat and shoes too, for they were all his size. Mr. Suggs was quite overcome. He told me that he would now be able to go to church, something he hadn't done in years because he had only the old overalls and shirts he wore in the fields.

The next Sunday Mr. Suggs went proudly by to walk three miles in the hot sun to attend the little meeting up in the chapel. Mr. and Mrs. Suggs were living in our little house down by the mailbox. I was talking to Harley as I pumped water to fill the big galvanized tub. He told me that he was glad his father could wear the good suit, but as for himself, he had never in his life worn anything but overalls and chambray shirts. "Not even when you were married," I asked, knowing better as soon as I had spoken the words. "No, not then, either. I can live right long like I been a-doin', without no fancy clothes, but Dad there, he sure is a proud man now."

As the early winter set in, with cold winds and freezing nights, I often heard the expression, "Just watch the graveyards in the bottoms fill up now." That meant that there would be many deaths from pneumonia and influenza and pleurisy, for bodies so poorly nourished, lacking the food necessary to build up defenses against cold and disease, and being constantly torn down by the ravages of malaria, were an easy prey to illnesses when the bad weather came.

It would be nice to say of those old cemeteries that I saw throughout the bottom lands, that the old headstones stood where the ivy twined amid the scattered rose bushes and honeysuckle; yes, nice and poetic— but untrue. The old rail fences, covered with sassafras shoots, sumac bushes and matted vines, often enclosed old decaying head-boards and rotting crosses, all awry. The newer graves were left, of course, after a funeral, with heaps of flowers or pathetic little bouquets from some home garden, and on Decoration Day one could see small groups of people pulling weeds and trimming bushes and redecorating the graves of their loved ones. A service of prayer and song was generally held, too, on that day, if the weather permitted.

It was in no way due to indifference or neglect that these graveyards were so fallen into decay. Too often the little loved child of a family was carried here with tears and heartbreak, to be left, the parents only too well knew, forever. Endeavors to ease the pain of their loss by revisiting the sacred spot or tending some little plant there, in token of the living memory, would never be theirs. They turned away, as the last sod was smoothed down, with tears blinding their eyes, knowing that in their shifting from place to place, as the wheel of fortune turned for them, they would never see this place again. This was the end.

I discovered a little old graveyard by the farther fence on our own farm. A little yard it was, with parts of an old fence still standing and old weathered markers with the names all worn away. And out towards the hills, too, there was an old graveyard at the corner of a farm. When I asked about this one I was told that it was very, very old, and most of the markers had long since disappeared. But it was there that a thieving rascal of a man hid the pigs he stole from a neighbor, feeding them in the graveyard and hoping to fatten and sell them. However he was caught and made to return the pigs and repair the fence and all the damage done to the grave stones where the pigs had rooted. No one seemed much disturbed by this. No one, it seemed to me, ever seemed to be much disturbed about anything.

42. Philosophy

Extracts from Letters to a Mother

I'M SITTING DOWN to write in a burst of enthusiasm. I've been reading that little book of Emerson's, Uses of Great Men, *containing essays on Napoleon, Shakespeare, Goethe, and others. Being rather familiar with Shakespeare's works, I could fully agree with Emerson about him, but Goethe—go and get this essay and read what Emerson says of Goethe. I looked up Goethe's life and habits, and studied Emerson's remarks on* Faust, *too.*

I got down the phonograph record book and turned to operas and found Faust, *in pictures and music titles. I remembered the song "Heart Bowed Down." I would like to be able to hear every record mentioned in the book, for after reading about Goethe, and the story of* Faust, *I'd doubly enjoy that music.*

I have been reading, lately, too, in your old scrapbook, and I'm sure that the emotions of the human heart have been the theme of every great poet, writer, and thinker through all the ages. It's perfectly fascinating. I don't know what I am going to do about it, but I feel just like a hound on the scent of something.

Recently I had the great pleasure of presenting the truth, as I see it, on the subject of death and resurrection, to Mrs. Varden. She had lost her oldest son by drowning in the Arkansas River some years ago. Grandma told me that Mrs. Varden had always felt comforted by the knowledge that they found her son's body before dark that day, so that it did not lie in the water all night. I am afraid that in opening her eyes to the hope her boy has in a future earthly life, instead of the Hell she believes in, I drew her from her own hope of heaven; but I managed at least to make her acknowledge that her honest longings were not for a harp and crown, but for a life among her children and friends, where there will be no parting, no sorrow. You see, she felt that her boy, not being a church member, was bound for the old Methodist Hell.

Your many magazines have come and are much enjoyed! Also the enclosed poems in your letters. By the way, I find in your old scrapbooks such wonderful poems. I can read your mind, your life, through them. I can't help but wonder after all what good our understanding of life does us. It is just as difficult as ever to paddle one's own canoe. Of course we can learn to "gently scan our brother man" and all that—and that's volumes—and when I imagine a violinist pouring out his soul on the strings of his violin, I can see and appreciate the good he can do to the souls in the audience that respond to his message. Yet I wonder if he can do anything for himself by his ability and understanding. Is he only the means used, the bridge over which the message goes to others?

You know I feel more and more every day that after all, our family, our blood relations, should mean more to us that they sometimes do. Looking back over my life, I am dumbfounded at the patience and love and help that have been given me. A friend once told me that she thought that whatever of life she experienced, whatever hardships or unpleasantness life held for her, she would not complain, for maybe she could have had her lot, her fate, fall on a lower scale. Who was she to complain or demand the best? Those are not her exact words, but something of her idea. But I don't feel that way. While I seem to be predestined to experience life to its fullest at first hand, though God knows what causes and reasons, still I know that I was not predestined to any plane. I can climb as high as any ability or ambition within me will enable me to. I accept no limits to my growth, only those that time brings. We all have our limitations, born in us, but I still try to feel most myself with my head high. I'll make my mistakes, and they are many, and swallow them like bitter medicine, for only I know who has suffered the most from them. But through them all I have gained a feeling that I know is growing. I can stand on the top of the earth and feel something within me that rises up to understand everything, and I would not exchange my spirit for the whole world. Sometimes I feel a power within me that gets so real that I could save the world through my own belief. As I told that woman who lost her boy, if she couldn't believe, I would believe for her.

After all, we are each of us nothing but a walking mind, a walking attitude toward life. We can go and stand before the people we know and touch them, but if we don't know their minds, we don't know them at all. Even the best religions known to mankind leave us to work and think it out for ourselves.

I can't stop thinking. I find few who can understand these things, but many who don't want to.

Yes, forget it all—and go to work. Sweep and wash and mend. Sew and teach and talk, and then pick up a book, let your eye glance over a bit of verse, look at a fine painting, go out under the stars at night, listen to good music, and it all comes trouping back again as alive and strong as ever, as full of meaning and question.

I don't know why, but lately I have been thinking of you so much. Not the you in the flesh, but your life as you've told it to me, and as I've felt and understood it. Your long sufferings and heartaches and longings. The passages I've read in books that you've marked for me; your old loved poems in the blue scrapbook; your letters, your courage, your beautiful music; your constant fight for the best that was in you.

And some day you will be gone, and I know now that I will be so alone! I have so much of your life in me, so much of your feelings, your understanding.

I feel that now is the time to tell you that when you are taken from this life, you'll know, I hope, that I, left behind, am not ever forgetting you, believe me.

I think we all avoid these subjects, but they are after all the very essence of our being. We feel them so strongly that we are afraid of them, and keep them buried deep down; but ever since my brother's death I have felt so strongly something I never knew before, a peculiar realization of life and death and all it means. And oh, it always happens so quickly, so unexpectedly; and through it all we are hedged in, and held back, hampered by this wall of flesh. Why does this understanding always come—too late?

I am trying hard to make you feel that these are my thoughts day after day and in the dark nights. They are a present to you. I love you, I value you, and I am filled with a fear that the day may come when I cannot say it, cannot let you know or understand how I feel. I know of nothing much

*materially that I can give you. You have given me so much. But somehow
I am so impressed with the deep realization that every one of us is so
hungry for love and sympathy and appreciation.*

*So think on all this, and believe me, I am always thinking of you. It's
the fellowship of kindred minds.*

Later:

*You'll remember these words of Othello in explanation of his love for
Desdemona: "She loved me for the dangers I had passed, and I loved her
that she did pity me." It came to me that these letters of ours, from the
"Land of Can Happen," seem so full of our trials and tribulations—the
dangers we have passed.*

*I'm beginning to believe that Longfellow is the only sane poet, and the
"Psalm of Life" the keynote to our existence. Charlie Chaplin earns his
salary because he makes us forget the hard facts of our lives for a while
as he lifts us up by mirth and hilarity; for we slump down again at the
fadeout, looking at each other and wondering why we laughed, and, seeing that we did, why don't we now?*

*Reading gets to be an opiate. If you read solid works, you're like Bun-
yan's Pilgrim, the slough only gets deeper. For instance, I sent for a copy
of* Elbert Hubbard's Scrap Book *when I was laid up with malaria, and
read and re-read it. It was full of quotations and extracts from Cicero
to Woodrow Wilson—masterpieces all, but causing one to realize how
many, many people have given their best thoughts to elucidating the ideal.
I closed the book and considered, and it seemed to me that the words
"Howe'er it be, it seems to me 'tis only noble to be good," would cover the
whole case.*

*I guess we're getting old—and were we ever young? I still get a great
pleasure out of anxiety over Kipling's health. Kipling, who wrote the
immortal "Chant-Pagan-Me."*

> *Me that have been what I've been,*
> *Me that have seen what I've seen.*

*We seem ever on the alert for some kindred soul that we might possibly
invite home to tea and a little music and much talk, and behold, there are
none; for if circumstances compel you to withdraw from its society, the
world ignores you, and if like Diogenes, you try to use the touchstone of*

just plain intelligence for a lantern, to find some poor soul like yourself, who enjoys Kipling, and the Bible, and poetry and tea and solitaire and cross-word puzzles—and music, and arguments and ideas, and plans— why, you find nothing. You're growing old. And you laugh and say, "Poor fools." And so it goes.

No news is good news, for my continual letters prove that our days are full of chance and disaster—sickness, trouble, and sorrow. No wonder my heart leaps up when I behold a rainbow in the sky—they are such rare occasions.

Wayne is in town buying corn from a carload that came in from Oklahoma today. Jimmy rode in the wagon with him. Donald is busy studying the New England states. I am trying to catch up on my letter writing. We are feeling well at present. The weather is very cold, mornings, and although the sun does come out during the day, it gets dark around five-thirty. Last week it was cold and cloudy and rainy. I haven't been to town for a long time. Maybe I can go tomorrow if the car can get through the deep muddy ruts in the road.

Three of four days of clear weather now will mean between five hundred and a thousand dollars on the hay. We have rented a gasoline baler and will begin the baling in the morning; but the sky is overcast and it may rain again.

The days have been cold, with heavy frosts and ice on the water outdoors. From the hottest of weather we seem to have plunged into real winter. All the little houses are full of cotton pickers now, new people from out in the hills. They are hard at work, but they have to gather the bolls with the cotton, because the wet weather rotted the seed and softened the bolls. So this year's seed-cotton is mostly a total loss, and the cotton itself is inferior. The bottom has dropped out of the market. Picking ranges from one dollar to a dollar and a quarter a hundred pounds. Selah! So much for our tribulations.

Everyone is looking for snow, for there is a cold wind blowing and whistling around the house, and the ground is frozen. It gets dark by six o'clock and the big heater is the center of attraction again. Wayne often plays solitaire to ease his mind. We have always contended that solitaire was a game for an empty mind or a troubled spirit, and now we are finding why so many hide their lives in cards or books—to keep from worrying or thinking.

I have put peanuts in the oven to roast and a big pumpkin on to cook, for Thanksgiving is but a few days away. Young roosters are running all over the place so we can have fried chicken whenever we want it.

Now the moon is full and sails in the northern sky. That is a sign of continued cold weather. It is ten degrees above zero. Since the latter part of November the weather has alternated between freezing and sudden changes to warmth, with cold rains and driving sleet. We are housebound, and although we burn lots of wood, we have lots to burn—a great blessing. The cotton is still in the fields and the picking goes slowly. I marvel that any of it can be picked in this weather, for the hands go out one day and get part of a day's work done and the next day it is raining or freezing and it is impossible to work.

It is not uncommon to see an old mule dragging a long narrow homemade sled up and down in the slush, while a group of pickers sit or stand on the sled, picking the cotton where the ground is so wet that they cannot walk between the rows. The poor mule will poke along, stopping so often that the cotton can be easily gathered from the rows on each side. On a very cold day an old bucket is often carried on the sled in which a tiny smouldering fire is kept burning so the pickers can warm their chilled hands occasionally.

We had a quiet time at Christmas. The deer sprouted oranges on his horns again, as he has done every Christmas, and the boys enjoyed their air rifles. They are getting too large for toys and games, and spend much time reading, or out with their ponies. They study regularly every day, after all the chores and daily tasks are done, and are trying hard to finish the fifth-grade work by fall.

Tonight is a bitter cold night with a cold wind blowing. We gather around the living room table, Jimmy playing with the cat, Donald reading sentences out of the old McGuffy's Spelling Book. The boys are as eager to learn as two magpies and I have to keep the big dictionary always handy. They will believe anything I tell them, so when I begin to answer all their questions I find it a hard matter to know just what to say. Today they told me that they must be real Christians because they liked everybody, they didn't smoke or drink or dip snuff, and they always did what they were told, brought in the wood, went for the mail, and studied hard. Really now, what could I say? Actually they are wonderful for their age, so good and so sincere; but maybe we were all like that once. Pity we couldn't stay young, isn't it?

Many of our neighbors have had the flu and pneumonia, and one has

died of pneumonia. Nobody goes anywhere these days. They only "aim"
to keep well and get the cotton picked as fast as possible. I notice that the
greatest struggle of all is just to get through the days and keep warm. Al-
though the days are clear and cold, I think it is getting ready to rain again.
I should say "fixin' to rain." Donald has just finished piling wood on the
back porch. Jimmy has been up at the barn milking his cow and feeding
the horses. Judging by outside appearances, we may have snow soon. It
seems a long time until summer.

I am reading Romeo and Juliet *through again in the evenings. The*
room is warm and comfortable and our oil lamps with their polished
chimneys really give a good light, but somehow we are always tired. Our
days are long and there is so much to do, and though we have been taking
quinine regularly, I think that malaria has begun to lay its hand upon us
all.

One night Wayne and I played our records and danced around the
room to the Merry Widow Waltz, wishing we could go to a real dance
with a good floor and good music. We have enjoyed the square dances
here, for they are good fun, but we miss the romantic atmosphere, the
orchestras, and our old friends that were so much a part of our lives at
the dances at home.

Generally we are able to keep comparisons from our minds, but then
something will happen that for a brief time will bring into sharp contrast
the old days and these present ones.

Wayne went up to Little Rock and did not get back until a week before
Christmas. He went to visit Frank, who lives near there, so we were late
in getting the gifts we wanted to send home. One evening we drove into
town, but just as we parked the car the town lights went out. We went
into the drug store, as we saw candles lighted there. The lights had been
going off and on all evening, we heard, but most of the storekeepers had
shut up shop and gone home in disgust, although generally they stayed
open for business until quite late during Christmas week. Apparently no
one in town could fix the lights, and the power plant was across the river.
Anyway, with the aid of the candles at the drug store we hunted around
and finally bought some things, and stood by while the clerk wrapped
and tied them.

We left the packages with the druggist to mail and went around the
block by the post office to mail our letters. It was pitch dark. We had to
strike a match to find the letter slot, and then we both fell over a pile

of wood that had been left out on the sidewalk in front of the bank. We stood and laughed like two fools, there in the dark, thinking how funny it all was. It takes so little to amuse us, for there is so little that breaks the monotony of our days.

The hardest part of the bitterness of home-sickness I have managed to keep in the back of my mind, for that makes the days seem easier to bear. Yet the cold weather and the bad roads keep us from going off the farm, and, lacking opportunity for pleasure or intelligent company, we are always remembering old songs, and music, and dance tunes; books we have read, places we've seen, or trips we've taken. It seems that not "half our hearts," but half our lives, are buried there in California.

Your loving daughter

43. Rain at Night

WHAT UTTER, UTTER COMFORT was the spring rain at night. To come gradually awake and lie stretched out between the soft blankets, slowly realizing that the softly repeated "thump-tap—thump-tap—splash" was the rain drops at the corner of the house.

Only occasionally the soft monotony was interrupted by a hurried watery scale of drops that broke through the steady drip-splash; drip, drip; as though the water had enjoyed hiding beneath some loose shingle until with a watery gurgle it could make the landing joyfully, each drop jumping to the earth below in measured succession.

A careful search the next morning, with the sun up and the water all soaked away into the sand, never revealed a reason for that particular corner being the one source of the running water tones. Hollow depressions were left, however, in the shallow trench on the ground below the eaves, where the water ran from the porch roof, and I always found several larger smooth hollows to show just where my musical drops had fallen.

Now, listening to the spring rain in the darkness, my mind flooded with childhood memories. Just the comfort, the old oft-remembered comfort, of being held tight and warm, while the old rocking chair swayed back and forth on the floor where a board squeaked always beneath the pressure of the rocker, with a peculiar human two-toned pitch that just, almost, made words of reason, but generally made the half-sense of a person doped with sleep. "I-yes," or "Now-no," high-low, high-low, "Squeak, crunch, squeak, crunch." The memory of my mother's arms and hands that held me so tightly even when my sleepy, relaxing body was half-slipping from the comfortable lap. Tightly, warmly, comfortingly held. No need to think of any hurt or loss, any more. Forget it all; let go. "Squeak, crunch," back and forth; eyes shut tight; sleep-sleep-sleep.

Oh, I wanted a baby of my own, again. A baby that I could hold and rock and sing to.

I dreamed often of a baby, a baby that leaned over my pillow and patted my cheek. It lay so warm in my arms. I never saw its face, only felt its soft body and tiny hands.

Maybe if one's hand or foot could be loosened, so that it could be picked up, whole, but separate from one's body—maybe that might give some idea of just how deep is the feeling of personal possession with a baby. So much is it one's very own flesh and blood; separate, yes, but so consciously part of one's very self.

44. Harley's Wife Dies

IT WAS ONE DAY in the spring, when the weather was alternating between days of freezing winds and warm spells that only made one feel the cold more keenly, that word was brought to me that Harley Suggs's wife had had a baby. They lived back behind Grandma's, on the farm behind ours—toward the bayou, in a little house next door to old Mr. and Mrs. Batson. The baby was a big nine-pound boy. Next I heard that the mother of the baby was quite ill with child-bed fever. Two weeks later I heard that the young mother had died.

The doctor told Grandma that it was purely a case of personal filth that had caused her death—but just how these poor people could be sanitary with little money to buy soap for laundering, or how they could ever have enough changes of bedding or night clothing, was beyond me. Besides, they didn't know what sanitation meant, anyway. They were clean, in a way, on the surface. But never having seen a germ, they didn't believe in them.

So Harley's young wife lay dead. I had been too ill during this time to visit her, and the doctor had sent orders for all of us to stay away from her place, for her infection might be carried to me, as it was very contagious, so all the news was brought to me. Under the doctor's orders, and much to the parents' dismay, her mattress and bedding were burned in the yard. Harley walked the six miles to town and bought his wife a pair of black patent-leather slippers. She was laid out in a white cotton nightgown and the new slippers, "for she wanted a pair of shoes right bad," Harley said, "and I know she'd think these right pretty." I found out that she had actually been going without any shoes and had even been picking cotton, in all weathers, in her bare feet.

The pine box was bought in town, but due to the doctor's warning, the neighbors were afraid to go near the house; yet Mrs. Barger went determinedly to the house to help Harley and his old parents to load the box onto a farm wagon.

Wayne, meanwhile, had phoned to town to try to get a preacher to go out to the burial ground, but the three who were called very bluntly refused to go because they didn't want to take their cars out in the snow, and because the weather was too cold, and anyway, these people were just "poor whites." There was about three inches of snow on the ground at the time. Old Mr. Batson went along to preach a sermon, I heard. Just Harley and his parents and Mr. Batson went out in the hills to a little graveyard. Later I learned that they all took turns digging the grave, then they put the box in the ground and covered it with the red earth. Old Mr. Batson made a prayer, and then they came home again.

Harley's parents took the baby to raise. They fed it from the start on cold cow's milk, in order to save trouble and get the baby used to that diet, for they would be unable to undertake any particular care regarding it. They had to get back to the cotton fields.

While the cotton was growing and was up about four or five inches, the hoeing had to be done promptly and with care. A good hoe-hand would step down the row quickly, swinging his broad-bladed hoe expertly, taking out at each stroke all the little cotton plants necessary to space the groups of three that were left together to make the stand. As the rain came and the sun shone and the weeds grew, the plants had to be hoed out many times. It was a good sight to see the little plants, that grew so quickly, cleanly hoed and evenly spaced, and all the field hands took great pride in their work and seldom needed overseeing. They liked to start down the rows together, always, and natural competition generally kept all the fields in first-class shape. The work was so much harder for them if they slacked in their efforts to rid the rows of weeds.

But this did not apply to Dove Crane. She dawdled along by herself, falling farther and farther behind. Wayne, out hoeing, too, would often help her, thinking that a little encouragement was all she needed. She was very sullen, though. He showed her over and over just how to hoe, and how to leave the three or four plants together in each group—taking out the weeds and the extra plants deftly. When his back was turned, however, Dove would walk on the rows and leave great skips—as she hoed along with bent head, as if in a trance. Her work became so bad that at last Wayne ordered her to the house. She was ruining too much cotton so he could not afford to let her work. She went from the field very reluctantly. Then followed a scene with her father, for Crane demanded to know why Dove had been refused work. Finally Crane became so abusive that Wayne and Henry ordered him to take his family and leave the place, and we felt we were well rid of them all.

After the Cranes left, that house stood empty until in the fall when the cotton pickers began again to come down into the bottom land.

Wayne met Crane at Cotton Town later in the fall. He had come in with a bale of cotton to gin. He said he was now working for a farmer down in the lower bottoms. Crane asked Wayne if he was still mad at him. Wayne said he thought of a hundred things all at once, and finally settled for a non-committal, "No, I'm not mad at you. Never was."

It was nearly time now to finish up the work and leave it until time for picking. The stalks were nearly four feet high, all ready to make bolls. Because it had been a wet spring, the heavy growth of weeds had been hard to dig out of the ground; they could not be cut easily with a deft swing of the hoe as was usually the case. Now the work should be finished immediately, the dirt scraped up against the young stalks, for a rain was coming and the cotton was unsupported.

Bunk told Wayne that he was going to take Mutt and Jeff and plow his cotton again. Wayne objected to this, and said that it was the cotton farther over in the fields that needed work badly. Barger's crop was all finished except for running out the middles where the pickers would drag the long cotton sacks. That work could be done anytime during the next thirty days, but Bunk wanted to finish his part of the crop completely and refused to do the two or three hours' work so necessary to the rest of the crop. He grew obstinate and insisted that he would do as he pleased and was going to take out the team. Finally Wayne ordered him to drop the neck-yoke which Bunk had picked up, and either go on and do as he was told or go to the house. Bunk refused to do either, and asked Wayne if he wanted to fight, advancing with the neck-yoke in his hand. So Wayne had to fight. They fought until they were worn out. Each was scratched and bruised and breathless, although in reality few blows were struck. Anger and resentment were surging so high that they soon exhausted themselves.

It was a very hot day, and the argument had started as Wayne and Bunk met on their way to the barn lot after dinner. Bunk finally reached for a monkey-wrench that lay on a wagon-tongue, and Wayne took out a file he carried in the hip pocket of his overalls, and flourishing these weapons, they advanced and retreated in the hot, loose sand.

Wayne had his scalp gashed so that several stitches had to be taken later—and Bunk had a swollen face and a black eye.

The upshot of it was that Bunk rushed away to town so as to beat Wayne there and report the fight to the sheriff. He wanted to put in his complaint first. As was to be expected, both Wayne and Bunk had to pay a fine.

Of course the next step was to get rid of both Bunk and the Bargers, so Wayne and Henry settled with them for their share of the crop and let them go. Before any agreement had been reached, however, each morning on the three successive days we were awakened early by Bunk coming to the front of the house and calling out, "Hello Wayne. Are you going to let me plow today? Can I take out the team?" Each time Wayne's answer was a cold refusal. Finally I noticed that while Bunk was doing this parleying, Barger was crouched just below the level of the bank holding a shotgun pointed at the house. Just what they planned or hoped to do we could never figure out.

Just before they moved from the farm a man called on Wayne one day, but refused to come into the house. He insisted upon sitting out on the ground below the front steps. That made me curious so I hung about to hear what the man had to say. He persuaded Wayne to sit down beside him so they could talk together, and as they settled themselves he pulled out his knife and looked around for a stick on which to whittle. The year before, he said, Bunk and Barger had worked for him, out in the hills, and when they had left him at last they were owing for groceries and supplies and the doctor's fee when the baby came. They owed him much more than their share of the crop had come to, and he wanted Wayne to pay him or give him a lien on their part of the crop now. It happened, however, that with the cash that they had drawn each they still owed us considerable, so there was nothing to be done about it. During all this conversation the man from the hills sat there aimlessly whittling on the stick. He had a vicious-looking knife, much too large for a pocket knife. I decided that I would stay right there until the man left, for I had often heard of the Southerners who would never engage in any argument or talk unless they had an open blade in their hands, so they would be ready in an instant should the discussion develop into a fight.

I heard Wayne explain the deal we had made with Barger over and over again and yet the man seemed dissatisfied. So finally Wayne said, rising abruptly, "You might as well put that knife up. It won't help you one way or another, and I have no more to say." The man looked baffled, and angry, and went away without another word.

45. Pink Lemonade

SEVERAL MILES AWAY, out toward the lower hills, there was to be a Fourth of July celebration. For weeks I had been hearing about it and was just as anxious to go as the children were.

A big open space in the woods at the edge of Centerville had been chosen for the site of the speakers' platform and the dance platform was nearby.

Big freezers of glaringly red ice cream, and barrels of pink lemonade, with the yellow peel floating on top, were all ready to serve to the gathering crowd.

At noon everyone settled with friends in scattered groups on the ground under the trees. We put our dinner with that of the Wood family, our former share-croppers who had moved to the hills, and lounged with them on quilts in the shade of the trees most of the day. Children ran everywhere. Dogs barked and chewed chicken bones and romped after the boys. Mothers with tiny babies and young children were having a hard day. There were many automobiles, but more farm wagons. The mules and horses had been led away to a nearby pasture, but the wagons were left under the trees, where the younger children could be put to sleep on the piles of home-made quilts.

Wayne and Jimmy and Donald were off together, and Grandma was deep in a discussion of Moody and Sanky's hymns with some old friends, so I curled up to rest on the back seat of the Ford, awakening now and then, from my deep sleep, to see the people all around me going aimlessly back and forth, or hearing the deep impassioned tones of the political orator, intoning and gesticulating as though the future of all the race lay in his fervent words. I watched the bright balloons in the hands of the smaller children, and heard the cries of the crowd that gathered around the horse-shoe pit, and the wild cheering from farther away, where an exciting game of baseball was going on.

As the afternoon wore away, it was a pleasure to see and hear the smaller groups that met and shook hands and settled on the green grass beneath the oak and hickory trees to talk in low, friendly voices. I was introduced to so many that I could not remember their names, realizing only that I had never met so many McKenzies and Georges before at one time.

Everything I saw and heard that day was so true to my memory of a big Sunday-school picnic I had attended as a child that I enjoyed it all doubly, surprised and glad to see even the red, white, and blue bunting tacked on the planks that formed the booths, and the boxes of cheap popcorn that were for sale, and the children's races that caused all the usual intense interest and jealous bickerings.

Even the weather played its part. The day began sunny and warm, with only a few scattered cumulus clouds that heaped themselves to the north. A breeze made the heat quite bearable all through the middle of the day, and kept the flies from settling on the food. The ants, only, proved a nuisance, getting into the boxes and baskets and making it impossible to rest long on the grass.

Before the baseball game was over, however, the clouds had thickened and blown across the blue sky, and a few drops fell.

We made haste to gather up our possessions and, after we were in the car and ready to start for home, we watched the unlucky ones who had rambled off from the crowd and now had to round up their teams and hitch them to the wagons in the pouring rain that grew heavier every minute. Loud peals of thunder and occasional lightning flashes added to the excitement. Such confusion! Wet quilts, lost children, and mothers screaming orders to youngsters who paid no attention. The men were soaking wet as they struggled with the frightened teams and loaded their families into the wagons, all facing the long cold drive in the storm, with no more protection than that afforded by the quilts that could be wrapped about them.

When we finally left, the baseball game had broken up, unfinished, the last of the wagons were pulling out, the small stands where the bunting had fluttered gayly in the summer breeze were soggy and forlorn, and wet paper lay everywhere.

"A sorry picnic," Matt Wood commented to us later. "Don't know when I've seen worse. No turn-out at all. No life to it. I declare I felt right ashamed. I saw you laughing and talking to the folks right well. And say, we did have a fine dinner, didn't we? Hettie sure can cook a good picnic dinner. Don't know as I've ever tasted better grub."

I had forgotten the part that transportation played in the daily life of the share-cropper until the next week when Mrs. Batson came to wash for me.

"I heard you-all went Fourth-of-Julying," she said with a grin. "Have a good time?"

"Yes, "I said. "I did. It was too bad though that the rain came and broke up the day. I think they were going to have a platform dance in the early evening. We would have stayed to see that."

"Well," she said, as she began to pump the tub full of water, "I ain't bin to ary picnic sence we lived across the river. That's bin five years now, I guess. Folks has got to have a way o' gettin' there. Dawson's folks all went, I see, but they wouldn't let us have a team. 'Fraid the old mules would be too tired for the plowin' today. I was glad to see the rain come. Knew he wouldn't be able to plow over in the black land today. Serves him right. No, I ain't bin nowheres 'cept to the camp meetin' last fall fer so long I forget when. Them Coatses had room to carry us along. I saw 'em goin' by early. A body sure needs a wagon and team, in this world. O, well, guess it ain't no use to fret. Them as ain't got jest ain't got, that's all, and nothin' you can do about it. Me, I took me a good sleep yesterday, anyhow. Been sorta chillin', and a little extra sleep don't do no harm."

Soon after the Fourth of July picnic we went to a dance "out in the sticks," at the home of Mr. and Mrs. Joe. They had two sons, both in their teens, and as several nieces lived nearby, their aunt was glad to supply entertainment for her own children and kinfolks, and so kept them all from wandering off on Saturday nights to seek amusement at other places.

Her house was a low, rambling log house, made doubly interesting to me because of the several fireplaces and the fact that I found myself going down three steps into the dining room and another two steps to the kitchen and another step down to an enclosed porch-room that was built to shelter a deep rock well of mountain water, cold as ice. The buckets were hauled up on a windlass.

The dance itself was in a small room at the end of another ell. As Wayne and I sat chatting with the family by the fireplace in the huge front room, that contained besides the living room furniture two huge poster beds, the other guests began to arrive. They threw aside their wraps on the beds and continued on through two other rooms that opened into the farthest room where someone was playing a phonograph. Usually there were violins and banjos and guitars to supply music, for many of the young men in the country played an instrument.

Several times the big front door, made of heavy planks, was thrown open and in would burst crowds of boys and girls, who, with only a passing glance at their host and hostess, kept right on through the house, saying only, "Where at is the dance?" When I expressed my surprise at this, Mrs. Joe assured me that it always happened so. It was impossible to invite young people to the house for a dance without others hearing of it, and to some people a dance was a dance, no matter where, and the formality of an invitation was not even considered. Mrs. Joe went on, "After all, what does it matter? It keeps my boys at home where they belong, and those who come unasked don't do any harm. Only thing is, I've learned to hide the refreshments. I always make a big cake and have coffee, but many times when we are talking here by the fireplace with our own friends who drop in for the evening, we forget to watch the young people. I found out that those outsiders would locate the cake and make off with it, so now I just leave a big basket of chips by the kitchen range, and all our regular crowd know what to do. They just wait on themselves. Some go down to the cellar and get into my crock of pickles, someone heats up the biscuits and makes coffee; several of the boys like ham and eggs. I never bother my head about it. They just go down to the kitchen and tend to things as they like."

It was an excellent idea, I found. We, ourselves, went later, and lifting off the clean tea towel that covered the dining table, found several fried pies and jelly and corn bread. Somehow the coffee made in a stew-pan lowered down onto the blazing chips had all the aroma and flavor of the best "camping-out coffee." We rummaged for some forks, and I found a tall old cupboard full of ancient dishes, all flowered in beautiful, deli-

cate colors, and old cups and big plates covered with interesting patterns.
There were huge covered tureens and cake stands and cruets and egg
cups and covered butter dishes that belonged to the family. They were so
much in the way that they had been placed here to be used only if necessary, but I thought them too rare for general use and wondered if some
of them were not valuable antiques.

Back to the dance room we went again, and round and round to the
strains of "Dardanella" and "Let Me Call You Sweetheart." The floor
boards were more than a foot wide, worn smooth and shiny by much
use. An old high-backed church pew was the only seat, but no one sat
down. As fast as one record was finished another was put on from the
stack on the little table. Years old, some of the records were cracked and
scratched, but anyway, it was music, and if the needle stuck obstinately
again and again, it made only a laughing diversion.

46. The Lady Moon

NOW AS I SAT out on the porch I was making little dresses and soft
gowns and flannel strips. There was a pink box by my side and I often
picked up an armful of soft baby-clothes and cuddled them close in my
arms, and tried to imagine what it would be like once again to hold a
baby.

I embroidered and sewed fine seams, then closed my eyes and relaxed.
I could hear the hum of the bees in the flower garden close by. From the
bodark thicket came the blissful cooing of the nesting mourning doves.
The breeze that caressed my face swept into the high pecan trees and
sang a lullaby to the tree tops. Every few minutes I heard the high, clear
call of the mockingbirds that was answered from the field behind me and
from the green meadow across the river. It was the fullness of summer
and the beautiful October moon was coming to her fullness above us.

I dozed a little and rocked back and forth in peaceful happiness.
Through the embroidered design that twined around the tiny neckline
I fashioned a pattern in minute stitches, to please my own fancy. Here
I embroidered several tiny star-fish, exact copies of the pink sprawling
shapes as I remembered them on the beach when the tide went out.
There I outlined a spider's web, sheer and fine, and grouped below were

three little birds, and two fat bees. It all represented what filled my mind. I was so happy, I was so contented. The very air around me was soothing and caressing and filled with a low hum of lazy summer life.

Jimmy and Donald stood by me and watched all my preparations with interest. They looked with deep satisfaction on every detail and were always eager to help. They took turns working the treadle of the sewing machine when I hemmed some sheets, sitting opposite me and very earnestly reminding me not to get too tired. They came and anxiously asked if they had not better stay home with me, when I knew they were eager to be off on a trip to town. I never ceased to be amazed at their never-ending thoughtfulness. When I was in that first period of extreme stomach irritation, and at the most inopportune times would crave things to eat that we did not have at home, they always insisted upon going at once on their horses to the little store a mile away to see if they could find something to please me.

It would be pickles, sardines, canned salmon, or olives that I craved, and so many times just a taste or smell of the new food was enough and I was satisfied. Many times though, after a day of intense craving for a certain food, when they finally were able to bring what I wanted, it turned out not to be the thing at all, and that was what pleased the boys, for they could then have the treat for themselves. Child-like, they often prompted me to consider whether there was not something I especially craved, hoping that they might share in these extra foods. They were at the same time deeply impressed, I could see, at Wayne's patience and helpfulness during those trying days. We all seemed to be waiting together for the arrival of the baby we so longed to have.

Old Zula came often to wash, for the crops were laid by now, and her smiling black face and kind words comforted me. Although she was the mother of ten children yet all her talks to me were filled with such a matter-of-fact acceptance of a woman's lot in life that at times I felt as though having my baby was a matter of only passing importance. Old Zula had none of the ghastly experiences to tell me that Mrs. Crane had insisted upon my hearing. All through the spring, as soon as Mrs. Crane had found out that I was expecting a baby, she had come often to the house to visit me. She was so grateful for the help I had given her that she wanted to repay me if she could, but I did not want her around. She worked every day now in the cotton, carrying her baby like an Indian on her back, but she managed to plod through the sand past our house on her way to Grandma's every day. It had become a custom to let her have some milk every day for her baby, so she carried two empty buckets and borrowed a few eggs and some peanuts and anything else she could cajole from Grandma.

One morning when Mrs. Crane had come to borrow some soap, she began telling me some of the harrowing experiences she had had, although I tried in every way to change the subject. I did not want to carry the thought of such dreadful things in my mind. I edged her from the living room and out into the dining room at last, while she still continued the account of one of her babies being born as she was riding horseback. By the time I maneuvered her to the kitchen I definitely ordered her to stop and go home as I felt I would surely faint. That only amused her, for she thought I was joking. She stood at last by the door telling me of another time that she suffered such agony while her husband lay asleep by the fireplace, the old doctor nearby, humped over, deaf to her agonized appeals, because he was drunk. That's all I can remember for I really did fall down in a faint on the kitchen floor.

The indelicacy of her expressions that I had overheard in some of her conversations with Grandma, and especially her references to me as "brooding," made me shrink from her. I pitied her, but I wanted none of my words carried back to Mr. Crane to be jeered at and made the subject of his scoffing and mockery. I wanted to avoid any chance of my learning of his possible crude remarks about my condition, for I knew that would cause me to droop and suppress this wonderful consciousness that pervaded me.

I visited my doctor and came away with a list of supplies which I was to have in readiness for the great day that was soon coming. We were almost sure that my mother would be with us, and I wanted to have my baby at home. The nearest hospital was about ten miles away, but somehow I clung to the desire to be in my own house. I had all the yearning of a little child that seeks comfort and warmth and safety in its accustomed bed at home.

The summer had passed away at last—a long hot summer, in which tedious strain gave way to patient endurance, and a happy sense of exultation came and stayed with me daily.

From the beginning I had relied implicitly, for relief from worry, on the advice given to me by friends in town regarding my choice of a doctor. I was told, "Just have *that* doctor. He'll give you ether, and the next thing you know, there'll be your baby." I heard this from several women and believed it literally, because that was what I wanted to believe.

I often conferred with the doctor, interpreting his assurances that everything would be all right and for me not to worry for he would take care of me, into terms of my own understanding. I did not tell him of the fearful, haunting memories of my past experiences which I had struggled so hard to be rid of. I simply obeyed all his instructions and spent my time in quiet domestic activity.

I felt so full of emotion at times. My thoughts seemed to seethe. I would sing for hours or laugh hilariously at trivial things. I felt that all the world must appreciate how wonderful it was that I was going to have a baby. I know my mind was more alert and keen than it had been. Malaria had been gradually wearing me down until I was losing all vivaciousness, but now again I seemed to be enveloped in a magic mantle. Every dear, sweet verse I had ever known stirred in my brain. Soft beautiful baby faces peered at me as I turned the pages of magazines. Music took on a deeper meaning. I was especially enjoying to the full, day after day, parts of *La Traviata* and the "Russian Valse."

I hardly knew sometimes whether I was living my own life or re-living the lives of others, but I did know that I was very, very happy, and I felt strong and able, and my thoughts tended to buoy me up and carry me through my days.

At last, one happy day my mother sent word that she was arriving. She was on her return trip from the East, where she had been visiting her own mother. How rejoiced we were. As she settled down into our daily living she soon saw how welcome she was and how much we really needed her, so she promised to stay until after the baby was born, and entered into all the plans we were making. Eventually we had everything in readiness. The doctor said I could not be better equipped even in the hospital, and that, since my mother and husband could be relied upon, he would prefer not to call in any of the neighbors. We would do better alone. Every precaution was being taken. The bed was set up on large blocks to raise it to hospital height. The rooms were scrupulously clean and everything ready for instant use.

One breakfast time, finally, I announced in a wild, hysterical manner that I was going for a walk. Upon being questioned I admitted that where didn't matter, I just wanted to go out alone for a walk. I wanted to get away from the careful scrutiny and concern of both my husband

and mother—away from everyone. So they let me go, out in the morning air, down the sandy road—alone. I started courageously, all excited, determined to walk far away to—some place. But I was able to go only a short distance when my strength gave out. Wayne brought me back. Late that afternoon I again felt suddenly that I must be alone. I must go somewhere—alone. The little house never seemed so small. It oppressed me. I wanted to hide. So I was again allowed to go. This time I went out to the hayshed and sat in some loose alfalfa hay and put my apron over my head and cried and cried as if my heart would break. Cried, because—a woman's reason—just because. Wayne again gently persuaded me to come in, and I knew now from my mother's anxious looks and unobtrusive activity that my waiting was nearly over.

Early the next morning Jimmy and Donald were sent up to their Grandma's to stay until sent for. Grandma was unable to be of any assistance but gladly welcomed Jimmy and Donald. She said little, but was very anxious, I knew, and she longed, too, to hold our baby in her arms.

The doctor came and talked kindly to me and stayed. I had only one request to make. I insisted that I be allowed to play my favorite record over and over and over while I waited, "and then," I said, smiling at the doctor, "it will be your turn."

So, all day long I tried to sleep, tried to rest, and played the overture of *La Traviata*, over and over, walking the floor with pain but with an inner sense of joyful expectancy. The doctor was amazed. As it grew dark he begged me to get into the high bed, but I would not. I wanted to do my part, fully and completely, I thought, and then, as I so foolishly believed, he would give me ether, and the next thing I knew, "there would be my baby—just like that." So the poor doctor waited, watching me anxiously, and wondering at my tenacity, and I waited too, the minutes stretching into hours. I succumbed at last. My mother stayed with me until she was on the verge of collapse and was sent from the room. I knew Wayne, at times. But the room seemed to become oddly dark and there was only myself, at last, only myself, not quite realizing where I was. The music was gone. There was only myself in a world that I so completely filled.

Finally when all time ceased to be, I came to the end—the end that seemed the beginning of unutterable glory. I was conscious and without pain during that most wonderful moment of life's greatest miracle, feeling and knowing that the firm, compact little body was actually making its final quick entrance into our living world. In one blinding flash of comprehension I was able to give all self to that rare sensation of ultimate accomplishment. Then I made the sudden drop into blessed deep oblivion.

To me, the most inspiring words in all the Bible are: "Eye hath not seen, nor ear heard, neither hath it entered into the heart of man, the things which the Lord hath prepared for those that love him." That statement seems to encompass and surpass all our knowledge and imagination.

While my little baby boy lay so soft and warm and sweet beside me, I used to whisper to him, "What makes your cheek like a warm white rose? I saw something better than anyone knows." I often thought of these words and the divinity of man—"Let us create man in our image." I felt that some of that magic baby knowledge does stay with us, although hidden and unrecognized, yet unconsciously creating the urge that builds upwards, ever, toward the ideals of mankind.

First the baby lay in a little pink-lined basket, a beautiful, soft, sweet-smelling shrine at which we worshipped daily. As he grew, we arranged a larger basket for a bed and carried this from room to room. The weather was cold now and the big heater was going all day and all night.

My mother said she was reliving the years of her own babies over again as she washed and dressed the beautiful little body.

I have no words to tell of the happiness that Jimmy and Donald found with their little brother, and Wayne never seemed to get enough of holding him.

We seemed to regain a sweet freedom in speech that we had been losing. Hourly we found ourselves searching for expressions of endearment that were fitting for this beloved child. The routine of our daily duties became pleasant and easy. Laughter and dancing and song and a general pleasantness had come to stay. It was as if that rare spirit that walks abroad through the world at Christmas time had come especially early to our little house.

We forgot our frustrations and disappointments, and the burden of our daily monotonous duties was lifted, as we gained joy and gladness in something perfect and beautiful and lovable that we could keep forever. Seldom does life yield such completely satisfying days as we enjoyed with our baby.

"The Lady Moon" was an old, old song that I often asked my mother to sing for me during my time of waiting.

THE LADY MOON

The lady moon came down last night,
She did—you needn't doubt it;
A lovely lady dressed in white,
I'll tell you all about it.

They put brother Len and me to bed,
And Auntie said, "Now maybe
That pretty moon up overhead
Will bring us down a baby.

If you can keep as quiet as can be,
When all the room is shady
She'll creep across the window sill,
A beautiful moon lady.

Across the sill, along the floor,
You'll see her shining brightly,
Until she comes to mama's door,
And there she'll vanish lightly.

And in the morning you will find,
If nothing happens, maybe,
She's left us something nice behind,
A beautiful star baby."

I didn't quite believe her then,
For auntie's always chaffing,
The tales she tells to me and Len
Would make you die a-laughing.

So when she stooped and kissed us both
And then went out a-humming,
Len said, "There isn't any moon,
Or any baby coming."

I thought myself it was a joke,
But still I wasn't certain,
So I lay quiet in my crib
And peeped behind the curtain.

And all at once, what do you think,
Without a minute's warning,
Both Len and I went right to sleep,
And didn't wake up till morning.

And there stood Auntie by the crib,
And when we jumped and kissed her,
She said, "You naughty sleepy heads,
You've got a little sister."

And sure enough, it was no joke,
In spite of Len's denying,
For at the very time she spoke,
We heard the baby crying.

Oh, then, we jumped and made a rush
For mama's room that minute,
But Auntie stopped us, saying, "Hush,
Or else you can't go in it."

I think it clear as clear can be,
As clear as running water,
Last night there was no baby here,
So something must have brought her.

At last I was up and around, but we needed help for a while, so I sent for Celesta, a young girl I heard about who lived down the road. Her family were share-croppers, and she had always worked in the fields, but it was only after a lot of persuasion that we got her to come up and work for me in the mornings. She said she preferred field work to housework, but finally consented to come, because I was so ill, and she wanted to be friendly. She ironed and swept and did some cooking.

So that we would not again be accused of disturbing the labor system in the community, Wayne agreed secretly with Celesta to pay her a dollar for the half-day's work. She refused, however, to do any washing, and as Mrs. Batson was now picking cotton each day, and Zula too, that daily task fell to Wayne's lot. He went at it with a will, however, and we could hear him outside whistling cheerfully as he pumped water, poked the driftwood tightly under the big black pot, stirred the boiling clothes and rubbed hard on the washboard. I never fully realized what a big job washing really was until I watched Wayne resolutely begin to sort the clothes into various piles, and when the flames were leaping up around the pot, saw him stand there seriously considering the task ahead of him.

Celesta wore old-fashioned buttoned shoes, with very high tops and high heels, a very tightly-laced corset, and a stiffly starched gingham dress that stood out from her body in all directions. She was a big pink-cheeked country girl, who had lived up in the Ozarks. So much of my time was spent in bed, under the doctor's directions, that I was not always up when she was doing the morning work, and I used to lie and watch her mince about, holding herself away from everything, working slowly. She often took two hours to wash the breakfast dishes for four people and sweep two rooms.

Of course, she spent some of this time leaning on her broom, talking. She had a beau, she told me, and was engaged to be married. Every day she would tell me what he said to her and what she said to him, and my mother was daily amazed and entertained by Celesta's stories.

Her beau had been married before, she told us, but his wife had taken their two children and left him some years ago, and now they were divorced. To offset the gossip of the neighborhood against him he spent much time, it seemed, explaining to Celesta just what their trouble had been and how well he had really treated his former wife. In proof of the fact that there were no hard feelings between him and the wife, he told Celesta that the previous summer, business taking him to the next town across the river, he had one day met the woman along the roadway as she was nearing her home. She had married again since leaving him, and had two young children by this second husband. He said he walked along the road a piece with her and inquired after her health and that of his two children and asked her if she were happy. She said she was and went on to tell how well she was getting along, turning at the gateway to say in the casual Southern manner, "Well, come and go home with me."

My mother said she'd remember all this to tell when she returned to

California, for at last we had to let her go. Again we found our emotions at the point of overflow. Four years now since we had left California, and still we could not even guess when we would be able to return, nor could we point to any more than the dogged determination that still possessed us to do all that lay within our power to raise excellent crops. This we had indeed done, over and over, yet we were conquered each year by the enemies of every farmer around us—the boll weevil, the army worm, sickness, the weather, and the low price of cotton.

We hid our sorrow under the guise of the enjoyment we were having with the baby whom we named Gareth, and my wise mother, understanding only too well, left us to our happiness and made no opportunity for the expression of either hopes or fears.

Again we took up our daily round of tasks. Again Jimmy and Donald settled to daily school work. But now our days revolved around Gareth and for the while we were content.

All during the fall and winter I watched Gareth with the greatest care. It was a problem to keep him warm, for late at night after the fire was out, the house grew chilled, and Gareth in his basket was cold, despite all the means we devised to keep him warm. The doctor advised me to let Gareth sleep in the bed beside me, and this I did. Contrary to my previous experience, and in opposition to the suggestions given by both my mother and Grandma, I decided to care for Gareth in my own way. I was saved all the work of caring for bottles and preparing milk formulas, for I nursed him myself. I found that the really old-fashioned way of feeding a baby when it was hungry, instead of waiting for the exact time set by a schedule, worked out amazingly well, for our baby never cried and screamed as I had heard so many do, but eagerly nursed whenever he was hungry, seemed perfectly satisfied, and went off to sleep with no trouble.

I disagreed with many modern ideas on baby care, and determined to enjoy my baby to the full. I rocked him every day in the big rocking chair, I sang to him, I cuddled him to me and talked to him as though he knew every word—as I pretended he did. Every night he lay warm and snug in my arms, his little feet braced against my side, and I would lean down and listen to his breathing that was so soft, so even and gentle, and almost inaudible. Surely he was that very baby for which I had yearned.

47. Day After Day

Extracts from Letters to a Mother

WAYNE HAS GONE *to Little Rock on business, but we are expecting him home daily, for he has been gone a week. A slight flurry of snow can be seen from the windows and a cold wind is blowing. My, but it is cold, cold! The porches are stacked with wood for the two stoves and there is a big pile out behind the house. Mrs. Joe sent us a lot of fine sausage. Always in the winter time the fires and the meals become of first importance.*

We are looking forward to our trip out to the hills again to get a fir tree for Christmas. We want to hang a little candle and make a little silver star for Garry. Jimmy and Donald feel that the oranges on the deer's horns and our "Mary Christmas" [sic] are all a vital part of the holiday. I have been rereading Gareth and Lynette *by Tennyson and finding myself more than pleased with the name I chose for our baby.*

There is another new family now in the house down by the mailbox, and Dale, a boy about the age of Jimmy and Donald, comes up every day to visit. Dale's only brother is much older than he, so Dale enjoys playing at our house and shares the games and toys, even helping cheerfully when wood is to be carried or water pumped. Although he was shy at first, we have finally convinced him that he is welcome to come as often as he likes. He loves to talk and we are a fine audience. Somehow he always seems to be in high spirits. We laughed with him as he told us how his uncle, who could "throw his voice," used to ride along with some farmer and suddenly call out as though from the bushes along the roadside, "Stop that team, stop that team!" Sometimes the uncle would call as though his voice came from the wagon bed where the cotton was piled high, "Let me out!" Dale would get up and act out the way in which the farmer would yell, "Whoa," and stop his mules and search in a bewildered manner for the hidden man. All our meals are very hilarious occasions when Dale is present.

While we were waiting for Wayne's return we opened your Christmas box and took out the Scotch records you wrote about. We all danced and pranced and sang, and put the chairs down the center of the room and danced and cake-walked between them. The favorite records of the children are "Bella McGraw" and "Love, Love, Love." But we love Harry Lauder anyway and thank you for the great pleasure we are having in this music. I've never felt so young before in my life. It's as though I were under a spell. Gareth looks on with amazement as we sing with the music.

Jimmy and Donald are doing a "John Gilpin" around the place on their ponies, and are picking up pecans. The days are still clear and sunny but with cold nights and mornings. All day the wild geese are flying south in great wedges.

With the sky cloudy and the wind changing, the stove smoking and the weather turned warmer we look for more rain or snow. Donald had a chill yesterday and is now sitting up in bed eating his breakfast. His only pleasure is to be out in my bed in the sitting room, where he is near Gareth and can watch all our activities. The doctor ordered the distasteful routine of quinine every three hours, day and night, with salts and soup and lemonade, over and over.

Later:

It's snowing steadily now, small, feathery flakes. Last night it rained and then froze, so the ground is covered with ice. The milk freezes in the milk pans in the kitchen every night, but we have so many uses for the lovely thick sweet cream.

Later:

Now, Christmas done come and gone. It is late at night as I write this, and beginning to sleet. We certainly enjoyed all the things you sent us for Christmas. I know that our box was late in getting to California but I can picture Wayne buying them all and getting them wrapped and tied for mailing with the weather ten degrees above zero. He did all the shopping by himself, for it was too cold for me to make the trip into town.

Wayne told us that on his return from Little Rock, he saw something that amazed him, and he kept thinking how amused and interested the boys would be when he told them. A farm woman boarded the train when it stopped at a small station. She was a big, fat woman, with a sunbonnet on her head and a blue gingham apron full gathered around her waist. On the crook of her left arm she carried a heavy market basket. Her right hand was closed into a fist, and there a young cockerel sat perched sleepily upon her thumb.

Now, with bath water to pump and heat on the cookstove and bread a-baking besides the regular household chores, I am as busy as the Little Red Hen.

Wayne said to tell you that he raised better than a bale of cotton to the acre, where half a bale was the top crop till he came, and he wants a word of praise, please; although he may have come down with a thump, he will try again. Jimmy has often done the work of weight-boss this fall and manages the field crew very well. Every day that is sunny means another bale of cotton.

This spring the weather is awful. I never saw the wind blow so hard and it is a hot, dry wind. We haven't had a real season in the ground since last September. Every day the clouds pile up but the winds blow them away. I'm afraid our flower garden will be ruined.

After a few thunder storms the days have been foggy and dreary, yet during sunny breaks we've started the vegetable garden again. I've been looking at pansies and violets in the seed catalogs until I can almost smell them. We are still staying close to home for the roads are impassable, and though the cold days alternate with the warm days yet there are too many that are bleak and dank and damp. It is these changes that cause so much sickness.

All the cotton is sold at last. Counting in everything we got an average of twenty-five cents a pound for it. We have rented the lower place and the tenant houses here on the farm to families who have no young children, thank goodness. I haven't forgotten Zeddie. One of the men is a carpenter so we will have part of the main roof reshingled and house jacked up again. The rains and winds keep weakening the shingles and the big rocks on which the house sits keep shifting in the sandy soil. Six men are planting cotton now. The corn is already up.

I just finished reading a Charlie Chan story from the Post. That line is good, "I have often been a witness when the impossible aroused itself and occurred." We are all well at present. Gareth has recovered from a spell of colic. It is hard to guard Gareth from colds but so far we have managed to keep people away from him. I've just finished bathing him. He is so lovely. We are certainly enjoying our baby. Wayne never tires of holding him or looking at him. I've had fever and headaches and been drowsy and blue, depressed and nervous, but I'm feeling better today. Malaria has a lasting effect, it seems, yet so many around here now have influenza and colds. I've been so miserable I've gone to bed early and have written wonderful letters in my mind.

In the evenings we have arithmetic and grammar. Wayne teaches most of the history then, too, for history was always of interest to him. What discussions we have! Often in the evenings, too, the boys get a crossword puzzle and start to work on it. Before long they are appealing to us for *help. We get interested, get stuck, and give up, vowing never to touch one again. But it happens over and over. The doctor happened along one evening just in time to give us the word, "Tyler," that was holding one puzzle from completion. The puzzles are fascinating, we find, and are a help with our vocabularies. The papers you send have so many in them, but there are none to share them with us. We seem to live our lives here as though we were on an island by ourselves, yet there are always people near us.*

I hear from my sister that she has a radio now and is enjoying opera with her dish-washing. How rejoiced I am that she finally got out of Alabama. Music surely is the world's comfort. We play our records every day and never tire of them.

We are now discussing the idea of leaving for California, yes, really. I have already done some sorting and packing and sewing. It looks like a big undertaking. I feel wildly excited yet terribly uncertain. We feel that we must go, for the sake of Jimmy and Donald.

Later:

The price of cotton is so low that our plans for California will have to be put off for the present. We had hoped and planned for a way out. I can only write these simple facts, for I dare not try to say all that I feel. I am trying to carry on as best I can, but it's a terrible struggle trying to adjust my mind to another year of this life. I read somewhere that if we could only see our troubles through other people's eyes we wouldn't be able to bear them. Think that over.

Your loving daughter

48. No Sloven Can Make Good Butter

"NO SLOVEN CAN make good butter." In my mind's eye I could see those words at the top of a page of my mother's old cookbook, under the chapter headed, "Butter Making." I stood and looked at Mrs. Suggs.

She sat out under the big cottonwood tree near her back door, lazily churning. The old split-bottom chair she sat on had lost one leg, and was propped up with a chunk of wood from the pile nearby. Her dress, of faded percale, was stiff with dirt. In one hand she held a torn and stained copy of *Capper's Weekly*, that someone had given her. Beside her on the ground was a leafy branch which she picked up and waved occasionally to drive away the flies, which kept settling on her face and legs, and around the top of the churn. They literally feasted on the smears of dried dirt on her dress.

The tall crock was held tightly between her dirty bare feet. Remnants of old dried buttermilk from the last churning still plastered the sides of the churn. The ground all around the back steps was wet and slimy from the pans of greasy dishwater and the tubs of washwater thrown out there. Thousands of flies, house flies and green flies, buzzed and circled everywhere. Even the ground was covered with them as they crawled over the putrid soil. The churn was immediately black with them as Mrs. Suggs laid down the branch she had been aimlessly waving. "Sure is pretty weather," she began, turning toward me. "You going for your mail, I see. You sure get a sight o' letters and papers. The crops is a-comin' along fine, ain't they? Jest look there at Harley's little boy. He's Grandma's darlin' now, you bet. Ain't he a bouncer? Tell me I can't raise a baby on cold cow's milk! I ain't had a speck o' trouble. 'Tain't no more trouble to raise a baby than a pig, I say. Jest look at him, ain't he fat?"

I looked. He was fat. He was more than a year old now. Dirty, and naked except for a little slip of filthy gray flannelette that passed for a dress, his nose running, his bare feet puddling in a muck of dirty water over by the pump. No more trouble to raise a baby like that than to raise a young pig.

"Notice your garden is kinda slow comin' up. Folks've sure left us a fine garden patch, here, all fenced, too. Ain't a chicken or a pig got in to ruin it yet. Let me show you my fine onions. I'd be proud to give you some garden truck, we got so much. Never had such a good garden before in all my life."

I walked with her around the end of the house, stepping over old cans and bottles and rags, and came to the garden gate. Matt Wood and Hettie had certainly had a garden spot of which to be proud. Mrs. Suggs had been glad to find that we not only allowed her, but encouraged her, to have a garden and keep her cow. I saw the beans already starting to run on the poles, and the heavy green pea vines by the wire fence, young turnips growing lush and green, and the long rows of big, well-developed spring onions.

Just as I was about to step over the board at the bottom of the gate, I happened to look down at the ground carefully. Where to step? The ground was filthy, polluted. Hardly a clean spot to set one's foot on could I see. Yet these people had an old out-house over behind the small barn.

It had been the only one on the farm when we moved down there, and the men had helped Matt Wood rebuild it again when a high wind blew it over that first year.

Mrs. Suggs saw my hesitation and said casually, "Here, I'll go and get you some truck, you seem kinda scared to go in. Maybe we should go farther away, but it's so handy like, and cain't a body see you from the big road."

However, I backed away, mumbling excuses, and once my back was turned on her, I fairly flew along the narrow sandy road, up toward the line of pecan trees and my own little haven. I just tried not to think. What was the use?

Once the doctor's wife, in talking about her experiences among the share-croppers—for she used to go with the doctor when he needed her help and could get no nurse—had said to me, "I think there should be a Social Worker to go among these people and teach the women and girls how to manage a home. Teach them to cook and sew and make improvements in their houses. Changes that might be done with very little expense."

How to manage a home, indeed!

49. The Storm Cellar

NOT A CLOUD appeared in the sky now except the smoky gray strips toward the western hills at sunset. The weather was settled for a while, with none of the signs of rain in evidence.

The pond began to dry up and Wayne and Henry thought it a good time to use dynamite to break up the subsoil so that all the water would sink and the plague of mosquitoes would be lessened. Our property line ran through the middle of the pond, and we had already found that it was useless to try to dictate or persuade our neighbors to help in keeping a light film of oil on that stagnant water to suffocate the mosquito larvae.

When the water disappeared at last, we found that there was an exodus of snakes up to the houses toward the wells in search of water.

It was a common thing for a man passing through the farm to stop and ask us for a hoe, so that he could kill a snake he had seen on the bank, and to tell us, as he brought the hoe back, that he had killed a big rattler and for us to keep a lookout for its mate, for snakes always went in pairs.

Out by the back steps near the pump we killed dozens of copperheads, a short mustard-colored snake with a triangular head, and deadly poison.

Close by the house under a large rock where the water dropped from the eaves, our pet toad sat in the smooth hole she had made in the damp earth. We used to lift the rock occasionally and look at her. In the cool of the day she would come out and hop around the house. Later we found five little baby toads with her. We felt as though this toad was an old family friend.

One day while I was washing in the shade on the back porch, I looked up to see a spreading-adder trying to swallow our toad. The toad was much too large for the snake to swallow, but part of it was in the snake's wide-stretched mouth.

I looked around for a stone to throw but could find none. I was afraid to go very close, so I threw drift-wood until the snake slithered away under the house, leaving our poor toad sitting there in a stupor.

Another morning I heard something moving and rattling up on the paper that was tacked over the tongue-and-groove ceiling of the dining room. It didn't sound like a mouse, and the paper moved in too large an area for it to be anything but what it was—a snake. I seized the dinner-horn and blew several blasts as loudly as I could. That was the signal we had agreed upon in case of fire or any necessity arising for the men to come to the house. Wayne and Henry came down in the Ford from the barn immediately. They tried to hit the snake through the paper, and finally ripped the paper open. Down dropped a long brown-mottled snake, not a rattler, just a water moccasin.

The next day another snake dropped from above the window frame with a flat sounding "plop" that I will never forget. It slithered away into a hole in the big oven of the coal-oil stove that stood in the corner of the room.

We got that one, too, but for weeks I examined our beds thoroughly each night and often awoke in the darkness stiff and cold with fear as I carefully drew up my bare feet thinking I had felt a snake. No more could the children play outdoors in their bare feet around the house where the sand was clean and cool, for they were afraid there would be more snakes, and I gave up hanging the covered pails of milk down in the cool well after the time I hauled up a pail of milk and found a snake coiled up in the lid.

Our big lounge that I had moved out into the kitchen between the windows to make a comfortable place in which to read or sew while the dinner was cooking was put back into the front room where it could stand across the corner away from the papered wall, for there was no pleasure in dozing off to sleep to awaken fearful lest another snake had crawled out from behind the loose paper, and might be curled behind one of the cushions.

Every time I had gone with the children and Grandma over to the big storm cellar of our neighbors, I was almost hysterical in my fear of the snakes that I knew would be in hiding there, and I was continually demanding that we build a storm cellar of our own. It was a long walk, too, up past the big house near Grandma's and over past the Batsons' house and up the road to the Dawsons'.

Our discussions were finally brought to a decision by the realization that lately when it stormed the skies looked more and more as though there might be a tornado in the offing. The great black funnel-shaped clouds would rise from the horizon and cover the whole sky to the northeast, and, although as yet no tornado had struck very close, still they were hitting all around us, uprooting trees, wrecking homes, laying small towns flat and killing and injuring many people.

Each time now that it looked as though a storm might develop, Jimmy and Donald would rush to the house to help me and to be near Gareth. Several times in a terrible windstorm, they dashed wildly to the house on their ponies, and Jimmy reached and took Gareth from my arms, and off they all went down the road to the house near the mailbox, for the man who lived there now had enlarged and cleaned out the small root cellar that had been boxed up on the slanting bank. He too, felt that it would be wise to have some protection in a case of a tornado. If we needed to take shelter in a hurry, should a tornado come, it would certainly be too far to attempt to reach the big cellar over at the Dawsons', and each time now the storms seemed to arise so quickly. I can remember panting and running in the loose sandy road, with my hand over my heart, trying my best to keep up, for Jimmy was ahead on his horse with Gareth in his arms. Donald had raced on his horse up the road to see what Grandma wanted to do. It was a long hard run down to the corner of our farm, and I used to wonder whether that desperate race was worth the strain and effort. In both cellars I felt afraid. Nothing but dire fear compelled me to grope into those dark pits.

Once I gave out, in my mad dash, and sat down by the roadside, for

I could run no farther. I was so exhausted that for a while I didn't care whether a tornado came or not.

Now Wayne and Henry and Harley Suggs went at the building of our storm cellar with a will. First they deepened the well-pit where the pipe came up from the ground for the pump. Then the well-pit was widened. It was about sixteen feet deep and wide enough so that our wash bench and army cot could both be placed there if needed—one on each side. The sides were now reinforced with split saplings, walled in behind large posts that were set securely in the corners. The top they covered with a criss-cross of split saplings and a huge mound of wet earth was piled on that and the earth well packed down. The water pipe now came up in one corner of the well-pit, through the mound of earth to the pump. Also through the top of the storm cellar was placed a long length of stove pipe which extended a foot into the air for ventilation. A low, heavy door was set at a slant at one side of the rounded mound. The door could be swung back, and steps descended down to the flooring of planks, which were laid across two-by-fours.

The old pipe had been pulled up and more pipe added so that it now went down twenty feet beneath the bottom of the storm cellar. A new filter point of copper wire was put on the end of the pipe before it was driven down into the ground to insure a good flow of water, as the fine sand was screened out, but we knew of course that after a few months the sediment would again settle in the point and again the pump would furnish only a trickle of water, so the job of putting on a new filter would have to be repeated.

Harley and Wayne screened in one half of the front porch, too, with fine wire netting, making a small room to be used as a bedroom or an outdoor sitting room. The floor we covered with carpet, and for fear that even a stray mosquito might get into this room, through the screen door from the living room or through the new one from the porch, we still hung a fine mosquito-netting from the rafters down around the bed that was placed there.

By constant vigilance and care we were able to guard Gareth from the malaria, and brought him through the summer. The doctor advised me to nurse him all that year if I could, and that I did gladly, to the surprise of many of the neighbors. Most of the farm women made no attempt to nurse their babies, for they had to work in the fields, and a bottle of cold milk was easily given to a young child in any place—especially when it was lying out in the cotton fields on an old quilt. The doctor told me that most of the women who lived in town did not want to nurse their babies, because they would be obliged to stay closely with the child or else take it with them whenever they left home for more than a short time, although there were a few, he admitted, who were physically unable to nurse their babies, owing to one cause or another. He seemed to regard all women as a farmer does his milk cows and talked to me with eloquence and vehemence on the subject. His general opinion was that most women were simply too selfish to nurse and care for their babies in the old-fashioned way, even in the cases where it was apparent that the baby would greatly benefit by having its mother's milk. He said that all his talks and pleas on the subject did little good, however.

In the late afternoons I enjoyed carrying Gareth in my arms out under the pecan trees, where he looked up at the swaying branches and seemed to be listening with me to the cicadas that whirred in the bushes and trees. When the locust trees were in bloom it seemed to me that he gazed at the big creamy blossoms with a look of definite pleasure on his face. Whenever the moon was full I would carry him outside and point to the stars and the great shining ball that floated in the clear sky above us. I could see that Gareth really noticed the moon, for when the sky was full of fleecy clouds and the moon appeared to weave in and out between them, he would turn his head and look and wait as the moon appeared and disappeared. Somehow this pleased me greatly. I think we all felt, regarding our baby, that we could truly say that he was our beloved, and in him we were certainly well pleased.

50. Beautiful Mountains

> *Oh, beautiful, beautiful mountains!*
> *The home of the pure crystal fountains!*
> *The home of the happy and free.*
> *Forever the streamlets are flowing, are flowing,*
> *Forever the fountains are going, are going,*
> *What lovelier place could there be?*

THE SINGING TEACHER in our community was a tenant farmer who lived down the road past the little store. About thirty boys and girls ranging in age from ten to twenty met every Sunday night in the schoolhouse to sing from the old Tewkesbury Hymnal. Many of the parents attended, too, so Wayne and I often drove down and took the children. In one of the hymn books they used, the old-fashioned shaped notes looked so peculiar to me, but everyone seemed quite able to sing all the parts by note—and with no one to play the old organ, they kept perfect time and made the building echo with the different hymns. A few favorites would be picked out by choice at first, and then many of the other songs were sung right through the book.

Even in the big room, the children's desks took up most of the space. A table and chair for the teacher stood on the small platform, with the small organ at one side. The windows were curtainless. Five or six coal oil lamps were set in brackets along the walls. The floor was of well-scuffed unpainted boards, and the gritty sand carried in from out of doors caused a loud terrible grating noise as the uneasy feet scraped and shuffled with the changing positions of the singers. During singing, although strict attention was paid to the teacher, one could hear the heavy clump, clump in the back of the room as the bigger boys tiptoed over the bare floor, trying to slip out unobserved to have a smoke in the darkness.

The seats were usually all filled. Most of the families who lived in the district were represented there, and almost every woman had a baby or a young child in her arms. The older boys and girls stood around the walls. How they could sing! This gathering was one of the few social pleasures they had.

One Sunday evening the teacher held up his hand for silence and said, slowly, "I've brought with me a few copies of another hymn-book. It contains a song we used to sing back home in the hills. I think some of you already know it; the others can take these books. Today is my birthday. I've been thinking a lot today of my old home in the mountains, and would take it kindly if you would sing 'Beautiful Mountains' for me.'"

It was a beautiful song, and they sang it beautifully. The tune was one that rippled beneath the words like a mountain brook singing over the rocks down a hillside, bringing to mind the tiny cataracts where mossy pools held up the crystal water only to send it on again, in and out among the ferns and grasses. The Ozarks are full of such tiny beautiful streamlets, clear crystal springs that break from every hillside, and tiny brooks that dally in every depression or run gayly across the roadbeds.

I happened to meet the singing teacher sometime later, down by the store one afternoon, and I asked him about that particular song. "It was written by a man I used to know," he said. "I played it as a boy by those streams. I get so lonesome for the mountains and the mountain life. We were so happy there. 'Tain't like thissen here. Here, we're always sick—malaria, chills, and fever. My wife ain't seen a well day since we hit the bottoms. I only hope to live to get out of here, and back home to the mountains where we belong."

"Had you always lived in the Ozarks?" I asked.

"Yes, my people have always been mountain people. Generations ago they settled in southern Missouri and northern Arkansas where they found what they were looking for. 'Tain't no wonder they tell me the state slogan was named Opportunity. It's opportunity, sure enough, for a good life.

"Yes, and the folks up there have got good common sense. They know they have got about all that life can offer. The best things aplenty, anyway, and, as far as I know, they are content.

"Oh, of course, there's some of the younger ones that's drifted out. New cars and the roads being fixed up, all helps. Been a heap of the young fellows from up there's been to Little Rock, and some has gone as far as Hot Springs and away into Springfield, looking for the greener pastures. But I don't see as they better themselves. Most of them I've kept track of have made a sorry bargain.

"You know, it seems like my mountain people are hidden away from all this drag of discontent.

"You might call 'em lackin' in ambition, maybe, but anyway, they ain't thrown in with real rich folks and don't see no one with much more than they got at home themselves. They ain't got much call to envy anyone.

"Then again, you take it, each man there's his own boss. Year to year, long as he sees how he can raise something to eat and a little over to trade for extras, he's his own boss. Can do his own figuring and planning, and sorta gets a pride out of bein' called in to give help and advice to his friends. Yes, they are all a pretty peaceful lot."

"Maybe," I ventured, smiling, "you're meaning that 'peace of mind, dearer than all,' that we were singing about last Sunday night in 'Home, Sweet Home.'"

"You said it," the teacher agreed, eagerly. "I don't mean they ain't given to a little fightin', if they're a mind to, but generally that ain't so harmful. No one don't bear no malice about it. We gave up feudin' a long time ago, up where I lived. There's just enough fightin' goes on to keep the folks interested.

"You see," he continued, "they own their own homes; everyone does. They got a real secure feelin'. No one goes without milk or eggs or smoked meat. Everyone has a root cellar, too, and you should see the wild berries. Don't know of a family that ain't picked four or five wash tubs of huckleberries every summer for jam. Ever had yuh hot biscuits piled with home-made butter and huckleberry jam?

"There's places, too, where tomato canning is comin' in. Small canneries here and there where lots of the women and girls work for extra money. And up here, too, we raise a sight of apples and good Concord grapes.

"You folks must have a wonderful climate in Californy. We done took notice of your clear skin and the boys' pink cheeks. They seem so full of life, too. You ain't been here long, and I hate to say a discouraging word, but you're all going to be sorry you ever saw this country. No health. There's more misery and poverty and dreary livin' in this bottom than is right.

"You might have seen that I could hardly bear up under the singin' of that mountain song. I'd been in bed all week, a-chillin'. Guess I was weak yet, for I mighty near broke down. But I wish you could see the ever-flowing springs of water, glorious water, everywhere in the Ozarks. All through the hot summer, when the things are ripenin', clear crystal steams and creeks and rivers—

"Guess you'll think me a little touched, maybe, but them mountains is my weakness. I've worked hard there, worked hard by day and slept good at night. Good times, too, a plenty. Knew everybody the country round. Had so many friends I was at home anywheres.

"There's where you find real home livin', with dances, and quiltings and bees, and fine singing going on all the time.

"Everywhere you turn there's a little white church. Nobody stays away. Sunday is a holy day like it ought to be. I just can't get away from my Bible training. My brother, he's a preacher, you know, and my father was, too. I can't help but feel that there ain't nothin' gives more meanin' to this life than a real belief in a future life after death. Else, I just can't see us all being so almighty good when we don't really have to be. 'Tain't in nature. I know some folks up there that got hold of an 'ism', once. Near as I could make out they held onto 'Root hog or die' as a motto, and that easily led to 'Every man for himself, and the devil take the hindmost.'

"No ma'am, I don't know of anything that gives more satisfaction than a religion that is clear and definite, that answers every question and don't leave no doubts. I've heard some as says that folks get too upset, sort of hysteric about it, especially at the camp meetings, but I done visited me some of these town churches and listened to a few of the folks that was picked out to do all the singing while the rest didn't pay no mind, and I figured right then that there didn't seem to be no heart in it. Nothin' you could warm to. Seems like they have a religion that could be easy laid away during the week without causin' no worry."

The singing teacher was so in earnest that I was completely carried away. We were standing there by the car, talking, near the little store. I knew it was getting late, but I wanted to hear all the man had to say.

"The changing seasons get to be part of us, too. Guess we just move in with nature, livin' so close and gettin' so mixed up in the weather and all. We have to do more than look at nature, we have to do our plannin' every day along with old Mother Nature. Hot or cold, snow or rain, spring or summer, we don't kick and we don't try to reason and argue about it. I've watched the birds come and go, saw the sun rise and set, and knew by the soil itself just when to get to plantin'. Never resented none of it and never felt I had enough of nature.

I always felt right well acquainted with the stars at night, and the moon was always a part of our lives. Courtin' days, as I recall, seems the moon was always a-shining, but I can call to mind too, the moon breakin' through the tree-tops over the hill back of our old home, or shinin' down on us at night when we were a-drivin' along home from a meetin'.

"I didn't aim to get to talkin' this way, but you seemed to want to know, and I've tried to tell you. It's a good thing my wife can't hear me. Life here is a hard row for her, and I try to keep her cheerful, but she's homesick, too, and sick all the time.

"Sure, it's true that this is the richest soil I ever see. Goes straight through to China, I guess. But I been aworkin' two years here now, livin' in that little shanty, side of the road. It's roast in summer, freeze in winter—mud, dust, sickness. I work hard, too, allus have, but I ain't got nothin' more than this pair of overalls I'm a-standin' in. I feel like gettin' up early some mornin' and pullin' out for the mountains where I belong. Might get some of my folks to come and fetch us. They'd be glad enough to, sure, and the wife would about die of joy.

"Wish'd I could make you-all see jest what I done swapped for my chance at a share-crop here in the bottom.

"We had a home of our own, a big log house in a clearin'. I was born there. All the neighbors round us was kind and good. Had me a big team and wagon and a buggy for the wife. My grandpappy and grandma lived there, too. They're buried there in the family graveyard, 'longside my sister and our little baby what died.

"Had a well of the coldest water you ever tasted, and a big spreadin' tree by the back door. Just down the road a piece was the little church where we was married. But it's all gone now. I done pulled up my stakes to come down here and raise cotton. Cotton! They won't even let us have a mite of ground for a onion patch. Only flowers we got is them zinnias that the hogs can't tromp in the muck.

"I don't often give way to all this, but somehow, come spring, the mountains draw me, and I know I'm goin' back.

"You ever get homesick?" he asked me, with something like a break in his voice. I nodded, turning my head away for he had brought the tears to my eyes. "Well. Then, you can sure know how we feel, my wife and me.

"All our friends are there, more than friends, folks we knew like we knew our animals. We was allus doin' somethin' for someone, and needin' somethin' done for us, too, I reckon.

"Nothin' there but what I long for, kinfolks and kindness. Kin you blame me, for longin' for my people, my fortunate, happy, people, and for aimin' to get out of these bottomlands?

"I call to mind my ma and pa, last time I seen 'em. I was all set to leave. We had all our goods up on the wagon, under the canvas, and aimed to go on down the mountain a piece and stay a day or so with some of our kinfolks. I'd been puttin' in that last day a-helpin' my brother, so only got packed up late in the afternoon.

"Pa and Ma was a settin' out on the porch of their little cabin. They had jest come in from workin' down in the creek bottom, where the garden patch was. I saw the two hoes leanin' up agin' the porch. The sun was just beginning to dip behind the hills, but everything was kinda rosy and warm. Real spring day. The kind I like to think on. The birds a-singin', the bushes leadin' out again, and the fruit trees all pink blossoms.

"The sun was shinin' on Ma's gray hair. I remember she had laid off her sunbonnet. Pa had been kinda bald for years, but in spring he allus went barefoot, same as he did when a boy. His bare feet was restin' up on the rail now. They was both of them tired from working, but they was happy, I knew.

"The old cedar bucket was on the top step where Pa'd left it, the gourd dipper floatin' on top. That was good water. Forty foot well, rocked up, and sealed in—best water I ever tasted.

"Old Make-Haste, the hound, was there, too, crouched beside them on the porch.

"I could hear the tinkle of the cow bell on old Bess, comin' up to the barn, and see the chickens slowly workin' toward the roosts.

"This was not our first home. It was jest the place Pa took after we all grew up and married.

"I jest sat in the wagon, lookin' and tryin' to not say nothin' that would set Ma to cryin' agin.

"I remember it all so well. I often stop and think of that quiet and peace and content. Yes, I'm shore a-goin' back."

What to say? What could I say? I felt as though I had been granted a vision, a look into the Better Land. Every vague idea and hackneyed expression that had been dinned into my ears for so many years rang empty in my own mind as I tried to think. "Ambition," "success," "fame," "getting somewhere," "amounting to something"—all were just words, meaningless. No—life, liberty and the pursuit of happiness, this poor singing teacher had fully explained. He knew. I didn't. His words stayed with me, and the more I thought of it all the more I was convinced that something in the whole situation was terribly wrong. I grew tense with a vague disturbing determination, but my mind groped in darkness seeking a solution. Surely he was not asking too much.

51. Clouds of Glory

ALL ABOUT US in the late summer we became more and more familiar with the creak of the wagons, the slow drag of the mules, white cotton in the fields and cotton piled high on wagons or lying in white patches along the roadside.

The hazy dust rose and floated on the stifling air. The monotonous katy-dids chirped and grasshoppers shrilled in answer, loudly and deafeningly. Day after day, the tiresome, dull sameness, in cotton picking time. Everywhere through the bottom land the scene was so dreary. Broken fences, rickety sheds, lopsided shanties, yards scattered with feathers and old rags and papers, black mucky barnlots and foul pig-pens.

Naturally, by contrast, the sky overhead was to us a great stage, where the restless clouds made an everchanging panoramic spectacle that compelled our attention daily, and it was indeed becoming a large part of our lives. The sky seemed to cover so much space, for there were no tall buildings to shut out the great sweep of the blue heavens.

How often we stopped our work to admire the flaky cloudlets floating high against an azure sky, shaping and reshaping themselves in the vast loneliness of the heavens. Daily our eyes unconsciously lifted to where the high clouds floated so pure and white against the blue, or to where the blazing sunset dazzled and awed us with its glory.

When the hot sun beat fiercely on the yellow cornfields and the sky burned with an ashen hue we knew how quickly the small clouds of plumy vapor, as yet only faintly visible, could thicken and blacken, and even though the hot beams might still be darting through, we knew that with hard gusts or a sudden roaring blast the sunlight could be quickly shrouded and deep-toned thunder echo overhead.

In the morning, before the sun arose, the colors varied from splendid purple to scarlet, flecked with gold. Then the clouds would become whiter and whiter in the early light, until they were splashed with rose tints or turned into flaming banners of orange that faded into tranquil shades of violet against an opal sky.

We knew full well that a rainbow early, arching its glory against the walls of heaven, presaged a wailing wind that would surely follow. The elms and cottonwoods would sway from side to side and the big oak trees groan as their limbs broke and crashed in the hard gusts. The sky would change to a greeny-blue hue, and the air would be thickened with dust and leaves.

Then back toward the mountains the rain would be seen gliding in a dark sheet over the hills and fields. The thunder would roll along the horizon, the wind would drop away, and the clouds blacken against the metallic blue of the sky, which now showed only in tiny patches.

Into the very heart of the storm we could look where the sky from horizon to zenith would be split by dazzling forked lightning, the crashes of thunder which deafened us seeming to increase in strength as torrents of rain fell.

Hail in summer! Hail and a shrieking wind. It rattled on the roof like bullets and crashed against the windows. The wind would be cold now and the rain would settle to a steady downpour, flooding the fields and roads. All day it would rain, fast and heavy, the roads becoming thick, black, slimy mud with the surface beaten to froth.

Suddenly, toward the west, a light streak in the sky would gradually appear, and grow, a contrast to the blackness in the east. The rain would lessen and the clouds lift above our heads, showing dark blue patches. Then the sunlight would stream down and the birds begin to sing again. The last faint rumble of thunder would die away. The giant thunderheads, white peaks that walled the sky to north and east, would be caught now in the splendor of the sunset, rising to the heights of heaven, monstrous walls of jasper with domes of gold, enthralling us in a vision of glory and grandeur, until they sank at last behind the hills.

Now near and far the shades of twilight would gather. The mockingbirds would trill. In the wet weeds the crickets would again begin their purring.

After a day like this, in the evening as we set the table and prepared our supper in the coziness of our little kitchen, we sang often that same old song of Grandmother's that we liked so well—"Tinkle on the Shingle." How beautiful our shining lamps seemed to us. How beautiful the firelight dancing on the hearth and on the floor. How the wood in the little stove crackled and snapped. The house would be filled with the odor of frying ham and coffee and freshly baked biscuits. Warmth and seclusion and comfort. It was home!

And afterwards when Jimmy and Donald undressed by the heater and put on their flannelette gowns and got ready to go into the North Pole bedroom, I would sing again the verse with the good lines:

> *What a joy to press the pillow*
> *Of a cottage chamber bed,*
> *And to listen to the patter*
> *Of the raindrops overhead.*

Often at night we would hear, down on the main road, the creak and rumble of a farm wagon that grew louder and louder and then fainter and fainter until it passed around the bend in the road. And a man's voice could be heard singing, the song rising and falling with the sway of the heavy cotton wagon, carrying a lament in rich syllables; loud and clear, then dropping lower and lower—just a man singing in the darkness, the deep, penetrating voice of some farmer returning late from the town or the gin, and singing to add cheer to his long, lonesome ride.

But in the darkness the tones had a special appeal. They had a throaty pitch, half yodel, half the doggie-drawl of the western cowboy. It was a haunting, beautiful tone that hung in the clear air, and we seemed to hear the sad melody long after the last note had died away, for like an echo it lingered on and left a feeling of loneliness behind.

Out on the screened porch we would lie at night and listen to the frogs. At first, in the early spring, only a few frogs croaked, but as the days lengthened more joined in the nightly concert.

First would come the repeated questioning voice of the one lone frog, then a run of three notes was dashed upward by a frog farther away; then a double croak, two-toned, "crack-ak, crack-ak." We thought that some of the frogs must keep the same positions at the pond night after night, for the deep voice that began the chorus and acted as master of ceremonies always seemed to come from the nearer end of the pond, while the hesitating thin tones of the second came from farther away. Soon so many were joining in that we stopped trying to distinguish among them, and gradually grew so accustomed to the pulsating tempo of their croaking that we were startled when it ceased, and only knew that sometime in the late hours, when the night winds had chilled the air, the frogs at last grew quiet, and the sibilation of all the insects fainter. When we awoke in the morning, we forgot all about them, and listened again to the mourning doves and the mockingbirds. How satisfying that coo-coo-a-coo, and the cheery chirp that soon changed to a whistle or a heavier hoarse note that was the mocker's imitation of another bird.

52. Bill Stafford

AGAIN THERE WAS a new family of cotton pickers down in our corner house, so one rainy afternoon I visited them. We gathered around the fireplace. There were only two chairs and the wash bench to sit on. I sat on the wash bench by the three girls while the two boys squatted on the floor. The mother and grandmother very reluctantly accepted the chairs, for they felt that as company I should not have to sit on the bench. The wet cottonwood steamed and sizzled and smoked. It seemed there was nothing to say. There was absolutely no furniture in the front room except the chairs and the wash bench—so I knew just about what the condition of the family was. There were five children here, and the parents and old grandmother. Finally one of the girls asked me if I would like to hear her granny sing. I was delighted. Granny, without any coaxing, began in a clear high treble to chant the words of an old English ballad, "Charlie Came A-Wooing." Then she sang "The Nut Brown Maid." When I recovered from my astonishment I asked where the family came from, and was told that they didn't rightly know but they had heard it was Tennessee, and Virginia before that, but couldn't recollect any family names.

Then one of the girls, about fifteen years old, sang for me the "Prisoner's Song," and "Titanic Was a Ship." These were both very popular songs at that time and as I seemed to enjoy them, one of the girls offered to give me the words to "Bill Stafford"—a favorite song of her father's. Just then the father came in. He was a big, handsome man, and he squatted down against the wall as though the whole situation was very pleasing to him. He was asked to sing "Bill Stafford," and immediately started in, and it was truly enjoyable, for his voice was another of those delightful Southern voices, that really thrilled me—throaty, nasal, very expressive and appealing, and full of a certain peculiar quality that quite charmed me while I groped in my mind for the source of the attraction.

After the song was well started, the man got up and leaned against the wall, tapping his foot with emphasis to the regular rhythm of the verses and gazing into space with a look of complete happiness and absorption. The song was not recited, it was drawled, a sing-song tune that was rather monotonous, but his evident enjoyment and the odd pronunciation of the words carried through to the end. This old song, I heard later, was considered to be full of wit and humor. Highly entertaining! It did amuse me while I listened, but later at home, listening to our good music, I was appalled at the comparison. It seemed to me to represent so much of the life around us that Jimmy and Donald would be feeling as they grew older and left our home. A way out had to be found for them.

BILL STAFFORD

> *My name it is Bill Stafford,*
> *I came from Buffalo Town,*
> *For nine long years and over*
> *I've roamed this country round.*
> *Of course some hard luck I have had,*
> *And better times I've saw,*
> *But I never knew what misery was*
> *'Till I came to Arkansas.*

I landed in St. Louis
With ten dollars and no more,
I read the morning papers
Until my eyes were sore;
I read the evening papers
Until at length I saw
Where a hundred men were wanted
In the state of Arkansas.

I wiped my eyes with great surprise
To read that joyful news
I bounded to an agent,
His name was Billy Hughes,
Said he, "Young man, for five dollars
A ticket you will draw,
That'll land you on the railway
In the state of Arkansas."

I handed him five dollars
And started pretty soon;
I landed in Van Buren
One sultry day in June.
Up stepped a walking skeleton
And offered me his paw,
He invited me to his hotel,
The best in Arkansas.

He fed me on corn-dodger!
His beef I couldn't chaw,
He charged me fifty cents for this,
In the state of Arkansas.
I rose up in the morning
To meet the early train,
Said he, "Young man, you'd better stay here
For I have some land to drain,
I'll pay you fifty cents a rod
Your board and lodging all,
You'll find yourself a different lad,
When you leave old Arkansas."

I stayed with the darned old blow six months.
Jess Hallings was his name,
He stood six-seven in his boots,
And slim as any crane.
His hair hung down in ringlets,
O'er his long and lantern jaw,
He's the photograph of all the gents
Who was raised in Arkansas.

He fed me on corn-dodger
As hard as any rock,
Till my teeth began to loosen,
And my knees began to knock.
I got so thin on sassafras tea
I could hide behind a straw;
You bet I was a different lad,
When I left old Arkansas.

The day I left I can't forget,
I dread the memory still,
I shook the boots clean off my feet
With a bloody blasted chill.
I staggered into a saloon
And called for whiskey raw,
I got as drunk as blazes
The day I left Arkansas.

So farewell to you, swamp angel,
To the canebreaks and the chills!
Likewise to sage and sassafras,
And them corn-dodger pills.
If ever I'm seen in this land again
I'll give to you my paw—
It'll be through a great big telescope,
From here to Arkansas.

53. Frankincense and Myrrh

CHRISTMAS HAD COME before the cotton was picked, and the weather had been so clear that it made everyone say it didn't seem a bit like Christmas. When I looked at the beautiful paper daffodils and sweet peas sent us by my mother, I fancied I could see them in reality, swaying in an April breeze with the sun on them, in California.

Our tree stood in the corner, and oranges again hung from the deer's horns. Gareth enjoyed the little tree. He sat down in front of it and smacked his lips, and then got up and smelled the branches. We had to watch to keep him from eating the decorations.

Mrs. Batson came to the door one day and asked me if I had "ary a button" that I could "lend" to her to sew on her coat. She was going to go to town and take the bus across the river to where her daughter, Lidy, now lived. Lidy was sick again and wanted her mother.

Mrs. Batson seated herself by the fire and I got out my button bag. When I took her coat and settled down in the rocking chair to sew I found that all the buttons were gone, and the coat lining hung in rags. So, I mended the coat and put good buttons on it. I looked at her shoes. They were high-topped button shoes, with but three buttons hanging on loose threads on each shoe. So, I got shoe buttons and sewed them on with good stout thread. Mrs. Batson was full of admiration for the beautiful paper flowers. She had never seen anything like them. They were well made and very natural looking, and I did not want to give them away, so I went into the bedroom to look for something to give her for a Christmas present. I picked out a new handkerchief, a tiny bottle of perfume, a bar of scented soap, and a pink boudoir cap—one of those hand-crocheted affairs that were always given and accepted as Christmas gifts, but so seldom worn.

Then I remembered that as she sat beside me while I sewed, I had looked at her carefully and decided that what she needed was a bath and clean clothes. Her scalp, where her graying hair was parted, was unbelievably dirty. I laid the things down and hunted up a comb, a new face cloth and a bath towel, and a bar of perfumed soap, but I paused and put them down too, for I began to remember. I remembered the day I had passed her house when out for a walk with the boys. It was just after a rain. Her yard, of black gumbo soil, was mucky. There were no vines, no flowers. Just the tiny house of two rooms, squatting at the edge of the

cotton fields. Mrs. Batson saw us and came to the door and invited us in.
I didn't want to go in but I didn't know how I could refuse her. No words
can possibly describe Mrs. Batson's situation so well as those of Thomas
Hood written back before 1845:

> *Work! Work! Work!*
> *My labor never flags;*
> *And what are its wages? A bed of straw,*
> *A crust of bread—and rags,*
> *That shattered roof—and this naked floor—*
> *A table—a broken chair—*
> *And a wall, so blank, my shadow I thank*
> *For sometimes falling there!*

I gave her the things I had first selected. To these I added a spray of
the sweet peas and a bag of candy and nuts and oranges. It was the right
decision. She was so surprised and pleased that it was pitiful. Her eyes
were large and blue in her small round face. Once, I could see, she had
been a pretty young girl. When she smiled she made me feel as though
she were completely filled with the pleasure of the occasion—no envy,
no jealousy—no self-pity—no comparisons—just the customary placid
acceptance of that position in life in which "it had seen fit for God to have
called her."

Never before, she told me, had she ever had such wonderful presents.
But still I looked at her dirty head, her dirty wrinkled neck and her soiled
dress, and then again at the real happiness shining in her eyes.

Cotton was selling at from eight to twelve cents a pound. It cost twenty
dollars a bale to pick it. It was worth from forty to sixty dollars a bale. The
seed went to pay the picking. Before the rains began we had fifteen thousand dollars worth, if we could have had it picked then. But we had six
weeks of rain, and the seeds sprouted in the fields. The cotton wouldn't
even pay for the cost of making and picking. The cotton pickers were
getting one dollar and twenty-five cents a hundred for picking, or rather
pulling, the bolls. The farmer was the loser, as usual.

Nobody had any extra money. Almost everyone was sending the bales
to the compress to hold for higher prices, so it was a poor time for re-rent-
ing and selling out the mules, wagons, tools and equipment. As Wayne
said, he had put in five years of time and money and wouldn't farm it
again, if he could possibly leave, but he hardly wanted to give it away,

when by waiting a little, maybe until spring opened up, he might deal better and get his bank accounts settled. In other words, he would know in several weeks what he could do financially, later than that as to renting, when the mules and farm equipment would surely rise in price with the spring work.

We talked sometimes of the possibility of my going ahead to California with the boys, and tried to figure out ways and means of living until Wayne could manage to leave and follow us. I was packing up, but I knew that it was leaving Wayne that counted. I realized this so strongly now that the time had really come to decide. Yet when I looked at Jimmy and Donald, I wondered why I hesitated. They were growing and developing fast, and needed something to inspire them, something better and bigger than they had. They didn't seem to realize that there was so much more to life than they could see around them. It wasn't that they needed a start. We didn't feel that their life or education had been entirely wasted up until this time. I thought they had gained some fine things, some good experiences, and had a solid foundation that cried aloud to be built upon, but sometimes it made me feel like crying just to look at them. I guess all parents come to the place where they must of necessity and nature give up their individual lives, needs and wants for the benefit of their children.

Both boys had fallen heir now to a form of malaria that came regularly, not just the occasional chills and fever. Donald often went to bed at night with a high fever, although we were always taking quinine and liver pills. The fever returned about every fourteen days. So there was their health—a big consideration.

At first, whenever we talked of the future, Wayne and I were positive we would both be leaving soon, together. Later, we hoped. Now we began to realize how things actually were and could only take the next step in the dark, though we wanted to do what was best.

54. New Year's Letter to a Mother

MONEY, IT SEEMS, still makes the mare go. This has been another 1920 year, with everyone down at the heel and full of hopelessness. Nearly all of last year's cotton and this year's crop, too, is stored out in the hayshed and in the compress in town, under insurance.

We have worlds of corn and hay and peanuts and sweet potatoes and

*Irish potatoes and onions, but we can hardly get anything for them when
we offer them for sale. We've raised the crops in abundance—in fact made
more to the acre than anyone for miles around, but we can't control the
prices.*

*No one has any money now, and we cannot rent the farm for cash, so it
means waiting on a renter's luck. The bank loans that we have made will
either be called in and the mortgage settled or the time extended. You see
how I am beginning to be steeped in a realization of the Southern share-
cropping system.*

*We aren't sure yet what we can do about leaving, nor when. I have
all my books packed, lots of things given away, and a large packing box
ready to nail up.*

*Lately, we have been seriously considering the idea of the boys and
myself going on to California, and of Wayne's staying for one more round.
So, at Wayne's urging, I am still half-heartedly packing up.*

*We have enjoyed "Riley Grannon's Last Adventure," printed in the
last* Adventure Magazine. *I wish you would get it and read it carefully.
Don't skim over it. Get the "flavor and the aroma." Wayne has been reading* Scottish Chiefs *aloud to us in the evenings, while Jimmy and Donald
and I crack black walnuts for the candy I promised you. I say, "Hurrah
for William Wallace."*

Donald is still weak from a spell of malaria, but will be up tomorrow.

Such freezing weather and cold winds as we are having. It's a regular
Capitola or the Hidden Hand *night; you remember when you read that
book to us, years ago? Great gusts of wind seem to lift the house from its
foundations and scream and bellow in the chimney.*

*Most of the time the sun shines and the skies are clear, but there was
a light fall of snow after Christmas, and the weather is not settled yet. It's
nine o'clock; a heavy frost and a weak, cold moon.*

*Old half-forgotten lines of New Year's songs, old tunes of the dying
year's vain regrets or hopeful aspirations have been coursing through my
brain all evening.*

> *Ring out, wild bells, to the wild sky,*
> *The year is dying in the night,*

or

> *Ring out the old, ring in the new,*

or

> *And the bells were ringing the old year out,*
> *And the new year in.*

Do you remember?

Your loving daughter

55. The Missionary Meeting

AS I BECAME ACCUSTOMED to driving the Ford over the deep-rutted roads I acted as chauffeur for Grandma many times. She still held her membership in the Home Missionary Society and went into town to visit her old friends as often as she could.

One day we drove in to attend a meeting of the Society held in the social hall of the Methodist Church. An elaborate luncheon was served at noon and the business meeting would come later.

Recognizing the fact that I was a stranger, one of the women, seated near me at the long table, asked me what I thought of the repast. The tables were beautifully decorated and many kinds of good food were in sight: ham, chicken, salads, cake, rolls, biscuits, relishes, jelly, coffee, thick cream, jello, and fruit.

Without hesitation I answered that it was really a fine banquet, but the wrong people were eating it. "Why what do you mean?" she said, in a bored, condescending voice. "What do I mean? Only this—that this food would represent Christmas and Thanksgiving and the Fourth of July all rolled into one, for so many families that live less than five miles from this church. In fact, they have never seen such food in all their lives. You can see for yourself that it is being wasted, really. No one here is actually hungry. Look down the table and see how the food is being picked at and tasted and mussed up, but not really relished. It seems rather peculiar to me that a Home Missionary Society should do this when there are so many hungry and ill-fed people around us."

Her only response to this was a cold stare and a polite withdrawing from me, as though I had the plague. But never mind. More was to come.

When all had settled down in one of the smaller side rooms, the business of the society was taken up. Among other correspondence was a well-written and appealing letter from a Negro woman in Tennessee asking for a donation for the home for aged and homeless Negro women. This letter was read by the president in a sarcastic tone of voice, and then followed a speech in which the whole matter was gone into; the president agreeing that the project was a worthy one that would have won her approval, but for the fact that the writer, a colored person, had had the presumption to sign herself as "Mrs."

My memory fails me as to the biting words that were used in defining the exact position the Southern women held against all such boldness on the part of the colored people, though, "Goodness knows, none could sympathize more deeply with, nor understand more fully, the poor colored people's problems."

So the entire matter was dropped on a note of determined self-righteousness.

As soon as the collection was taken up, and the minister, who was visiting from another part of the state, had breathed his last supplication to the Lord for blessings on this missionary work, I edged my way through the milling women, and asked him a few questions. My questions all dealt with such points as why the town churches couldn't send down young men from their Bible classes to take turns as teachers in the bottom Sunday-school, where often no school at all was ever held over long periods, because the itinerant preacher had not come, and there was no one to substitute. I even suggested that some of the preachers from town might come down to the bottom of a Sunday evening and preach in the tiny chapel, so as to hold together those who would attend so willingly.

This talk led to the preacher questioning me as to where I lived and how I happened to know of such needs. My answers were such that he was quite taken aback, for I told him—oh, how I told him, of how the young wife of Harley Suggs had died when her little boy was born, because there was no one there to take care of her, and how she had been buried with no one in attendance except her young husband and his aged parents, as no preacher would answer the begging request for a funeral service, for there was snow on the ground and it was a bitter cold day.

I told him of the poor little children I had seen, picking cotton, barefooted on the cold, frosty ground, their feet and legs wrapped only in strips of sacking. I told him of the awful conditions under which Lidy's baby had been born, and the total lack of nurses at all times. I mentioned the fact that the doctors only came when their fee was guaranteed in advance—and told him of the many deaths all through the bottoms from malaria and dysentery and pneumonia—and that even pellagra was a common thing in many places, even though few people would admit that. So many things came to me that I did not lack information or words, and he, poor man, stood speechless, not even able to enjoy the release of beating his breast, for now I had fixed him with my glittering eye, and he could not break away.

So the dear town women, who would have lionized the visitor, had to stand back and endure all this. Every question he asked cut deeper and deeper into actual conditions, and every answer only awoke his interest anew.

I don't remember how it ended. I was exhilarated, I know. I hoped I had done some good, but I drove the old Ford home in a daze, and later when I recounted my visit to the missionary meeting to my astounded husband and his brother, they were a little shocked at my unlady-like behavior, especially when they learned that the wives of the leading merchants and of a certain judge and colonel were among those who heard me. I didn't care, but they seemed to.

No, nothing every came of it, and nothing ever will.

56. Aulo Acquaintance

AT LAST I HEARD that a new family had come from the hills to live in the bottom and I was very eager to meet them. Mrs. and Mrs. Dole and their son, the age of Jimmy, were living where Mrs. Randall had lived on the Cossey place. Grandma told me that Mrs. Dole had a piano and played guitar and sang. So one day I decided to take Gareth in my arms and wade the hot sand to her house.

We had a wonderful visit. She played the guitar and sang for me, and we found so much to talk about. I could share with her all my petty woes and trials. All the harshness of the life around seemed to soften and become more bearable as we laughed together at the things that before seemed so wearisome.

Mrs. Dole had lived in Oklahoma and in Kansas and also out in the Ozark hills. I delighted in her delicate beauty, her dainty ways of house-keeping, and her happy spirit.

The Doles were share-croppers, and she worked every day that the weather permitted in the crop—but she also planted beans and corn and onions, and worked hard to train vines around her house and plant little plots of sweet petunias.

But, alas, she was far from strong. She had already spent many months in a mountain sanitorium, trying to get cured of tuberculosis, yet she was no better. Grandma persuaded Mr. Dole to take her to one of the older doctors in town, who had been practicing years ago when Grandma was younger. Mrs. Dole saw the old doctor and he strongly advised her to go to Mayo Brothers in Minnesota. At last her husband borrowed the money and Mrs. Dole left. At the Mayo Clinic it was decided that she was in a very critical condition with malaria, and she had never had tuberculosis. After a most strenuous and complete course of treatment she came back, almost entirely cured. It amused all the bottom folk, however, to see her going to the trouble of buying celery and keeping it fresh by burying it. She ate lots of onions and tomatoes, as she had been advised, and learned to like green salads and soups. She had been brought back from death's door, she told us, and had learned at last the part mosquitoes had played in her illness, as well as the value of a proper diet.

When I felt an urge one day to blacken my face and dress like a Negro mammy and have fun, it was straight to Mrs. Dole's home I went, sur-prised that for a while I succeeded in fooling her, and I danced and sang and felt almost like a California girl again.

On weekends the Doles usually went out in the hills to visit with old friends, but she came back to the farm again with more anticipation than she had felt in years, she told me, knowing that she would again be near me and little Gareth and her son would be happy, too, with Jimmy and Donald.

One of Mrs. Joe's nephews brought us three duck eggs and we hatched them out under an old Plymouth Rock hen behind a bale of hay in the hayshed. As soon as the little ducks were old enough to be left alone we put them in a coop near the house so we could give them special atten-tion. They grew fast, lost their downy yellow coats, and soon were covered with feathers. Jimmy and Donald sometimes pumped a big tub full of water and put the ducks in it and watched them swim around and around.

These ducks were indeed our pets. They consumed great quantities of wet bran and waddled in a lop-sided manner around the yard. Whenever it rained we had to be careful to find them and put them safely back in the coop or they would wander helplessly in the cotton fields and stand there crying piteously in the drenching storm. By accident two of them were killed, for they got under the big box to which they had been transferred and were crushed to death. So that left only Molly Duck.

Molly Duck grew very rapidly, and changed daily from a sprawling duckling to a glossy-backed young duck with a beautiful soft gray breast and long wings with green and blue shining fathers. We all loved that little duck. She would run wildly after us out to the garden and come quickly to us the minute she was called, making no attempt to fly. She seemed quite contented and had a look of real intelligence in her bright black eyes.

At last one day in the fall, the boy who had given us the eggs asked if he could have Molly Duck to use for a decoy, and we gave up our pet. We knew that we must leave Molly behind us anyway, when we went to California, and it made us feel better to leave her with someone who would really care for her when we were gone. But soon after that, the boy told us, one day, down by the bayou where he used to hunt, Molly rose from the water and flew away with the wild ducks. Molly, he told us, was really a drake, and would have soon left us, too, unless we had kept him a prisoner.

— · —

It seemed that in a short period of time we were to suffer the loss of more than our pet.

Only a few days after we had heard how Molly Duck flew away, Wayne came in looking sick at heart, and I knew something dreadful must have happened.

It was all he could do to restrain himself in telling me that our mule, Joe, was dead. Dead—killed by a stupid, ignorant fool.

Joe and Jerry had been taken with Belle and Rhoddy down to the lower place and up onto the plateau of the Petit Jean Mountain where Wayne and Henry had leased a small farm with a stretch of fine cotton ground. The mules were left in the care of old Hames, the brother of Zeddie's father, who was working the place. In spite of all Wayne's explanations, Hames simply could not or would not believe that Joe and Jerry, like all mule teams, absolutely must be worked together.

Wayne had tried to change Joe's name to Tom, so he and Jerry would be "Tom and Jerry"—but Joe refused to respond to the word "Tom," and so remained Joe, as he had been named.

Hames took Jerry and Belle and left Rhoddy and Joe behind in the little barn lot. Joe was so determined to follow and be with Jerry that he ran about wildly and at last jumped the fence. Hames turned back and got a heavy log chain and tied Joe to a big hickory tree. He put the chain around Joe's neck and fastened it by slipping one end of the chain through the big link at the other end. Then he tied the chain to the tree. Several hours later Hames returned to the lot and found old Joe, dead. Joe had thrown his head around and jumped and leaped to free himself so he could follow Jerry, the chain had tightened on his neck, and he had choked to death. Old Hames had learned nothing from the loss of his cow, even though Wayne had carefully explained to him the mistake he had made in tying the cow so that it choked to death. Now, this was our valuable mule, Joe.

— · —

Spot, the brother of the collie Looie, used to run along the country road inside of our fence, barking at passing teams. The drivers usually urged Spot on by yelling and cracking their whips. But one day Spot was killed, right at the corner of the farm. Frank Potts shot him, just for the fun of it, we were told. Frank was one of those men that I saw at the Stringtown dance, who used to saunter onto the floor, whenever a dance was going on, and start his bullying by demanding of some fellow that he relinquish his girl, if she were very attractive, so that he could dance with her. He always carried a gun and a big knife and he kept the younger boys cowed by his overbearing ways. He was a big handsome man, but ignorant and mean.

— · —

Looie took sick later on when there was an epidemic of black tongue among the dogs of the neighborhood. We tried to cure Looie, working under a veterinarian's orders, but our efforts were of no use, and so Looie died, too.

Then one hot day the news was phoned to us that a mad dog was on the rampage and had been seen turning in at our gate by the barn. Wayne got out his forty-five and the boys and I went into the house. Soon we heard several shots. Later Wayne told me that the dog appeared with saliva streaming from its mouth. Several men were following along, yelling and waving sticks. Under the urging of the men Wayne took a shot at the hound, hitting it somewhere, and the dog crawled beneath the bushes on the bank. Wayne followed and shot it again, for once having started he had to finish the job and not just wound the animal. At last the dog crawled out and came trotting over, straight toward Wayne—where he stood by the well. Wayne had one bullet left, so he knelt and shot again and this time the dog fell dead. He told me later that he was convinced then, as the dog walked so slowly to him, wagging his tail and apparently asking for help, that it had only the black-tongue, and not hydrophobia. Wayne said the look of patient pleading that he saw in the dog's eyes as he took that last fatal shot was something he'd like to forget.

These losses seemed heavy burdens indeed and it was hard to realize that such things must be borne bravely and forgotten if possible. Yet in our lives our animals were like friends.

57. The Waters of Life

THE FLOWER GARDEN, under our constant care, was growing well, and we enjoyed it every day. In one corner stood the brown velvet cannas. The sapphire-blue morning glories and the delicate tangled sprays of the wild grape made a hedge of the wire fencing, and enclosed the clumps of four-o'clocks, the circle of dazzling white sweet-alyssum, the petunias, and the heavy yellow and orange zinnias. Friends in town often gave me roses from their gardens in the summer, and these I carefully treasured, but so many times through the long winter months when all the flowers in their gardens were dead, and their potted geraniums and begonias were stored safely in their cellars, I would almost ache with longing to see, to touch, to smell—a rose. So acute was this feeling that I would visualize roses, pink, and red, and creamy tea-roses, and in imagination touch the silky petals or hold the tight buds in my fingers. I recalled yellow and pink moss-roses I had known, with their sticky, mossy stems. It seemed as though I could feel within me an actual straining to capture in reality the dainty rose odor, so clean, so spicy—so like a rose.

Somehow roses didn't grow well in the bottom lands. The soil was too sandy, but even in other yards where there was more clay, or more of the black gumbo soil, it was the same—zinnias. They flourished. Their hardy stems, so stiff and straight, held up brilliant orange and yellow and maroon blossoms, but as they had no pleasant odor, their regular shapes and firm petal displeased me. I resented the fact that they grew so well when I yearned for delicate violets and pansies and roses.

Our special pride now was the one big sunflower with a face more than a foot across. This we watched anxiously, lest it fail some day to be a sunflower and forget to turn its huge flat, yellow-petaled face to the west before the sun went down. Every morning it faced the east, and Jimmy and Donald wondered just how it managed this feat, for they reminded me that Grandmother had sung them a song about the sunflower turning to her God when he set the same look that she gave when he rose.

Through the middle of the day, all these flowers and vines hung in a soft wilt under the oppressive heat, but with the sun dropping lower and lower over the far blue mountains, they began to revive, and then we watched the many butterflies that came flitting with gem-like wings, and the big red dragonflies that darted back and forth above the blossoms.

Along the bank the twilight stillness settled and the musk-like wild odor from the weeds scented the air. There, where the long looped vines of wild grape hung, and sheltered in the heavy grasses, the cicadas would begin their evening whirring—a grinding rasp that filled the air and that could be stilled for a while only when we beat on the tree trunks with a stick.

In the scented dusk in the early summer when the fireflies flitted in swarms, their lights flashing here and there, and when the mosquitoes gathered with their monotonous lazy drone, there would arise an incense from the bed of petunias and an over-powering sweetness from the flowering locust nearby. Many times Jimmy and Donald and I carried a small table and four chairs out to the little garden and set our supper table there, surprising Wayne when he came in from his work. We gathered extra long twining grape vines from the bank nearby to fill in the vacant spaces in the wire netting around the flower garden. We were so proud to have created this tiny enclosed spot of beauty, and I, like the boys, would childishly demand admiration from Wayne for each one of the plants, for they were the result of the hard work that we had willingly undertaken, pumping and carrying the heavy tubs of water around to the garden, and tending each plant with loving care.

We would stay out in the garden as long as we dared, enjoying the coolness that we could feel descending as the sun sank lower behind the hills, but we were always driven indoors at last by the gnats and mosquitoes that swarmed about us.

Especially after a rain, breathing at night became a sensuous pleasure when the breeze swept over the unseen blooms in our garden. Often when the moon was full, we would go out and walk about in the little plot, feeling as though we were in fairyland. Those were the nights when the air was so soft and pleasantly warm, and the moonlight so clear that we could sit out on the porch and read. It was indeed "our" October moon, big and round and shining like burnished gold.

During the warm summer showers when Gareth was pattering about the house in his little bare feet, he would often ask for my umbrella so he could go out in the rain and sit on the large flat rock that was half buried in the sand by the front steps.

There he loved to sit, his bare feet tucked in beneath him, wearing the little soft white dress with the starfish and bees embroidered around the little yoke that I had shortened and adjusted to his growing body. I would stand inside the screened porch and watch his complete absorption in the curtain of water that fell around him.

Out to the flower garden he loved to go, and I was pleased to see that young as he was, he stepped carefully on the narrow paths and touched the blossoms and smelled them and laid his little fingers upon them softly and did not seem inclined to break or destroy anything. He seemed to know by instinct that this little place of beauty was being cared for and loved.

Jimmy and Donald and Wayne, too, noticed that Gareth had a unique way of doing everything. He seemed to touch life with appreciative hands, lightly and lovingly, already finding such pleasure in the things that had always given me supreme satisfaction. And when the shadow of the house fell on the four-o'clocks, I would show him the blossoms opening one by one. That pleased him so.

— · —

"Dysentery—an epidemic of infant diarrhea, a contact infection caused by a virus. An acute medical emergency, the bacteria multiplying so rapidly in the baby's intestines that the child is not capable of combating the disease successfully. The body loses large quantities of salts and water and a severe dehydration occurs which often results in death."

Yes, Gareth was very ill, and those were the essential facts that the doctor read to me from one of the books in his office when I insisted upon knowing the cause of my baby's ailment.

Dysentery and typhoid were the chief "fly" diseases, and I felt very resentful that all our efforts at cleanliness and sanitation, our careful

screening and constant watchfulness had not prevented the severe illness of our little boy. Yet I should have known that with conditions as I
had seen them all through the bottoms, it was indeed a miracle if any of
the young children escaped an epidemic. Wayne and Henry were very
serious and thoughtful and so depressed by Gareth's illness that I began
to suspect that they were keeping something from me. That was true.
Wayne finally told me that already eight young children and babies had
died down in the lower bottom lands.

Jimmy and Donald spent their days now trying to amuse and interest Gareth. They carried him around, rode him about in their wagon
and swung him in the hammock. We boiled all the water we used and
tended Gareth faithfully. At times he seemed improved, but at last it
became very evident that he was growing weaker and whiter by the day.
He was entirely indifferent to food and water. The doctor had prescribed
a medicine which he hoped would cure him, but I saw no improve-
ment. Mrs. Batson suggested that I dig up dewberry roots and make a
strong tea from them. Grandma advised flour, baked to a deep brown
and made into a gruel. We all felt as though a dark cloud had passed over
the sun.

At last, early one morning when Gareth lay white and limp and quiet,
we sent in haste for the doctor. His orders were brief but imperative. We
must take Gareth as high up in the Ozark Mountains as we could get by
nightfall. He told us of several good places where we could stay.

Long before noon we were well on our way. Grandma and Jimmy and
Donald went with us. We had quickly gathered together bedding and
food and cooking utensils. I had hastily ironed some needed clothes for
we were resolved to stay until Gareth was well.

By nightfall we were camped at Freeman Springs where we rented a
small house. It was nearly dark when we prepared our supper. The doctor
had said not to give Gareth anything to eat or drink, until we were up at
the springs, then we were to let him have all the water he would take. He
said he felt like a witch-doctor for recommending so strongly the pure
mountain water, but he knew it was highly beneficial, and in Gareth's case
the sudden change of water was the only hope left. We gave Gareth so
many drinks that we began to think he would refuse the water altogether,
but he drank eagerly and when his gruel was prepared he reached for the
dish and ate hungrily.

We were all greatly surprised and happy, but not jubilant, for it took some time for us to realize that he was actually better, but before the next sun went down our thankfulness and relief knew no bounds. We stayed at the springs for more than a week and Gareth improved rapidly.

Each day we drove over the mountain roads, seeing rough country that was still thickly timbered and houses built in little clearings where the trees had been felled, and many places where the gardens were planted in fields where the trees had been deadened—waiting to be cut down later.

Never could I have believed that there were such places as we saw here. Away off the beaten path one day, we met several families driving along with their wagon-beds filled with tubs of huckleberries which they had gathered. We talked to these people a long time, and I went myself and lifted up the quilts that were stretched over the tubs and stared unbelieving at that great amount of wild berries. It just didn't seem possible. Once we came across a man driving a dozen razor-back hogs. He was having a lot of trouble keeping them on the road. He told us he was taking them along further to where they would be left to eat the oak mast. The ground in places was thickly covered with acorns. They really were pigs, I decided, as I stared and stared. Their back went up into a peak and their bristles were rough and course.

Many of the mountain homes were built of logs. They all had big porches and were built in regular house style, without the open hallway making a division between two parts of the house, as most of the houses were built down where we lived.

This was indeed a dog's paradise. Every family owned several dogs. Most of them were hounds, with long, flapping ears and long, whip-like tails. Of course all the men and boys went hunting here in the fall and winter and so the dogs were almost a necessity. Coons, opossums, and squirrels were plentiful and were the only meat many of the mountain people had.

Near the springs was a big wooden platform with a railing around it and branches at the sides. One night when we were preparing for bed we were surprised to hear music. Someone was playing an accordion. Although Grandma was already in bed, she dressed too and went with us over to the platform where we found a group of campers gathered there in the moonlight, singing and dancing. One man was teaching the group to do the square dances, and he entertained us all with his antics. A young girl endeavored to teach me to do the Charleston, but my efforts were very unsatisfactory. She explained that it all depended on just the right way of rotating the knees. Her own exhibition of that dance was all part of the fun. It was a real old hill-billy dance we were having, and a most spontaneous pleasure.

It was both surprising and delightful that we, as strangers, were so quickly and pleasantly taken into the group and made to feel so welcome. Everyone laughed and talked and joked and had a very happy time. Even the older men and women, gray-haired, got up on the rough old floor and danced as well as the rest to the gay music. Between dances the conversation was general, and all introduced themselves to everyone else. It seemed to me that such carefree laughter was something very strange. I think that life had indeed become a very serious affair with us all. Wayne, especially, seemed very happy. It was an occasion that brought out the best in each one. None of the gaiety seemed forced, yet it was evident that each one intended to have a good time, and they were surely enjoying themselves. These people were all city people, vacationing through the mountains. The next evening I watched when they were preparing their supper at a camping spot across the road from us, and saw that they set their large folding table with a white cloth and with gleaming silver and glassware. They had three cars, and this change to the mountains was only an extension of their usual manner of living.

Perhaps the warm nights, the mountain breezes, the scent of the pines and the deep woods that surrounded us, created a magic spell, for we all seemed warmly united here, with all the rest of the world forgotten.

When it came time at last for us to leave the mountains and return to our farm in the bottom land, we suddenly realized as never before the life to which we were returning. It was hard to realize that Gareth had come so close to dying, he was now so completely recovered. But although I said nothing, I was firmly resolved to be away from the conditions that made our life so unbearable before another summer came. We tried to enjoy each minute of our return journey. We still passed many beautiful places, saw ideal mountain homes that we thought were all anyone could ask for, and loitered by small lakes and streams where hundreds of people were out enjoying the freedom of travel and the beauty and change on every hand.

One afternoon we stopped to eat our lunch by a creek. While Grandma slept under a tree with Gareth snugly beside her, Wayne and Jimmy and Donald and I all went in swimming. This was surely the place that every poet has described—warm water, trees overhanging the pool, a warm breeze, the water deep enough for a good swim, the sun still high in the heavens, a sense of peace and happiness, and, best of all, a sense of unlimited time, time in which to loiter and enjoy all this to the full.

58. The Holy Rollers

A HOLY ROLLER camp meeting was being held across the creek toward the hills, and as the crops were laid by, we drove over one Sunday evening. Bea and Herman went with us. Grandma told me what I would see, and said she wanted none of it. She was a staunch Methodist and declared that "talking in tongues" and "laying on of hands" was a part of the Methodist religion that had long ago been given up. I could go if I wanted to, but she knew what I would think of it all. One trip would be enough.

She was right. I never saw anything like it. Under a huge oak tree a small platform had been built. Below this, in a semi-circle, the benches were grouped, with a few chairs close to the platform. Sawdust covered the ground. From the branches of the tree hung several lanterns, and at one side, lower down, hung two cedar buckets with gourd-dippers floating on the top of the water.

About a hundred people—men, women and children—filled the benches. All around, behind the seats, automobiles were parked, and dozens of farm wagons were waiting, the mules tied to nearby trees. We had to park our car some distance away so we walked over to the lights and finally found places to sit. The chairs placed up toward the front were reserved for the "mourners." The evangelist was a woman who was traveling through the South holding revival meetings and healing by the laying on of hands.

Hymnbooks were passed around and the congregation all joined heartily in the singing. The preacher would read a verse, and the crowd would sing it; then another verse was read and sung, and so on until the hymn was finished. There was a small organ up on the platform, and there a woman sat all evening, pedaling with all her might and waving her hands and shouting "Amen" as she accompanied the songs.

At last the preacher announced that she would discuss some of the sins that were becoming too flagrant. Her speech was very rapid, her voice high and shrill, and the most I could make of it was a jumble of accusation against everybody, delivered in a voice of doom—the threat of hellfire and damnation held over their heads, and a frequent beseeching to all the audience to come forward and give up their souls to be saved. She jumped up and down, waved her arms, clapped her hands, knelt to the crowd with arms outstretched, paced back and forth, and stamped her feet. Her voice, sharp and penetrating, rose higher and higher, and the words came faster and faster, intoning with a regular cadence or falling dramatically by degrees into such a low pitch that she startled everyone by her hissing whispers. Then up again with a scream would go her voice, calling on the Lord to witness the sin-ridden people whose hearts were so hardened that she could not move them. I whispered once to Bea and asked her what the preacher was trying to say. Bea, who was trying to keep her baby quiet, said she had missed some of it, but that she thought she was preaching against the short sleeves and low necks of the women and the new fashion of bobbing hair.

The preacher herself wore a dark dress with long sleeves and a long skirt. Her hair was twisted into a tight knot. She was not pretty—nor yet ugly—just severely plain, with large dark eyes and a very high forehead.

We shifted our places several times, for the seats were hard, the children restless, and the mosquitoes kept buzzing around constantly. At last we got in the Ford and found a place to park it so we could see and hear well.

Now the preacher began a long chant in a monotonous tone that went on and on in a sustained ardor of supplication and intense excitement. Every now and then she burst forth, without warning, into a childish chanting of happy rejoicing, with much hand-clapping and feet-stamping and body-swaying. Suddenly, after a dramatic stillness, with beseeching arms extended and eyes closed, she began to sway, to stamp her feet, and clap her hands, and to moan and mutter; beginning in a low voice and increasing in pitch and tempo, with added motions, until each burst ended in a downward, hopeless tone of impending doom. Finally she began a long shrieking dirge, the first words of which, falling on my critical and attentive ears, were like nothing else than, "Holy Mackinaw." The words that followed, in an awful jumble, sounded like nothing so much as the gibberish we used to call "pig-Latin," with no resemblance whatever to any foreign language I had ever heard.

Noting a commotion down toward the front of the platform, I got out of the car and stood up on a bench to get a better view. About a dozen men and women were gathered on the mourners' seats, and beside each one knelt a woman, praying loudly and wildly. These women were helping the preacher. The noise of their voices, that seemed to be a mixture of crying and praying and shouting "Amen" and "Praise the Lord" all together, drowned out everything else. Throughout the crowd I saw men and women gather to urge some man or boy forward. Then I noticed right down in the aisle close to the platform five men and two women stretched out in a row. These, Bea told us later, had finally fallen into a trance. They would get up later and cry and pray and declare themselves saved. Meanwhile the preacher was going from group to group. I could hear her high voice above the others, begging, laughing, praying and shouting.

Throughout all this excitement mothers sat and nursed their babies, children made endless trips to the hanging buckets to get drinks of water, dogs barked, old Fords were cranked up for an early start home, and little children fussed and cried, or left their parents and ran up and down the benches, playing tag and getting in everyone's way. The swinging lanterns gave an odd brilliance to the scene and attracted millions of gnats and mosquitoes.

As we drove home Herman told us that he did not hold with such goings-on, for the men we saw there who prayed the loudest, and even some of those we had seen stretched out "with the Spirit on them," were men he knew who would be out in the fields the next day cursing and beating their mules. "'Tain't a religion that lasts," Herman said. "They get too much pleasure out of it. As for the laying on of hands, they're the ones like as was up by the chapel where they let that girl die because they wouldn't call no doctor."

I doubted no one's sincerity at the camp meeting, although we had witnessed an extremely hysterical demonstration of religious fervor. Some actually believed it all, I knew, and tried to live their lives accordingly. They seemed to get a lot of satisfaction out of it. To see their shining faces and hear their joyous songs of praise and their earnest testimonials of the Lord's personal love for them, was to realize that, although here there was a total lack of dignity, yet they were emotionally transported to the seventh heaven and believed they were truly on the road to Beulah Land. I knew that better minds than mine had tried to weigh the true value of such a religious demonstration.

Later, at home, as we discussed the amusing incidents we remembered, Wayne quoted, to put an end to my expressions of astonishment and opposition:

> *It is not ours to mediate*
> *The tangled skein of will and fate,*
> *Or say what metes and bounds shall stand*
> *Upon each soul's debatable land.*

59. The Slough of Despond

I DID NOT DO any canning that fall. We still had some fruit left from the year before, and Wayne told me it would be enough, for we would certainly be leaving for California.

I felt that by this definite decision, in some way a chain of events was being set in motion that must surely result in our dreams coming true at last.

Faith and hope are wonderful powers, although I found it very hard to fool myself. In my actions I carried on with courage and resolution and doggedness, yet I felt doubt and fear, and a sense of grimness hanging over our lives.

Wayne seemed to take little real interest in the packing, leaving it all to me, yet when I appealed to him time and again for assurance that this would indeed be our last year in Arkansas, I got it readily enough. Each day I went at the task with renewed vigor.

Grandma was not to be told of our plans until everything was actually settled, for disappointment would be too hard on her, too upsetting. In all this time I had seldom discussed with Grandma my feelings about our life here in the South, nor had I often voiced the strong longing that I had to get back to California. The shock and suffering I had undergone during Gareth's illness made me grimly determined now to carry through to the end. It seemed a great and grave undertaking, but I felt so compelled, so driven to it. We had come so close to sacrificing Gareth to these Southern conditions that I could hardly bear the thoughts and emotions that filled me when I thought about it. The days passed slowly—and then at last I came to know in truth what it really meant to have a shadow over my heart, for finally, toward the last of the cotton picking, Wayne came in late one night, and with little talk ate his supper and settled to some reading. This was the day when an estimate was to have been made of the crops' valuation, so that definite steps could be taken for the next year's program. I wasn't smart or clever. I had no head for figures. It seemed that Jimmy and Donald and I, too, were only children, wanting to know, "Is everything settled? Will we have enough money? Are we really going this time?"

The answer to my questions was a quiet, "No, I guess we'll have to make another crop."

Years have passed, but I will never forget that bitterly disappointed woman that was myself. My hopes were dashed—my courage gone. Mentally I had been through here, finished—and ready for a new life. I knew that Wayne was doing all he could—I acknowledged with my mind all the facts, but my heart actually ached.

When Donald and Jimmy went to their bed that night they asked me if it was really true that we were not going back to California yet. I sat on the side of the bed and looked at them, my two fine sons—so young, so eager, sharing my hopes every day, through every long year. Hope deferred!

I lay down beside Jimmy and he put his arms around me and hugged me, and begged me not to care. I cried myself to sleep in his arms. I felt that I had touched bottom and did not know where to turn now for the strength to take up my life again and go on.

60. Cotton Town

I DIDN'T REMEMBER seeing Cotton Town as we drove down to the lower farm, and I was not sure where it could be, so when Wayne asked me if I wanted to go with him in the Ford down to Cotton Town to sell a load of melons, I was full of anticipation. I thought it was a little community further down the river, and I was very anxious to see it and perhaps do some shopping at the general store there. I was amazed when I found that it was only a small store with about a dozen houses grouped around it within sight.

It was a glorious fall day, when all the world seemed bursting, like the melons in the fields, with sweet ripeness. We drove down the river road. All along the fences the big black clusters of elderberries hung on the bushes, and the sumac leaves were red and shining with a lacquered brilliance. Although it had rained recently the roads were passable enough, and I scanned the countryside with interest as we drove along.

Wayne pointed out the place where Mill Creek cut across from the bayou to the Arkansas River. The acres here, around the mouth of the big creek, lay in low ground, so that whenever the river rose to any great extent the crops would be under water.

A levee district had formed, and the people were taxed to pay for one levee after another. But all the levees were built far too close to the river, apparently to protect the seventy-five acres or so of some special land owners, and then when the usual spring and fall rises came, these levees, which supplemented the high banks where the creek mouth lowered, were usually washed away as the softening sands of the banks became soaked with the rising water, and they fell in. Then the river water poured onto the land in a long strip parallel with the main current and water of course backed up, as it so well could, until even we, too, on our farm up the river, received part of the overflow. Even when Mill Creek was dammed up at the flood gates, and the torrent of water from the hills and bayou was held back for a while, it has to be released eventually, and the respite was only a short lessening of the damage.

Poor management and some favoritism, I heard, kept these lesser levees where they were, for year after year they were in a perpetual state of breaking down and being rebuilt. At last, however, the costly lesson seemed to be learned, and a large high levee was built much farther back toward the hills; then, excepting when an extremely high rise came in the Arkansas River, the land adjacent to the low places would flood, surely, yet the great stretches of bottom farms would be saved. But if ever a big flood should come, the river might still break through wherever a low place allowed the water to cut across to the hills and the bayou, and it could flow far across the bottom lands, for Mill Creek would afford no great relief, and the big new levee might be of no avail.

As we drove past the little shacks along the road, we saw the usual patch-work quilts a-sunning on the porch railings and out along the fences.

Every family had home-made quilts. It seemed to me that a recognized badge of social standing was given according to the amount and kind of quilts that a family could display. Most of the quilts I saw were beautifully sewn and quilted, with stitches fine and even, and were not to be compared to the ones made of old pieces cut from worn clothes. Generally the quilts were backed with gray or brown chambray or brighter colors of percale and gingham.

These quilts were a vital necessity, for often, with little wood for the fires, the heavy home-made quilts were their only barrier against the cold. When there was a large family of children it took a lot of quilts to cover them. Sheets were not as generally used as one might suppose, here in this land of cotton, for they had to be bought in the stores and there was little money for such luxuries. Light-weight blankets took their place sometimes, but often quilts formed the entire bedding.

Old quilts were used, too, to pad the wagon seats when the mothers and children went on the long trips to town to visit friends. If a storm came up, extra quilts were wrapped about the babies and younger children as they huddled from the rain on the wagon seats. There are the "haves" and the "have-nots" in every condition of life. I found homes where a pile of extra quilts, well-made and beautifully clean, were always handy for the quick making of the extra pallets, in this land of where "you-all come and go home with me" was the invitation so heartily given and just as heartily accepted. In too many homes, however, it was evident that even the necessary amount of quilts for reasonable warmth was sadly lacking.

None of these people seemed to own more in the way of furniture than their cookstoves, beds, chairs, and a kitchen safe. Most of their beds were empty ticks which they filled with straw and which could be laid on a floor or placed on a crudely made bedstead, a thing of slats and no springs. Those who had feather beds or maybe a heater or an old dresser with a tiny mirror or perhaps a phonograph or sometimes an enlarged picture with bulging glass in an oval of gilt, were considered to be way ahead financially, for they had either happened to be fortunately situated in a year when cotton prices were up and the crops good, or they had started housekeeping with an outfit from their parents.

I always noticed the beautiful calendars that were saved year after year. These were given away by the mercantile companies all through the South. Most families treasured the highly colored prints of Scriptural scenes and the bright motto cards from the Sunday-school, but there were seldom any other pictures.

The floors were always bare and on nails around the walls hung the overalls and shirts of the men, the gingham dresses of the women, and the inevitable snuff-sticks for the adult members of the family, each stick dangling from its own nail.

The plowing and hoeing were finished now, and the green cotton fields were deserted until the fall picking began. I knew what the women in all these little homes could be found doing. They would be spending all their spare time mending overalls and making more quilts.

When we got to the little store at last we found that the place was crowded with idle men. These were hoe-hands and share-croppers, resting now, since the crops were "laid by." The men seldom took to themselves any odd jobs when their regular work was finished. They simply sat around home or sat around the stores talking. Rarely did they even "get up" wood in advance of the coming need, or make repairs on their houses or fences. "Sufficient unto the day" seemed indeed their motto, and since they suffered from no sense of omission or failure they frankly enjoyed the idle days with lazy tolerance and calm good nature.

If they had malaria, they could be seen sprawling on pallets on the tiny porches, or they would lie inside on the beds, tossing and muttering with pain and fever. Some of these people, I found out, never had a real malaria chill. They were gaunt and ailing and complaining, but they apparently became immune, although they not only carried the unmistakable evidences of the ravages of malaria in their bodies and minds but were potential sources of infection always.

To me, these men all had the same general appearance. They were thin and sunken-eyed. They looked tired, and they were tired, showing a lassitude in their every movement. When they walked they slouched. Not one ever stepped out with a firm elastic step showing a feeling of well-being or of possessing a keen zest in living. I decided that even their drawling voices, so slow and easy, were also the outcome of general debility. I even thought that their tolerance and easy-going ways, for these qualities were so evident in every least business transaction or conversation with them, were more the result of a desire to avoid an active part, either in thinking or in doing, than an expression of an acquired philosophy.

I felt that I myself was slowly succumbing to the same conditions that affected everyone around me. It even seemed strange to me to think how at home in California, and during my first year here, I used to walk briskly or even rush about in the house or out in the street with a dash and vivacity that was now gone entirely.

I made Mrs. Joe's nieces laugh until they cried one day when I put on my blue tailored suit, and taking my umbrella in my hand showed them how we walked on the streets in California. They could not believe that anyone could walk so fast and so jauntily along. When they tried to imitate me they were unable to get out of the slow languid tempo of their own strolling steps. My tailored suit, which I had always liked, was too hot and close-fitting for the warm weather here, and too light for the cold weather. Heavy underwear and heavy dresses were worn under heavy coats in winter, but no suits were seen, so it was a treat to the girls to try on the suit, take the umbrella, and try to imitate my lively performance. That umbrella of mine was of no real use to me. When it rained, it poured—no weak drizzle here; and generally when it rained the wind blew. No umbrella could withstand the gusts of wind. If I were on the street when a rain started, like all the others, I was forced to run to shelter.

I was realizing more and more that the intense heat in the summer that kept everyone in a wash of perspiration also seemed to drain their vitality, and this, combined with the malaria in their systems, gave them their gentle calmness, their tranquil acceptance of life day after day. But gradually those sweet soft voices of the Southern women awoke in me an unnamed annoyance. They were so docile, so uninspiring. I observed the quiet deftness with which they washed and wiped their dishes and set their kitchens to rights, hurrying to sit down to talk that was mainly doleful matter-of-fact statements about the daily work or their health, or to pointless discussions of trivial things; using the half-endearment of "little old" in reference to anyone they had a fondness for, and ceaselessly "aiming." They aimed to have, get, go, do—everything, always.

How I tired of the expressionless faces, the twisted mouths turned down at the corners, indicating a seeming soul-sadness that enveloped both mind and body. They were so lacking in hearty good humor. Light-hearted gaiety—never! How often in my youth I had been admonished when sent on an errand, "Now don't let the grass grow under your feet." That remembrance often made me smile, here.

The men, too, were so lacking in larger affairs to worry about, and they could do so little to control their immediate situations that they seemed to be forced to take a petty interest in small details. They could only be carried away occasionally to larger horizons, when they talked of the lives of others or grandly referred to world situations they "had heard tell of."

With Mrs. Joe's nieces I tried to have some discussions that might help me to understand the women better. I asked the girls just what they did, what hobbies they had, or in what they took an interest when their housework was done. I noticed that their houses were very clean, always; the routine of regular housework never slacked. Monday was the unvarying wash day followed by a dutiful ironing and mending. One could walk through their rooms and find everything very simple but very pleasing. Yet, the minute one stepped out of doors a big difference could be noted. Trash and sweepings, empty pails, and bits of wood and old glass jars were left carelessly around the yards. Apparently the women felt that their domain ended literally at their doorsteps. My questions were usually answered with a rather condescending remark, "Oh, we manage to get by." That answer aggravated me. I knew that if I ever accepted their easygoing attitude, I would not only be drawn down into the quicksands of peaceful endurance, I would sink right down to the bottom. Though they might all manage to keep their heads afloat, I knew that I would be completely submerged. My Scotch-Irish ancestry was such a strong current in me that in spite of everything I still rose each day with a feeling of hope and ambition and great expectancy. I craved excitement and change, and desired above all some overpowering motive that would provide a goal for each day's efforts.

Yet year by year my sense of frustration and defeat was deepening. There was nothing tangible of which I could accuse these friends, nothing definitely offensive, but I felt strongly in opposition to this tranquil acceptance of life. Sometimes I wondered what impelled them, having so little stimulation or incentive, to arise each day and live so calmly, so serenely.

Some of the characteristics of the Southerners that I noted both in town and country I never definitely understood. Almost without exception I found that the smooth, cool attitude they presented to the world was something behind which they seemed to hide. It baffled me. If an argument became too concrete and direct, they would always find some quiet and unobtrusive way of escape. A man would be seen getting to a nearby door where he would stand to add a few words as a parting shot in the strengthening of his side of the argument at hand, but before any rebuttal could be made, he would have disappeared. Seldom could one get a straightforward, sincere "yes" or "no" to a question, or feel that the answer given was the one backed by the man's own convictions.

Yet I wondered just how these poor share-croppers must feel to be forced into submission to an intolerable, hopeless economic life, to be so surrounded by custom and routine and habit, both of body and mind, that they were effectually trapped for all time. I read once of a man who trained a starling to talk. It sat in its cage and called, "I can't get out—I can't get out." These people were not starlings, wanting out. They seemed not to know that they were "in."

I used to wonder whether the terrible economic burden laid upon the Southerners after the Civil War had been so heavy, so curtailing to their efforts at release, that they had at last in self-defense learned the craftiness of apparent submission. I wondered if the time element under dominance was longer and harder than their forebears had foreseen, so that these descendants had gradually acquired a feeling of smouldering rebellion, of suspicion and artfulness, inheriting an ability in the use of cunning, stratagem, insidiousness and guile that found its best release only in the uses of diplomacy and politics.

Anyway, we managed to get rid of the melons, and while Wayne was talking to the storekeeper I examined the place with interest. I found just what I expected—the usual supplies that were supposed to fill the needs of the poor white people. Flour, corn meal, soda, baking powder, lard, salt, sugar, sorghum molasses in gallon buckets, harness, hoes, tobacco, the one brand of poor cheap coffee, straw hats, cheap shoes, and a few bolts of dreary-looking cloth.

One old man was going from counter to counter peering at the shelves and lifting up various articles for closer inspection. Finally the storekeeper became worried when the man went behind the counter and began to examine the plugs of tobacco, and asked him to move out to the center of the store with the other men. The old man was busy pushing plugs of tobacco into his jumper pockets, and he turned testily on the storekeeper, demanding, "This is one of them touch-and-take places, ain't it? I was just only waiting on myself." He paid for the tobacco, but only after the storekeeper insisted upon dragging out the last plug from his pocket.

We were late in getting home, for a heavy rain and a terrible wind that came up suddenly kept us, with all the others, inside the store for more than two hours. But at last the rain stopped, and the sun came out. We found the road blocked in several places by large branches, but we managed to drag these to one side so that we could drive on.

As we passed one of the little one-room shacks that stood at the side

of the road I saw Lidy Cowder standing there in the door with seven children in sight. All the children looked half-starved as they stopped their
play to stare at the passing car. Lidy seemed like an old woman now. Her
hair hung in strings around her face. Her clothes were torn and her feet
were still bare. They had moved here soon after they left our neighborhood.

Yes, Lidy had lived, to have another baby each year.

The cotton land stretched away as far as I could see, back toward the
willows where Mill Creek ran. The cotton crop was in excellent condition,
fully five feet high here, the leaves large and glistening, newly washed
from the rain.

I returned home with a heavy heart. As we drove up the road, around
bend after bend, we began to catch glimpses of the familiar line of western hills, with the sun fast sinking behind them. Against the deepening
blue of the sky floated bird-like clouds with wide-spread wings. Below
them, in snowy heaped softness, the rounded puffs were made glorious
with the gold of the fading sunlight. Surely, "Every prospect pleases, and
only man is vile."

61. Last Letters to a Mother

NOW YOU MIGHT ASK: "Sister Ann, sister Ann, what do you see?"

*I see Wayne with the cotton and corn laid by. We are all feeling much
better, but are still dull and listless, a result of the typhoid shots we took
and the malaria in our systems. Jimmy and Donald have been hoeing a
little. They haven't felt like studying, and would rather be busy at something. They each had a spell of dysentery with fever, as most of the children here in the bottom have had. Castor oil, quinine and iron tonic! Will
I ever forget those medicine bottles?*

*The mosquitoes are more troublesome than ever before as there is so
much standing water everywhere. We have been able to get extra good
nettings for the beds, however, the type used in the army, bound with tape
and of a fine mesh.*

*The magazines have only touched the high spots, as usual, regard-
ing the flood. It isn't half what they could tell. We have had nothing but
weather lately. So much rain. Yet we know that we have been exception-*

*ally lucky when we hear of the floods and repeated floods in other parts
of the country.*

*We take extra doses of quinine every three days, so we will not be sick
as soon as we change climates. There are lots of cases of pellagra among
the poor people living in army tents lower down on the river bottom. Of
course, they have been making a crop with no garden, no cow, no chickens, just the usual diet of beans, fat meat, corn bread, coffee and sorghum.
But I see that the Foreign Missionary Societies are still flourishing. I think
it was an Alabama editor who said, "They are going to fill Heaven with
Negroes, Chinese, Japs, and Malays, and Hell with citizens of the U.S.A."*

*I'm planning to be in California by September. I will be soon seeing
California in all the glory of October. I can't believe it after nearly seven
years of waiting. My thoughts keep running ahead of me.*

*I see Gareth out playing ball, and I am corralling the last of our things,
getting ready to leave. Yes, really. Jimmy and Donald will look so nice in
their new suits. Jimmy is about as tall as I am now. The time is so right for
them to be given something better—a chance for a better education and
a better environment. The fall term of school in California will begin in
September. I am sure they are ready for seventh grade work.*

*I feel that if we don't make the move now it will never be made. It is so
necessary. What awful conditions are here for children.*

*I am not worrying about the future for I have been through the slough
of despond here. Nobody knows but myself what this change will mean to
me. I find myself tense and determined, but with my mind rather vacant. I
do all the necessary things as though I were in a trance. I am afraid, even
now, that I will weaken at the last, and lack the courage to go on ahead
without Wayne. How can I leave him behind? There is nothing I shall
be sorry to leave here but Wayne, and that won't be for long. Grandma
and Henry will be leaving, too, but later, with Wayne. They are all sorry
that I am going ahead but can see the many reasons why it seems a wise
thing to do.*

*The weather is very peculiar for July, with lots of sudden rains. But
the crops look good, where there are any, with prospects of twenty-cent
cotton at least. Much good soil is now a stretch of sand where nothing will
grow and it has been difficult to get enough hands to work the crop, so
much of it is in the grass. But there is too much rain, and the outcome is
something that cannot really be predicted. Wayne feels satisfied that he'll
break even, anyway.*

Yes, the time is actually getting short, yet I'm not excited or nervous now. At the last we find that there is so much to do. I am not sorry to leave, for Wayne promises to follow soon, as soon as the crop is gathered, he says, and the place is re-rented and the deck re-shuffled for other tenants. I do realize his feelings about returning to California with nothing to show for his years of hard work here. I do acknowledge his right to pride and self-respect, and his reluctance to accept defeat. I do admit that I am torn between my love for him and my intense anxiety about our children. But in this he agrees with me, and knows that year has followed year with no indication of better conditions. No hope.

I am so anxious to see Wayne living in such a way that he will get rid of his thin face, and will learn to smile again and not take reading as a sedative. I want to know him again when he will be able to keep his worries out of his mind and will remember how to laugh and live again before he gets too old. Wayne has the loveliest character of any man on earth, and beneath any faults he's a real human. Little Gareth takes after his daddy. God Bless him, he's so sweet. No other words fit. Oh, you'll think I'm daft.

Your loving daughter

62. October's Bright Blue Weather

ON THE TRAIN, speeding west, Wayne left behind for a short time, for only a short time, surely, we settled down to watching the great stretches of Oklahoma and Texas flash past the windows. I was greatly impressed by the vastness of the farm lands, and, in too many instances, the insignificance of the farm houses.

The routine incidents of train travel, the stops made at way points, our hurried lunches, all made the time pass quickly. The children were interested in everything and eager to arrive in California, and it made me feel proud when the conductor paused by our seats to tell me that my boys were exceptionally well behaved.

I slept but little, for my thoughts would fly ahead, full of plans, although I was often overwhelmed with sadness for I kept thinking of Wayne as he had stood waving good-bye to us at the last, promising to bring Grandma and follow us soon, and then hurrying back to the Ford just as a terrific storm broke. Loose papers were blown high in the air and forked lightning cut sharp against the dark clouds. I imagined Wayne's return to the farm and pictured him walking through the house, alone, and the tears flooded my eyes.

Once the train passed close by a farm and we watched a farmer driving his milk cows into a barn lot and a woman standing out in the wind feeding the chickens. We had such a sense of knowledge and judgment about these glimpses of a life we had known so well, but we agreed that the rich bottom lands we had left were the best lands of all for farming.

The sun was about down when I noticed that Jimmy and Donald, who had been chatting happily, suddenly became quiet and thoughtful. I saw that they were watching a young boy driving a team of mules through a bit of meadow-land where a creek flowed under a little bridge. Donald poked Jimmy, and seemed about to say something, but turned back to the window and kept his face tight against the pane for a long time. Somehow, that, too, made me sad, and my mind was again in a tumult of questioning and wondering. I was conscious of the miles between us and our little home. The long drawn whistle of the train, the twilight settling down over the great barren expanse of land, the first stars showing over the low hills, filled me with fear and uncertainty and made me wish that we were back again with Wayne, all together, safe and secure.

Yet I could not forget how, as year had followed year, I had weighed the advantages and disadvantages of this very step we had at last taken. I could not forget the sleepless nights when I had sat beside Jimmy and Donald where they lay tossing and feverish, enduring another onslaught of malaria. I remembered the nights when I had arisen and walked out around the house on the cool sand in the moonlight, thinking of the boys lying asleep under the mosquito nettings, and trying to decide what to do for their future. They were so eager, so trusting, so sincere, so young, looking to me to make the big decision, hoping that everything would turn out all right. I felt again that bitterness of our yearly disappointments, and my heart ached anew as I sat and looked at the three children. They needed so much to be guided wisely and safely now.

We had lived in a little house, I thought, and we had outgrown it. I remembered the lines from the Chambered Nautilus:

> *Build thee more stately mansions, oh, my soul,*
> *As the swift seasons roll;*
> *Leave thy low-vaulted past ...*

And then I was swept by a deep realization of all we were leaving behind us, of all the life we had known since that day long ago when we had first moved down to the farm. As the train rushed on, carrying us toward a new life, an unknown future, I composed these lines:

THE LITTLE HOUSE

We thought it was a big house.
We were within. Our minds pushed back the walls.
Our thoughts went outward as our lives expanded,
'Till never did we realize or feel
We lived in a small house.

We lived in a little house;
And yet, in all those years, I know now,
We never felt that we were anything but free,
With spaciousness and grace.
I think about this, now,
As we speed by these little houses.

We are all together now, in California, under sunny skies, in October's bright blue weather. But I know that down South, near the Arkansas River, there's a little house still standing by the pecan trees facing the Bermuda meadow. Across that meadow, of a late summer afternoon, comes the low, sobbing moan of the mourning dove, and in the morning, with the sun fast rising over the cottonwoods, the cheery, "Kutie, kutie, kutie" of the mockingbirds still arouses to the day's work.

One mockingbird came every day and sang "Kutie, kutie," and then settled on our roof and jumped up and down for pure joy of living. I'm lonesome, at times, for it all.